NATO ASI Series

Advanced Science Institutes Series

A series presenting the results of activities sponsored by the NATO Science Committee, which aims at the dissemination of advanced scientific and technological knowledge, with a view to strengthening links between scientific communities.

The Series is published by an international board of publishers in conjunction with the NATO Scientific Affairs Division

A Life Sciences B Physics	Plenum Publishing Corporation London and New York
C Mathematical and Physical Sciences D Behavioural and Social Sciences E Applied Sciences	Kluwer Academic Publishers Dordrecht, Boston and London
F Computer and Systems Sciences G Ecological Sciences H Cell Biology I Global Environmental Change	Springer-Verlag Berlin Heidelberg New York London Paris Tokyo Hong Kong Barcelona Budapest

PARTNERSHIP SUB-SERIES

1. Disarmament Technologies	Kluwer Academic Publishers
2. Environment	Springer-Verlag/Kluwer Academic Publishers
3. High Technology	Kluwer Academic Publishers
4. Science and Technology Policy	Kluwer Academic Publishers
5. Computer Networking	Kluwer Academic Publishers

The Partnership Sub-Series incorporates activities undertaken in collaboration with NATO's Cooperation Partners, the countries of the CIS and Central and Eastern Europe, in Priority Areas of concern to those countries.

NATO-PCO DATABASE

The electronic index to the NATO ASI Series provides full bibliographical references (with keywords and/or abstracts) to about 50000 contributions from international scientists published in all sections of the NATO ASI Series. Access to the NATO-PCO DATABASE is possible via a CD-ROM "NATO Science & Technology Disk" with user-friendly retrieval software in English, French and German (© WTV GmbH and DATAWARE Technologies Inc. 1992).

The CD-ROM can be ordered through any member of the Board of Publishers or through NATO-PCO, Overijse, Belgium.

Series H: Cell Biology, Vol. 105

Springer
Berlin
Heidelberg
New York
Barcelona
Budapest
Hong Kong
London
Milan
Paris
Santa Clara
Singapore
Tokyo

Gene Therapy

Edited by

Kleanthis G. Xanthopoulos

NHGRI, National Institutes of Health
10 Center Drive, Bldg. 10, Rm. 10C103
Bethesda, MD 20892-1852, USA

Current address:

Aurora Biosciences Inc.
11010 Torreyana Road
San Diego, CA 92121, USA

With 32 Figures

Springer

Published in cooperation with NATO Scientific Affairs Division

Proceedings of the NATO Study Institute "Gene Therapy", held at Spetsai, Greece, August 17–28, 1997.

Library of Congress Cataloging-in-Publication Data

NATO Study Institute "Gene Therapy" (1997: Spetsai, Greece)
Gene therapy / edited by Kleanthis Xanthopoulos.
p. cm. – (NATO ASI series. Series H, Cell biology; vol. 105)
Proceedings of the NATO Study Institute "Gene Therapy", held in Spetsai, Greece, August 17–28–1997 – T.p. verso.
"Published in cooperation with NATO Scientific Affairs Division."
Includes bibliographical references (p.).
ISBN 3-540-64112-2 (hardcover)
1. Gene therapy–Congresses. I. Xanthopoulos, Leanthis, 1958– . II. Title. III. Series.
RB155.8.N38 1997 616'.042–dc21 98-2583 CIP

ISBN-13: 978-3-642-72162-5 e-ISBN-13: 978-3-642-72160-1
DOI: 10.1007/978-3-642-72160-1

Softcover reprint of the hardcover 1st edition 198

Typesetting: Camera ready by authors/editors
Printed on acid-free paper
SPIN 10571817 31/3137 - 5 4 3 2 1 0

PREFACE

Gene therapy has generated enormous expectations as it holds a great promise, namely to revolutionize clinical and molecular medicine. This is because the successful introduction of a gene sequence in somatic sells followed by long-term expression at therapeutic levels can fundamentally correct a variety of human disease such as inherited and acquired disorders, cancer and AIDS.

The proceedings of the first ASI on Gene Therapy that took place from August 17-28, 1997 on the island of Spetsai, Greece, summarizes recent advances in the field as presented elegantly by the internationally-recognized experts that served as faculty at the meeting. Their contributions constitute brief but comprehensive reviews presented as independent chapters in this book and emphasize the influence of chromatin structure on gene expression, describe the principal mechanisms of gene expression, followed by articles on epigenic elements, viral vectors, non-viral vectors as gene delivery systems, animal models of human diseases, inducible regulatory systems, gene therapy of cancer, AIDS, monogenic disorders and ongoing clinical trials.

Gene therapy is the logical progression to human health management because it offers a potential low cost and efficient treatment. Gene therapy can reach it's full potential and satisfy all expectations. However, a number of critical technology advancements need to be completed. These include the development of vectors capable of safe and efficient gene transfer, the control of cell type targeting and regulation of transgene expression levels. I sincerely believe that by maintaining a delicate balance between the basic science of gene therapy and technology development and a careful design of clinical trials with well-defined end points, gene therapy will be validated. This delicate balance is elegantly illustrated in this book.

Kleanthis G. Xanthopoulos

Table of Contents

A Few Glances at the Function of Chromatin and Nuclear Higher-Order Structure in Transcription Regulation................. 1
J. Mirkovitch

Transcriptional Regulation and Gene Expression in the Liver.. 17
M. Tomizawa, J. Lekstrom-Himes, K.G. Xanthopoulos

Modeling Genetic Diseases in the Mouse.............................. 37
A. Wynshaw-Boris, C. Barlow, A. Chen, M. Gambello, L. Garrett, T. Hernandez, S. Hirotsune, W. Kimber, D. Larson, N. Lijam, G. Ryan, Z. Weaver

The Use of Genetic Switches for the Generation of Conditional Mutants at the Level of Cells and Animals 45
H. Bujard

Application of Tetracycline Regulatable Systems for Gene Therapy... 53
D. Bohl and J. Heard

Alphavirus-Retrovirus Vectors 61
H. Garoff and K. Li

Integration of Retroviruses into a Predetermined Site............... 71
K. Davies and G. Kalpana

Amphotropic Envelope/Receptor Interactions........................ 77
P. Rodrigues and J. Heard

Transduction of CD34+ PBSC Mobilized in Cancer Patients Using a Novel Fr-MuLV Retrovirus Vector Derived from FB29 Highly Titrating Strain.. 83
L. Marina Restrepo, M. Masset, J. Bayer, L. dal Cortivo, J. P. Marolleau, M. Benbunan, M. Marty, O. Cohen-Haguenauer

HIV Gene Therapy: Current Status and Its Role in Therapy........ 91
E. Bohnlein

Adeno- Associated Viral Vectors: Principles and In Vivo Use.... 103
O. Danos

Genetic Correction for Gene Therapy..............................111
E. B. Kmiec, A. Cole-Strauss, M. C. Rice and P. Havre

Retroviral Vectors for Human Gene Therapy 119
G. Ferrari and F. Mavilio

Gene Therapy for Cancer..133
S. A. Zullo, N. J. Caplen, R. M. Blaese

Feasibility Study Intended for In Vivo Retroviral Mediated
Gene Transfer of Bladder Urothelium 147
N. Dumey, M. Masset, P. Devauchelle, M. Marty, O. Cohen-Haguenauer

A Gene Therapy Approach for the Treatment of ALS 157
P. Aebischer and D. Braguglia

Modular Fusion Proteins for Receptor-Mediated Gene
Delivery..167
C. Uherek, J. Fominaya, W. Wels

Intracellular Combinatorial Chemistry with Peptides in
Selection of Caspase-like Inhibitors...............................171
S. M. Rothenberg, J. Fisher, D. Zapol, D Anderson, Y. Hitoshi,
P. Achacoso, G. P. Nolan

Lipid Gene Transfer, a Story of Simplicity and Complexity......185
N. J. Caplen

Theory and Practice of Using Polycationic Amphiphiles
and Polymers for In Vitro and In Vivo Gene Transfer..............195
B. A. Demeneix, D. Goula, C. Benoist, J. S. Rémy, J. P. Behr

Lipid Gene Transfer and Clinical Gene Therapy 205
N. J. Caplen

Human Dystrophin Gene Expression In mdx Muscles
After In Vivo Ballistic Transfection, Application of Synthetic
Oligopeptide Complexes and Cationic Liposomes................219
V. Baranov, A. Zelenin, O. Tarasenko, V. Kolesnikov,
V. Mikhailov, T. Ivaschenko, A. Kiselev, O. Artemyeva,
O. Evgrafov, I. Zelenina, R. Shafei, T. Kascheeva, G. Dickson,
A. Baranov

A Few Glances at the Function of Chromatin and Nuclear Higher-Order Structure in Transcription Regulation

Jovan Mirkovitch

Swiss Institute for Experimental Cancer Research (ISREC)
Chemin des Boveresses 155
CH-1066 Epalinges
Switzerland

Keywords: gene therapy, nucleus, chromatin, transcription, insulator, chromatin domain, nucleosome, chromatin remodeling, histone acetylation

Gene therapy necessitates the accurate expression of a transferred gene; that is the gene has to be expressed at precise levels in the right cell at the appropriate time. Much before "Gene Therapists" started to show up in laboratories throughout the world, cells had devised a large number of mechanisms to reach that goal, that is to tightly regulate the expression of about 100,000 genes. The most common mechanism utilized by cells for controlling gene expression occurs at the transcriptional level. In this case, deciding if a gene will be transcribed or not, and adjusting the level of transcription initiation and elongation, determine the amounts of gene products. Although the cell has found it advantageous to control the expression of many genes at many other downstream steps, control at the transcriptional level prevents the synthesis of useless molecules and the utilization of further control levels.

Mammalian transcription takes place in a complex environment, still poorly understood, the cell nucleus. The goal of this minireview is to briefly summarize the different structural levels that organize the genome in the nucleus as these structures are implicated in the

NATO ASI Series, Vol. H 105
Gene Therapy
Edited by Kleanthis G. Xanthopoulos

regulation of gene expression. Due to the complexity of the field, only selected topics will be discussed. As the number of original publications is extremely large, most citations throughout this article will be reviews.

1. Genomic DNA compaction levels

Each human somatic cell contains about 2 meters of DNA which has to be packaged into nuclei of the order of 10 micron in diameter. This packaging involves different levels of compaction as depicted in the figure below. The genomic DNA is first wrapped into nucleosomes by the histone proteins. These nucleosomes are made out of an octamer of histone molecules stabilized by the DNA wrapped around the histones octamer (21, 34). Nucleosomes assemble along the DNA and form a structure often called "beads on a string". This open and relatively relaxed structure is probably adopted by actively transcribed genes, those which serve as templates for RNA polymerases and various regulatory and structural molecules.

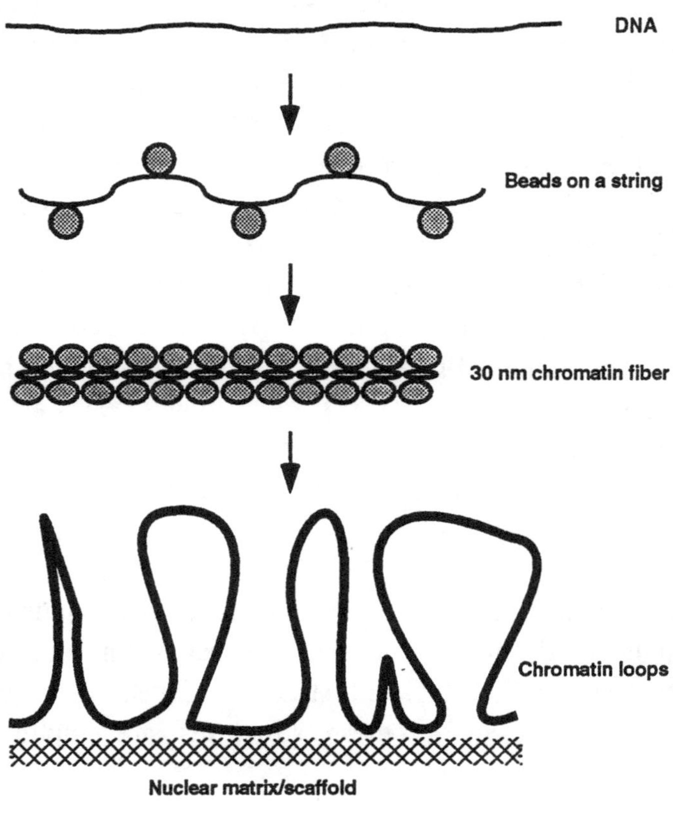

In a second level of compaction, the beads on a string fiber is wrapped into a compact solenoid. This structure is stabilized by histone H1 which is found in the axis of the fiber. This compact structure called the 30 nanometer fiber is thought to be adopted by most of the genomic DNA (30). These two levels of organization, the beads on a string and the 30-nm fiber are usually referred to as chromatin. However, chromatin can also be defined as a more complex structure which consists of the cellular structure formed by the genomic DNA and it's associated proteins and RNA. This vague definition, which does not restrict the structure just to DNA and histones, also includes the thousands of different DNA binding proteins which regulate the expression and maintenance of the genome.

The 30-nm fiber is further compacted into chromatin loops, which range from ten to a few hundred kilobases of DNA. These loops, which are most easily seen in particular forms of chromosomes such as polytene or lampbrush chromosomes, are believed to be stabilized by a structure called nuclear matrix or nuclear scaffold. This last structure is still poorly characterized, but by organizing chromatin into loops, it has been proposed to create chromatin domains. These domains could be units of regulation as will be discussed below. Finally, the chromatin loops are themselves organized into solenoids which form the chromatids of metaphase chromosomes (24).

2. Chromatin domains

Evidence the higher order structures such as chromatin loops and nuclear matrix are essentially obtained by microscopy and cell fractionation. These are difficult approaches, subject to many artifacts and to the natural mefiance of investigators. There is a methodological gap between these structural studies and the results obtained with molecular genetics. However, in the last years some functional assays have been devised to test the existence of chromatin domains.

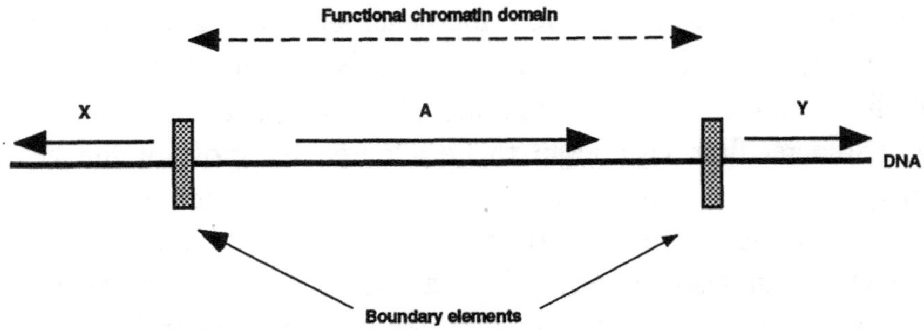

A chromatin domain can be defined as a stretch of the genome where the expression of a gene is not subject to the influence of neighboring sequences. In the above drawing, the expression of gene A is not influenced by the nature of the adjacent sequences or by the expression of genes X and Y. This is possible only if some type of DNA sequences plays the role of boundaries, or insulators; these sequences insulate a domain from its neighbor sequences.

The effects of insulators on the expression of a transferred gene could be expressed as follow:

- The expression level of the introduced gene should be independent of site of integration
- The expression level should be directly proportional to the number of integrated copies
- If all control elements are included, the expression level per copy should be similar to the expression of the native allele
- The expression should be stable and inheritable

It is clear from these properties that inclusion of insulators in constructs that are to be transferred into cells for gene therapy should result in a precisely controlled and stable expression of the gene.

In the last years, a large number of studies characterized various sequences which could present insulator functions (1, 6, 9, 14). Sequences such as locus control regions (LCR), specialized chromatin structure (SCS) and matrix attachment regions (MAR) have been shown to play a role in creating chromatin domains in various ways. LCRs have been described about 10 years ago and were shown to provide insulator effect. However, the few LCRs characterized up to now are tissue-specific and can provide insulating function only in some cell types. In contrast, *Drosophila* SCSs present insulator effects in all tissues. They do not seem necessary for the regulation of promoters except for providing insulation. MARs or SARs present different functions. Some seem to provide insulation when others only increase transcription without providing insulator function. MARs effects do not appear to be tissue specific. Recent data suggests that MARs may be the entry point on DNA to either activating proteins, or repressing factors such as histone H1 (7, 11, 38).

From a gene therapy point of view, there is still no ubiquitous mammalian insulator such as *Drosophila* SCS elements. At this time it seems necessary to characterize each gene's proper LCR, if it has any, in order to obtain vectors which could express a gene independently of the site of integration in the appropriate tissue.

3. Sensitivity of chromatin to nucleases

Boundaries/insulators provide a functional definition of chromatin domains. However, there is some structural evidence, in addition to the previously mentioned microscopy and cell fractionation studies, that suggest the existence of chromosomal domains. Active and inactive chromatin have been shown to present different sensitivities to nucleases.

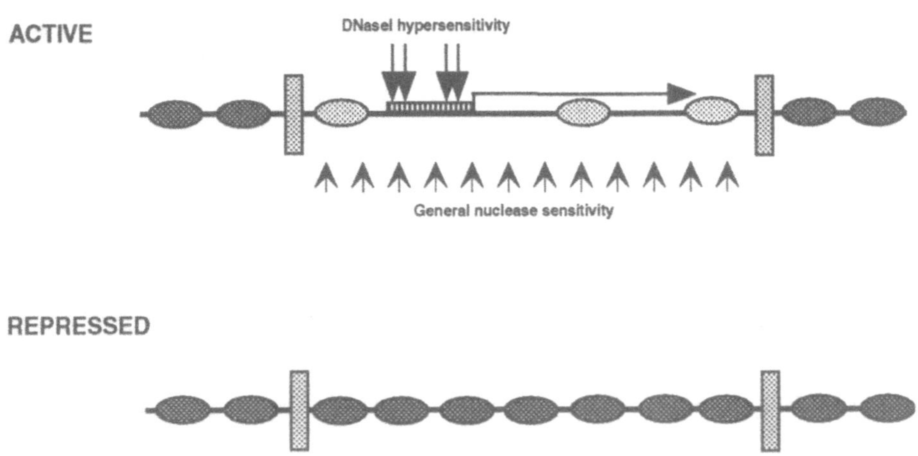

Repressed chromatin can be viewed as a compact structure covered with nucleosomes, protected from the action of nucleases added to isolated chromatin. However, Weintraub and Groudine (35) showed that active genes reside into domains which are accessible to nucleases which present a general sensitivity to nucleases. This sensitivity is another way to define a chromatin domain, the portion of DNA accessible to nucleases in an actively transcribing tissue.

In contrast to this general sensitivity to nucleases, there are precise regions ranging from 100 to 500 bp which present a hypersensitivity to DNaseI. Although DNaseI produces single-stranded breaks in DNA, in these DNaseI hypersensitive sites (DHS) the density of cleavage is such that they result into double-strand breaks, and are physically mapped as such. Typically, promoters, enhancers and LCRs are DHS in active tissues (10). The mapping of tissue-specific DNaseI hypersensitive sites is a convenient way to define candidate sequences which regulate a gene (26).

4. Nucleosomes as transcriptional repressors

If the description of general DNase hypersensitivity as well as the presence of DHSs suggest a precise organization of active chromatin structures, it is only recently that the molecular mechanisms that control the assembly of such actively transcribing regions are coming to light. Nucleosomes make about two thirds of the chromatin proteins. The role of this major constituent of chromatin during the assembly and maintenance of structures directing active transcription has been a major field of investigation in recent years.

In vitro assays of transcription on nucleosomal templates exemplify the problems the cell has to solve (17, 18). When nucleosomes are assembled on a plasmid, they form a structure which can not be transcribed as nucleosomes prevent the assembly of transcription complexes on the template. If the plasmid is however first preincubated with general transcription factors, of the resulting nucleosomal template can be transcribed as nucleosomes in this case do not assemble on the critical DNA elements. This *in vitro* repression of transcription by chromatin illustrates how a gene can be more stringently regulated than appearing from *in vitro* transcription assays in the absence of nucleosomes. In these *in vitro* transcription reactions, the leaky basal transcription is abolished by the presence of nucleosomes. However, as the presence of nucleosomes along the transcribed template does not impede transcription in a dramatic way, templates which were preincubated with general factors are still efficiently transcribed.

Therefore nucleosomes have been proposed to participate in transcription repression by assembling on transcriptional regulatory elements to stably shut down the basal transcription of a gene.

One example of transcription repression mediated by nucleosomes *in vivo* may occur over the human Alu repetitive sequences. These 285 bp sequences form about 5 % of the genome in primates. These elements can be very efficiently transcribed by RNA polymerase III *in vitro*, and therefore could titrate most of the RNA polymerase III in the cell. However, most of the Alu sequences, even those found in open chromatin regions, are not transcribed. Recent work showed that a positioned nucleosome may be responsible for Alu repression (4). The Alu sequence is sufficient for precise nucleosomal positioning and these nucleosomes adopt a specific position in a way that makes a regulatory element and start site inaccessible to transcription factors, resulting in repression of transcription. A naturally occurring mutant, found at the neurofibromatosis locus, contains a $T_{14}A_{11}$ insertion. This short sequence creates a bend in the DNA which hinders nucleosome assembly, and probably results in transcription of the Alu repeat as the RNA polymerase III control elements now are accessible. The precise mechanism by which insertion of that Alu repeat in the NF1 locus shuts down the expression of the tumor-suppressor gene is unclear (33), but may possibly result from RNA polymerase III antisense transcription driven by the Alu element.

A well described system where nucleosomes are present over control elements in the repressed gene but leave the template after activation is found at the *PHO5* locus of *S. cerevisiae* (27). *PHO5* encodes yeast major acid phosphatase. In the presence of phosphate the gene is transcriptionally inactive and the promoter is covered by precisely positioned nucleosomes. Upon phosphate starvation the gene is activated, transcription is induced and both binding sites for the positive PHO4 transcription factor are accessible and bind the PHO4 protein. In the active state, four nucleosomes seem to have abandoned the promoter. How does this chromatin remodeling occur? Does nucleosome release precede transcription activation, or does transcription activation result in nucleosome displacement? Phosphate starvation results in increased PHO4 binding activity in cells. However, PHO4 seems to be able to bind the DNA even when it is assembled into a nucleosome. It seems then that nucleosome disruption is the consequence of PHO4 activated transcription and not a prerequisite for it. Recent work has provided evidence for that mechanism (8). Recruitment of the RNA polymerase II holoenzyme to the promoter is sufficient to remodel chromatin at the PH05 locus. The RNA polymerase II holoenzyme is a large complex which consists of RNA polymerase associated with many basal transcription factors and coactivators. Some of these coactivators exhibit chromatin remodeling activities as further described below.

5. Nucleosomes as transcriptional activators

If nucleosomes participate in transcription repression, in some instances the transcription machinery seems to have taken advantage of the structural properties of nucleosomes to stabilize the interactions between activating factors. When DNA is wrapped around a histone octamer, it brings distant sequences together in two different ways. First, as DNA "enters" and "leaves" the nucleosome on the same side of the nucleosome, sequences separated by more than 146 bp, the length of DNA wrapped around the octamer, can be brought close to each other. A second, and a less restrained mode to bring sequences together, is present at the surface of the nucleosomes where sequences which are normally separated by 73 base pairs are found spaced by about 2.8 nanometer. Both ways to bring closer DNA sequences seem to participate in transcription regulation as described below.

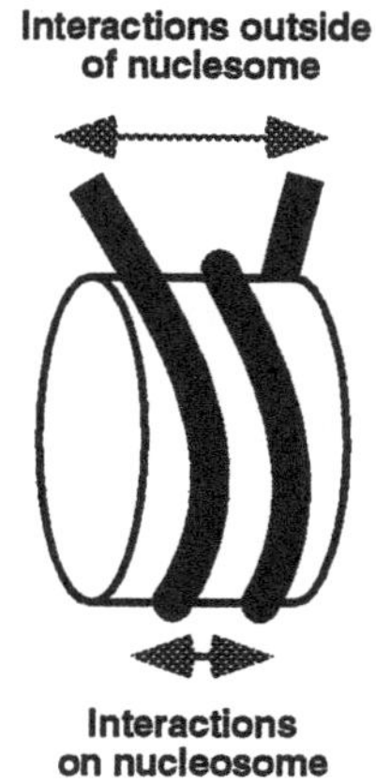

The laboratories of Sarah Elgin (13, 28) and Peter Becker (22) have studied the nucleosomal organization of the *hsp26* and *hsp27* genes, respectively, of *D. melanogaster*. Both genes have a similar regulation by heat shock, and their control elements are made of two sets of DNA binding sites separated by a spacer DNA. In both cases, this spacer has been shown to be assembled into a nucleosome. This situation is represented on the figure below either in a usual view when DNA is represented as a line, or as a "wrapped" view where the length of DNA wrapped around a nucleosome is represented by a circle which circumference represents 146 bp. As the DNA "enter" and "leaves" the nucleosome on the same side, the two sets of transcription factor binding sites may be brought together and could synergize transcriptional induction by heat shock.

LINEAR VIEW

WRAPPED VIEW

This nice model has not yet been strictly demonstrated *in vivo*. However, *in vitro* studies on a similar system conducted in collaboration by Walter Wahli and Alan Wolffe on the *Xenopus* vitellogenin gene showed that the presence of a nucleosome between transcription factor binding sites resulted in a synergistic activation in the presence of these factors. In this case, distal estrogen response elements are brought close to NF-1 binding sites (25).

Regulation of the mouse mammary tumor virus (MMTV) promoter also seems to involve the interaction of two transcription factors brought close together by a nucleosome. In this case however, the binding sites for the transcription factor NF1 and the glucocorticoid receptor are found on the surface of the nucleosome (3). This proximity of the two binding sites results in synergistic activation of transcription.

6. Histone acetylation

It is clear from the above examples that the cell has organized it's transcription apparatus so to work with chromatin. The molecular mechanisms that result in chromatin remodeling have provided some of the major new breakthroughs in the field. Two aspects have attracted much interest lately: The role of histone acetylation on one hand, and the identification of ATP-dependent chromatin remodeling activities as described in the next section.

One structure that turns out to be associated with transcriptionally active chromatin is histone acetylation (23, 32, 37). . It has been observed for many years that histones can be acetylated at their N-termini. Core histones have their N-termini on the outside of the nucleosome. The termini are lysine-rich, and these positive charges may interact with the negatively charged DNA. Acetylation is believed to loosen the nucleosome structure by neutralizing these positive charges, making it easier for transcription and replication to displace nucleosomes. The precise structural transitions are still to be described, but accessibility studies point towards a more unstable nucleosome complex. Histones are first acetylated after their synthesis in the cytoplasm, and this modification is necessary for the assembly of chromatin during replication. Once integrated into chromatin, most histones are deacetylated and remain so in repressed chromatin. However, active genes seem to be enriched in acetylated histones

Histone acetylation is mediated by two counteracting types of enzymes, histone acetyltransferases (HAT) and histone deacetylases (HDAC). HATs responsible for histone acetylation in the cytoplasm and import into the nucleus have been cloned recently and have a defined protein domain which turns out to be present on some nuclear proteins. Interestingly, some of these proteins are activators of transcription. Examples of these transcription regulators that show HAT activity are GCN5 (23), P/CAF (16) and TAFII250 (15).

The yeast GCN5 mediates activation of GAL4-VP16 in yeast while P/CAF is involved in cAMP activation by the CREB protein. In this way, acetylation of nucleosomes can be restricted to precise genes. However, a more general transcription factor such as TAFII250, a TBP binding protein, also presents HAT activity, suggesting that the transcription apparatus mobilizes its own tool for histone acetylation. Still many questions remain on how these regulators assist transcription by acetylating histones. Transcription activators are supposedly restricted to the promoter region. Do they acetylate only a few critical nucleosomes at the promoter, or is there a spreading effect, mediated possibly by other HATs?

Conversely, if histone acetylation is associated with transcription activation, histone deacetylation is associated with transcription repression (20, 36). Recently a histone deacetylase activity has been purified and cloned. This has identified the HDAC domain which was found in another protein, HDAC2. These histone deacetylases have been shown to associate with a number of repression processes. Repression through the Mad:Max complex or unliganded nuclear receptors implicate HDAC, which seems to be associated through docking proteins such as mSin3 or corepressors such as N-CoR. The YY1 protein, a DNA-binding protein which bends DNA and can either activate or repress transcription depending of the context, tightly binds HDAC2. The RbAp48 protein was isolated as a HDAC1 and Rb-binding protein, and was shown to be part of the chromatin assembly complex (CAC) which participates in the assembly of chromatin in a replication-dependent manner. As mentioned above, histones acetylated after synthesis in the cytoplasm are assembled into chromatin during replication. The HDAC activity found in CAC may be responsible for stabilizing the nucleosomes after assembly into replicated chromatin.

These discoveries show that control of histone acetylation is tightly coupled to transcription. Although the precise role in this nucleosome destabilization remains to be understood, on the overall acetylation correlates with expression and deacetylation with repression. However, the precise residues which are acetylated on histones may play a critical role. Similarly to phosphorylation, the balance of these activities at specific residues may control the regulation of specific genes by different mechanisms.

7. Chromatin remodeling activities

Nucleosomes are very stable structures which participate in stable repression of transcription. In order to activate transcription of specific genes, for example during tissue differentiation, nucleosomes have to be displaced from key regulatory sequences or positioned at places where they would not interfere with the buildup of the RNA polymerase complexes. One major opportunity to redistribute nucleosomes, either for transcription activation or repression, is DNA replication. However, a large number of genes, such as the yeast PH05 and mouse MMTV LTR can be activated without DNA replication and present a redistribution of nucleosomes. If acetylation may decrease nucleosome stability, specific activities that require energy to displace nucleosomes have been recently identified (5, 12, 19).

Genetic approaches in *S. cerevisiae* have identified the SWI/SNF complex which alters chromatin structure as the first member of a new class of ATP-dependent nucleosome destabilizers. This huge complex of about 2 MDa has a mammalian homologue. By copurification the SWI/SNF complex has been proposed to be part of the RNA polymerase II holoenzyme as further suggested by the finding that mobilization of the holoenzyme is sufficient to alter nucleosomes at the PH05 locus (8). The study of the 11 subunits of this complex identified related proteins from different organisms involved in transcription regulation. For example, the yeast SWI2/SNF2 component of the complex has two human homologues, Brg1 and hBrm. Brg1 contains a DNA helicase domain and is a member of the trxG-like proteins homologues of the *Drosophila* homeotic regulators. This family of proteins assist in the stable expression of homeotic genes during *Drosophila* development.

Another DNA-stimulated ATPase with nucleosome destabilizing properties has been isolated from *S. cervisiae* on the basis of it's homology to the SWI/SNF complex (2). This activity called RSC (remodel chromatin structure) is about 1 MDa and made of 15 subunits, some which share similarity to SWI/SNF subunits. RSC is about 10 times more abundant than SW/SNF and may therefore play a wider role.

Chromatin remodeling activities have also been identified in *Drosophila*. The nucleosome remodeling factor (NURF) was found as an activity required to alter nucleosomes after binding of the GAGA transcription factor (29). In contrast to SWI/SNF and RSC, the ATPase activity of NURF is activated by complete nucleosomes and not naked DNA. NURF is a 500 KDa complex of four polypeptides. The ISWI subunit of NURF is related to the SWI2/SNF2 subunit of the SWI/SNF complex. Recently another activity called CHRAC (chromatin accessibility complex) has been isolated from *Drosophila* embryo extracts (31). Interestingly, the five subunits of CHRAC include the ISWI subunit of NURF as well as topoisomerase II. In addition to a nucleosome destabilizing activity, CHRAC is able to regularly space nucleosomes along a DNA template.

The exact structural changes that occur on a nucleosome under the action of chromatin remodeling factors are still unknown. One can expect that different complexes act in various ways on the nucleosome. Possibly some activities are required for specific genes, or for remodeling during replication or transcription elongation. The complex subunit structure of these activities suggest that they interact with the multiple structures involved in gene expression in the nucleus.

8. Conclusions

Transcription takes place in a complex environment, the cell nucleus, which is organized into many structural levels. Recent evidence points to the participation of these structures in the control of transcription. Higher-order structures define chromatin domains which are responsible for stable and precise control of expression. Various activities which modulate the properties of the basic element of chromatin, the nucleosome, have recently been identified and implicated in the dynamic control of transcription. The cell has devised it's transcription machinery to work on a complex template, chromatin. At the moment just about every week brings its new share of exciting discoveries. Some of the major findings, not discussed in this review, concern the control of epigenetic traits during development by chromatin. A future challenge will be to make these findings useful for gene therapy.

9. Acknowledgments

I wish to thank Melanie Price and Hubert Renauld for critical reading of the manuscript. Research in my lab is supported by grants from the Swiss Federal Office for Public Health, the Swiss National Foundation for Scientific Research and the Swiss Research against Cancer.

10. References

1. Bonifer, C., M. C. Huber, U. Jagle, N. Faust, and A. E. Sippel. 1996. Prerequisites for tissue specific and position independent expression of a gene locus in transgenic mice. Journal of Molecular Medicine 74:663-71.
2. Cairns, B. R., Y. Lorch, Y. Li, M. C. Zhang, L. Lacomis, H. Erdjumentbromage, P. Tempst, J. Du, B. Laurent, and R. D. Kornberg. 1996. RSC, an essential, abundant chromatin-remodeling complex. Cell 87:1249-1260.
3. Chavez, S., and M. Beato. 1997. Nucleosome-mediated synergism between transcription factors on the mouse mammary tumor virus promoter. Proceedings of the National Academy of Sciences of the United States of America 94:2885-90.
4. Englander, E. W., and B. H. Howard. 1996. A naturally occurring $T_{14}A_{11}$ tract blocks nucleosome formation over the human neurofibromatosis type 1 (NF1)-Alu element. The journal of Biological Chemistry 271:5819-5823.
5. Felsenfeld, G. 1996. Chromatin unfolds. Cell 86:13-19.
6. Felsenfeld, G., J. Boyes, J. Chung, D. Clark, and V. Studitsky. 1996. Chromatin structure and gene expression. Proceedings of the National Academy of Sciences of the United States of America (USA) 93:9384938-8.
7. Festenstein, R., M. Tolaini, P. Corbella, C. Mamalaki, J. Parrington, M. Fox, A. Miliou, M. Jones, and D. Kioussis. 1996. Locus control region function and heterochromatin-induced position effect variegation. Science 271:1123-5.
8. Gaudreau, L., A. Schmid, D. Blaschke, M. Ptashne, and W. Horz. 1997. RNA polymerase II holoenzyme recruitment is sufficient to remodel chromatin at the yeast PHO5 promoter. Cell 89:55-62.
9. Gerasimova, T. I., and V. G. Corces. 1996. Boundary and insulator elements in chromosomes. Current Opinion in Genetics & Development 6:185-92.
10. Gross, D. S., and W. T. Garrard. 1988. Nuclease hypersensitive sites in chromatin. Annual Review of Biochemistry 57:159-197.

11. Jenuwein, T., W. C. Forrester, H. L. Fernandez, G. Laible, M. Dull, and R. Grosschedl. 1997. Extension of chromatin accessibility by nuclear matrix attachment regions. Nature 385:269-72.
12. Kingston, R. E., C. A. Bunker, and A. N. Imbalzano. 1996. Repression and activation by multiprotein complexes that alter chromatin structure. Genes & Development 10:905-920.
13. Lu, Q., L. L. Wallrath, and S. C. Elgin. 1995. The role of a positioned nucleosome at the Drosophila melanogaster hsp26 promoter. EMBO Journal 14:4738-46.
14. Milot, E., P. Fraser, and F. Grosveld. 1996. Position effects and genetic disease. Trends in Genetics 12:123-6.
15. Mizzen, C. A., X. J. Yang, T. Kokubo, J. E. Brownell, A. J. Bannister, T. Owenhughes, J. Workman, L. Wang, S. L. Berger, T. Kouzarides, Y. Nakatani, and C. D. Allis. 1996. The TAF(II)250 subunit of TFIID has histone acetyltransferase activity. Cell 87:1261-1270.
16. Ogryzko, V. V., R. L. Schiltz, V. Russanova, B. H. Howard, and Y. Nakatani. 1996. The transcriptional coactivators p300 and CBP are histone acetyltransferases. Cell 87:953-959.
17. Owen-Highes, T. A., and J. L. Workman. 1994. Experimental analysis of chromatin function in transcription control. Critical Reviews in Eukaryotic Gene Expression 4:403-441.
18. Paranjape, S. M., R. T. Kamakaka, and J. T. Kadonaga. 1994. Role of chromatin structure in the regulation of transcription by RNA polymerase II. Annual Review of Biochemistry 63:265-97.
19. Pazin, M. J., and J. T. Kadonaga. 1997. SWI2/SNF2 and related proteins: ATP-driven motors that disrupt protein-DNA interactions?. [Review] [20 refs]. Cell 88:737-40.
20. Pazin, M. J., and J. T. Kadonaga. 1997. What's up and down with histone deacetylation and transcription? Cell 89:325-8.
21. Pruss, D., B. Bartholomew, J. Persinger, J. Hayes, G. Arents, E. N. Moudrianakis, and A. P. Wolffe. 1996. An asymmetric model for the nucleosome: a binding site for linker histones inside the DNA gyres. Science 274:614-7.
22. Quivy, J. P., and P. B. Becker. 1996. The architecture of the heat-inducible Drosophila hsp27 promoter in nuclei. Journal of Molecular Biology 256:249-63.
23. Roth, S. Y., and C. D. Allis. 1996. The subunit-exchange model of histone acetylation. Trends in Cell Biology 6:371-375.
24. Saitoh, Y., and U. K. Laemmli. 1994. Metaphase chromosome structure: bands arise from a differential folding path of the highly AT-rich scaffold. Cell 76:609-22.
25. Schild, C., F. X. Claret, W. Wahli, and A. P. Wolffe. 1993. A nucleosome-dependent static loop potentiates estrogen-regulated

transcription from the Xenopus vitellogenin B1 promoter in vitro. EMBO Journal 12:423-33.
26. Sippel, A. E., H. Saueressig, M. C. Huber, H. C. Hoefer, A. Stief, U. Borgmeyer, and C. Bonifer. 1996. Identification of cis-acting elements as DNase I hypersensitive sites in lysozyme gene chromatin. Methods in Enzymology 274:233-46.
27. Svaren, J., and W. Horz. 1997. Transcription factors vs nucleosomes: regulation of the PHO5 promoter in yeast. Trends in Biochemical Sciences 22:93-7.
28. Thomas, G. H., and S. C. Elgin. 1988. Protein/DNA architecture of the DNase I hypersensitive region of the *Drosophila hsp26* promoter. EMBO Journal 7:2191-2201.
29. Tsukiyama, T., and C. Wu. 1995. Purification and properties of an ATP-dependent nucleosome remodeling factor. Cell 83:1011-20.
30. van Holde, K., and J. Zlatanova. 1996. What determines the folding of the chromatin fiber? Proceedings of the National Academy of Sciences of the United States of America 93:10548-55.
31. Vargaweisz, P. D., M. Wilm, E. Bonte, K. Dumas, M. Mann, and P. B. Becker. 1997. Chromatin-remodelling factor CHRAC contains the ATPases ISWI and topoisomerase II. Nature 388:598-602.
32. Wade, P. A., D. Pruss, and A. P. Wolffe. 1997. Histone acetylation: chromatin in action. Trends in Biochemical Sciences 22:128-32.
33. Wallace, M. R., L. B. Andersen, A. M. Saulino, P. E. Gregory, T. W. Glover, and F. S. Collins. 1991. A de novo Alu insertion results in neurofibromatosis type 1. Nature 353:864-6.
34. Wang, B. C., J. Rose, G. Arents, and E. N. Moudrianakis. 1994. The octameric histone core of the nucleosome. Structural issues resolved. Journal of Molecular Biology 236:179-88.
35. Weintraub, H., and M. Groudine. 1976. Chromosomal subunits in active genes have an altered conformation. Science 193:848-56.
36. Wolffe, A. P. 1996. Histone deacetylase: a regulator of transcription. Science 272:371-2.
37. Wolffe, A. P., and D. Pruss. 1996. Targeting chromatin disruption: Transcription regulators that acetylate histones. Cell 84:817-9.
38. Zhao, K., E. Kas, E. Gonzalez, and U. K. Laemmli. 1993. SAR-dependent mobilization of histone H1 by HMG-I/Y in vitro: HMG-I/Y is enriched in H1-depleted chromatin. EMBO Journal 12:3237-47.

Transcriptional Regulation and Gene Expression in the Liver

Minoru Tomizawa, Julie Lekstrom-Himes, and Kleanthis G. Xanthopoulos

Clinical Gene Therapy Branch, National Human Genome Research Institute, National Institutes of Health, Bethesda, MD 20892, USA

Keywords: transcription, gene expression, targeted mutations

Abstract

Temporal and spatial regulation of gene expression is essential for the evolution of multicellular organisms. Eukaroytic cells regulate gene expression at the transcriptional, post transcriptional and translational level. Transcriptional mechanisms that control differential expression of RNA polymerase II genes include modulation of the stability and speed of assembly of the transcriptional apparatus via general and tissue-enriched transcription factors, transcriptional pausing, and alternative mRNA splicing and stabilization. Furthermore, leaky ribosomal scanning, post-translational modifications, and locus control regions also contribute to the regulation of gene expression following the generation of mRNA. The CCAAT/enhancer binding proteins (C/EBPs) family of transcription factors encompass a variety of these regulatory methods and will be discussed as a model for differential gene activation that is critical for normal cellular differentiation and function in a variety of tissues. The prototypic C/EBP is a modular protein, consisting of an activation domain, a dimerization bZIP region and a DNA binding domain. All family members share the highly conserved dimerization domain, required for DNA binding, by which they form homo- and heterodimerizes with other family members. C/EBPs are least conserved in the activation domains that vary from strong activators to dominant negative repressors. The pleiotropic effects of C/EBPs are in part due to tissue-specific and stage-specific expression. Dimerization of different C/EBP proteins precisely modulates transcriptional activity of target genes. Recent work with mice deficient in specific C/EBPs underscores the effects of these factors in tissue development, function, and response to injury.

NATO ASI Series, Vol. H 105
Gene Therapy
Edited by Kleanthis G. Xanthopoulos

Introduction

The processes of cellular differentiation and proliferation involve many molecular regulatory events. Control of gene expression is often regulated at the transcriptional level by a variety of mechanisms, including modulation of the assembly of the transcriptional machinery, use of multiple promoters, stability of the transcript, and alternative mRNA splicing. Transcription factors are a heterogeneous group of proteins that share the capacity to interact with the transcriptional machinery and effect the rate at which a given gene is transcribed. Many transcription factors bind DNA directly, either near the promoter elements of a gene or distant to these elements, binding enhancer or silencer sites. Cofactors are sometimes necessary, to facilitate DNA binding or alter the tertiary structure of the protein, revealing critical regulatory domains. Protein dimerization is another mechanism employed by transcription factors; the versatility of transcriptional regulation is increased by modulating transcriptional activity in response to the levels of other transcription factors present in the cell. Protein dimerization also allows for interactions between factors from different families of transcription factors promoting regulatory "crosstalk."

The liver is an ideal model to study the role of transcriptional regulation in directing proliferation and differentiation. Gene expression in hepatocytes is most commonly regulated at the transcriptional level. Additionally, the liver is capable of regeneration in a well-organized manner following partial hepatectomy, allowing careful study in non-embryonic tissues (1).

In this chapter, we will first review general aspects of gene control and the machanisms of action of transcription factors in regard to signal transduction and initiation of transcription and then we will review liver specific gene expression.

Gene Expression

Regulated gene expression comprises several steps: extracellular signaling, intracellular signal transduction, and gene transcription (Figure 1). Untransformed, eukaryotic cells are responsive to changes in their extracellular environment such as the local concentration of hormones (2, 3, 4, 5) and growth factors, and cell-cell & cell-matrix interactions (Fig. 1). Extracellular factors modulate the

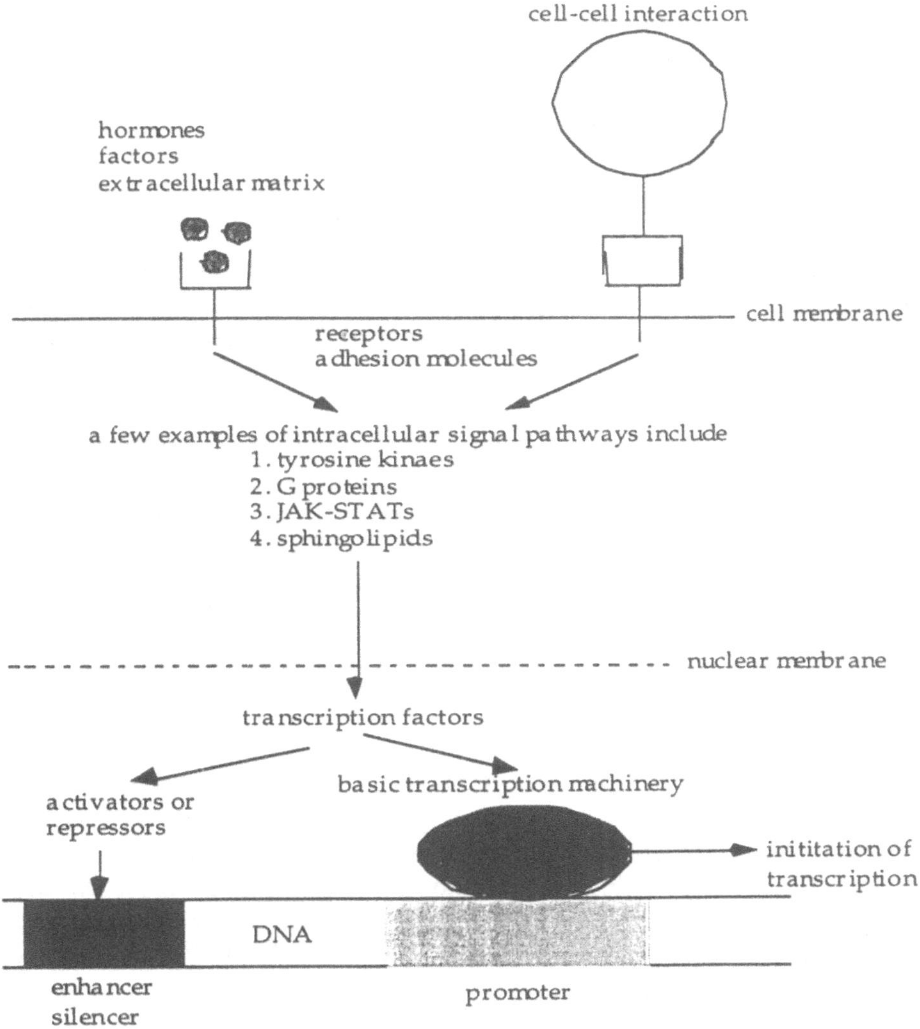

Figure 1. Multiple signals may bind to receptors or adhesion molecules, followed by intracellular signal pathways. These signals are often transmitted by transcription factors. Transcription factors act as activators or repressors binding to regulatory elements that are located either upstream or downstream of a promoter region.

activities of membrane-bound or intracellular receptors (6). For example, steroid hormones interact with cytoplasmic glucocorticoid receptors. Signal transduction occurs via the interaction of the steroid hormone with its receptor, facilitating dimerization of these receptors and subsequent DNA binding. Other extracellular signals such as cytokine proteins are transduced through various cytoplasmic signaling cascades including tyrosine kinase, G protein, JAK-STAT, and sphingolipid pathways (6, 7). Finally, signals are transmitted to a nucleus as transcription factors, whose levels are, in turn, regulated by a variety of mechanisms including differential promoters, alternative splicing, transcriptional stability and post translational modifications.

Gene expression is regulated by a variety of mechanisms and at multiple levels, such as genomic imprinting, locus control regions (LCRs), and anti-repressional activities. Genomic imprinting, by which paternity of the allele determines its availability, is probably mediated by DNA methylation (8). Chromatin structure modified by LCRs and interactions with histones and other DNA binding proteins also regulates the access of transcriptions factors to DNA. Finally, repressor transcription factors which interfere with DNA binding directly, or indirectly by altering protein-DNA binding domains regulate access of the transcriptional machinery. Once a gene is made accessible the level of transcriptional activity is greatly affected by sequences termed enhancers, silencers, and promoters (9). Enhancers contain DNA binding sites for proteins that are able to activate transcription of a gene (Fig. 1). Enhancers and silencers are cis-acting regions that are located either upstream or downstream of the gene and are able to exert their influence from a distance (10, 11, 12). Promoter sequences are always located at the 5′ end of the gene, directly adjacent to the protein-coding region that is transcribed into mRNA, and typically extend 150-300 nucleotides upstream of the transcriptional start site (Fig. 1). Chromatin structure also modulates promoter and enhancer function (13, 14, 15). Additionally, not all promoters or enhancer-binding proteins are transcriptional activators (16). For example the Sp3 factor, related to Sp1, occupies the same site as Sp1, however lacks transactivating function, effectively inhibiting Sp1 activity. Also, other proteins known as mediators or adapters may be needed to form bridges between a transcription factor and a component in the basal transcription machinery (17, 18).

Commonly, transcription factors binding to the promoter sequences facilitate the assembly of the RNA polymerase complex. Once a functional transcription complex has been formed, the RNA polymerase ll will transcribe the gene, making a heterogeneous nuclear RNA (hnRNA) (19). The transcript is capped in the 5′ end with a methyl-guanidine, cleaved at a specific 3′ site and then polyadenylated. Most eukaryotic genes contain intronic sequences and are subject to splicing, resulting in heterogeneous mRNAs from the same gene. Many genes, including transcription factor C/EBPε, are alternatively spliced in a temporal, tissue specific or developmental fashion (20). Once the RNA has been spliced, it is transported out of the nucleus accompanied by RNA-binding proteins (21).

Regulation of gene expression at the translational level occurs by several mechanisms. Repressor proteins binding to the 5′ end of the mRNA can block translation. Also, leaky ribosomal scanning, can result in translation of multiple proteins from a single RNA transcript. Usually the ribosome begins translation of the mRNA at the first initiation AUG codon, however, in some cases an upstream AUG might be neglected and the translation starts at a downstream AUG (22). Certain transcription factors utilitize leaky ribosomal scanning, generating different protein isoforms with variable transactivating capacities. Additional regulation of mRNA stability, localization, and translation allows cellular responses to environmental changes to occur more quickly than de novo (23, 24, 25). After translation the protein product may be subject to post translational modifications such as glycosylation, myristilation, and phosphorylation. For nuclear proteins, translocation to effector sites within the nucleus is regulated by phosphorylation, interaction with cytoplasmic proteins, and modulation of the import machinery associated with the nuclear envelope (26). Finally, regulation of protein stability depends on such factors as proteases and the accessibility of cleavage sites on target proteins.

Cell Specific Gene Expression

The existance of multicellular organisms drives the complexity of cell-specific gene expression and the variety of gene regulatory mechanisms. Many of these mechanisms operate at the level of transcriptional initiation, and are well modeled in the regenerating liver (27, 28). The set of factors binding to a gene promoter differs between genes and also in a temporal manner for the same gene, allowing changes in expression levels over time. Importantly, the specific site of transcription factors binding to the promoter can determine the spatial expression pattern, thus being the basis for tissue specific gene expression at the transcriptional initiation level. It is fundamentally important to learn more about how transcription factors act and how they are regulated in order to understand the basis of these regulatory mechanisms.

Transcription Factors

Transcription factors comprise a functionally and structurally diverse group of nuclear proteins. They share a modular structure, with motifs specific for DNA binding and domains responsible for transactivating or repressing abilities. Most transcription factors are classified according to their protein structural motifs used for DNA recognition (29). The three dimentional structures have been determined for a number of DNA-binding proteins. Based on these structures, transcription factors are ordered into groups of closely related proteins. Large important groups of transcription factors in higher eukaryotes are the homeodomain, zinc finger, basic helix-loop-helix and the leucine zipper proteins.

Homeodomain proteins bind as monomers via a sequence-specific motif called the homeodomain. The homeodomain is a folded structure including an α-helix, a β--turn and a second α-helix forming a helix-turn-helix motif. The helix-turn-helix motif in the homeobox protein facilitates DNA binding of the protein, controlling the exact DNA sequence that is recognized.

DNA binding motifs of zinc finger transcription factor proteins utilize finger-like structures by coordinate binding of zinc atoms via cysteine and histidine residues. A phenylalanine or tyrosine residue and a leucine residue occur at nearly constant positions in the loops, and are also required for DNA binding.

The basic helix-loop-helix (bHLH) transcription factors contact DNA via a basic region. Adjecent to the DNA binding domain is the dimerization motif, a prerequisite for efficient DNA-binding. This domain consists of two α-helices with an intervening loop sequence.

The basic leucine zipper (bZIP) proteins are structurally similar to those in the bHLH group. DNA-binding is mediated via a basic region and dimerization via a leucine rich region (leucine zipper) close to the basic sequence (Fig. 2). Dimerization occurs by hydrophobic interactions between the leucine residues on two molecules, forming a molecular zipper. The leucines are repeated every seventh residue in the zipper region, aligning them on a common face of the helix and facilitating interactions with the leucines of the dimerization partner. Several important transcription factor families belong to the bZIP group. The first protein of this group to be studied extensively was isolated from rat liver cell extracts. This protein was called C/EBP because of its ability to act as a viral enhancer binding protein and because it was also thought initially bind to CCAAT boxes. Several C/EBP proteins have been discovered. The original C/EBP protein is now called C/EBPα.

Regulation of Transcription Factors

Similarly to the regulation of other proteins, transcription factors are regulated in several ways: regulation of synthesis and regulation of activity.

Transcription factor expression is regulated so that specific genes become activated only at the appropriate time and place, particularly in cell type-specific or developmentally regulated genes. This cell type-specific expression of the transcription factor mRNA may result from regulated transcription of the gene encoding the transcription factor (30, 31). Differential mRNA splicing yields varying mRNA isoforms encoding different proteins. Other cell-specific regulation occurs at the translational level.

Preformed transcription factors which exist in an inactive form prior to activation permit rapid induction of activity (32). Many transcription factors are modified extensively following translation by adding 0-linked monosaccharide residues (33) or by phosphorylation (10, 34) representing obvious targets for agents that induce gene expression.

Regulatory Molecules Involved in Hepatocyte-Specific Gene Expression

A fundamental understanding of the mechanisms of normal gene regulation in proliferating, developing and differentiated hepatocytes is necessary in order to comprehend many liver diseases. Numerous studies have identified, isolated, characterized, and cloned hepatocyte nuclear factors, now classified as HNF-1, HNF-3, and HNF-4, and C/EBP families, each comprising of several members. We will discuss these families in detail.

The HNF family

HNF-1 was originally detected by its ability to bind to the β-fibrinogen gene promoter (35), and has since been described by several groups (30, 36, 37). The regulatory regions of many genes expressed in hepatocytes contain HNF-1 DNA binding sites, including albumin, transthyretin and α1-antitrypsin (38). The consensus binding site, 5′ ATTAAC 3′, is a palindrome and it is suggested that protein dimerization is necessary for DNA binding (39).

HNF-3 was identified as a nuclear protein interacting with the promoters of the mouse transthyretin and α1-antitrypsin genes (40). The consensus HNF-3 binding site, 5′ TATTGANTTANC 3′, was used as a probe to isolate and clone members of the family, HNF-3α, HNF-3β and HNF-3γ (41). HNF-3 proteins are proline rich with no distinct DNA binding domains. Family members also show extensive homology with the protein product of the Drosophilia, forkhead gene, suggesting a broader developmental role.

HNF-4 was identified by a variety of *in vitro* (40) and *in vivo* methods with the consensus DNA binding site as 5′ GGCAAAGGTCCAT 3′. The protein was consequently purified from rat liver nuclear extracts and the corresponding cDNA was isolated (28). HNF-4 family members share homology with the steroid-thyroid receptor superfamily and dimerize via a zinc finger motif. HNF-4 is expressed in limited tissues, including liver, kidney and intestine. Evidence suggests HNF-1 and HNF-4 play a role in energy metabolism. In albino deletion mice that die after birth due to failure to induce gluconeogenic enzymes, levels of HNF-1 and HNF-4 are strongly reduced (42). In mouse embryos, HNF-1 and HNF-4 are expressed differentially according to embryonic stage (43). These factors may also play a role in liver development (44).

HNF-5 was recently idenftified using DNAase I footprinting (45) and is located in the promoter regions of human MDR3 P-glycoprotein (46) and human histidase (47).

HNF-6 was identified as a third liver enriched factor that recognizes transcriptionally important sequences in the HNF-3β (-138 to -127) bp and TTR (-94 to -106) promoter regions (48).

CCAAT/Enhancer Binding Protein (C/EBP) Family

The C/EBP family, to date, consists of six members. C/EBPα was cloned from rat (49) and mouse liver (50). C/EBPβ was originally cloned from human glioblastoma cells (termed NF-IL6) and subsequently from rat, mouse liver and mouse adipocyte cell lines (termed IL-6-DBP, LAP, AGP/EBP, C/EBP(CRP ll, and rNF IL-6: 51, 52, 53, 54, 55, 56). Four additional members of the family include C/EBPγ (57), C/EBPδ (51, 56), C/EBPε (56, 58) and C/EBPζ.

C/EBPα is a critical transcription factor in the regulation of terminal differentiation and function of the liver and adipose tissue (50, 59, 60). Full length C/EBPα possesses the DNA binding and dimerization domains and two activation domains. Truncated, 32kDa C/EBPα is generated by leaky ribosomal scanning. It possesses only one of the two activation domains, and is predictably a less potent transactivator. C/EBPα expression is very low in early embryonic tissues, however, increases with temporal maturity (61). Likewise, C/EBPα protein levels are drastically reduced in regenerating, proliferating hepatocytes and are highest in terminally differentiated hepatocytes (62). Ectopic expression of C/EBPα drives terminal differentiation of preadipose cell lines despite other signals which prevent differentiation. Thus, C/EBPα may play an important role in liver development and hepatocyte proliferation and adipose differentiation.

C/EBPα and C/EBPβ can activate the transcription of several liver-specific genes associated with the differentiated state: albumin (63), α1-antitrypsin (59) and phosphoenolpyruvate carboxykinase (PEPCK) (63, 64). C/EBPα and C/EBPβ may work differently in liver. For example, after partial hepatectomy, an inverse regulation of C/EBPα and C/EBPβ results in up to a sevenfold increase in the β/α DNA binding ratio between 3 and 24 h while total C/EBP binding activity in nuclear extracts remains relatively constant (65). As another example, C/EBPβ binds to the albumin gene enhancer in embryo liver (66).

Signal transduction and transcription factors involved in liver acute-phase response

Partial hepatectomy activates nuclear factor-κB (NF-κB) and signal transducers and activators of transcription 3 (STAT3). These factors preexist in normal liver in an inactive form, becoming activated as a part of the initial response of the remaining liver (67, 68, 69, 70).

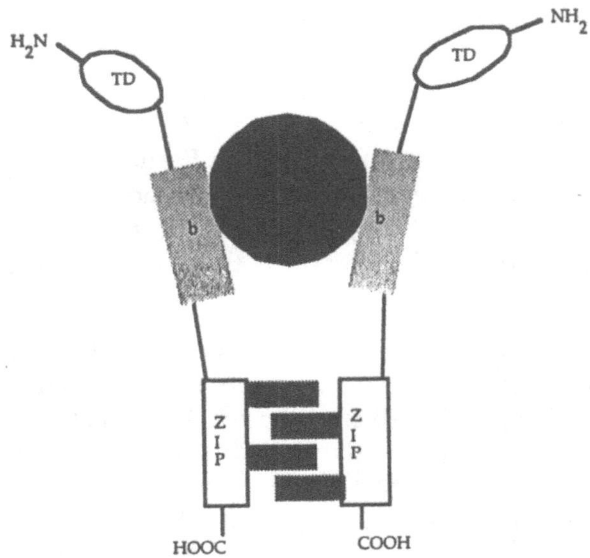

Figure 2. Schematic model of a C/EBP dimer. The two molecules are dimerized via their leucine zipper domains (ZIP) and bind DNA via the basic domains (b). The N-terminal part of the proteins contains the transcription activation domains (TD).

NF-κB is a family of homo- and heterodimeric proteins characterized by an NH2-terminal-300 amino acid Rel homology domain and proteolytic processed NF-κB1 and NF-κB2 subunits, as well as the Rel A, c-Rel, and Rel B subunits (71). Dimerization of NF-κB proteins produce complexes with a variety of intrinsic DNA-binding specificities, transactivation potentials, and subcellular localization. NF-κB is inactivated in cytoplasm by binding to IκBα. Upon stimulation of the cell, such as in partial hepatectomy, IκBα undergoes phosphorylation and rapid degradation resulting in release of the dimer and activation of NF-κB (72, 73).

Stat proteins consist of Stat1α, Stat1β, Stat2 (74), Stat3, Stat4, Stat5, and Stat6 (75, 76, 77, 78, 79, 80). They have both signal transduction domains and a DNA-binding region. Once activated by cytokines and growth factors via Stat tyrosine phosphorylation (81), Stat proteins translocate into the nucleus, where they bind to specific DNA-sequence elements and activate the transcription of target genes. Cytokines such as IL-6, EGF, TNF-α, and IL-1 induce both NF-κB and JAK-STAT pathways with activation of transcription factor complexes. Interestingly IL-6 deficient mice show impaired liver regeneration leading to liver failure (82). TNFα produces a biphasic Rel A•NF-κB1 nuclear translocation (83).

Targetd Inactivation of Representative Transcription Factor Genes in Mice

In recent years, several colonies of mice deficient in several transcription factor genes with a role in liver development have been established, showing a variety of phenotypes (Table 1). Many such mutations are embryonic lethal (85, 86, 87), suggesting that these transcription factors are important for critical developmental stages. C/EBPα deficient mice die perinatally, demonstrating increased hepatocyte proliferation and severe hypoglycemia (88, 89). C/EBPβ deficient mice develop normally however are immunocompromised and susceptible to discrete pathogens including *Listeria monocytogenes* and mycobacteria. Differentiation of myeloid cells is blocked in both C/EBPα and C/EBPε deficient mice. (90; Yamanaka et al. in press). Stat5β deficient mice lose multiple sexually differentiated responses (80). HNF-1 nullizygotic mice may be a model of human phenylketonuria and renal Fanconi syndrome (84).

Table. Knockout mice of transcription factors

Transcription Factor	Phenotype
HNF-1	hepatic dysfunction
HNF-4	abnormal gastrulation
C/EBPα	lung and liver abnormality, impairment of energy metabolism
C/EBPβ	lymphoproliferative disorder
C/EBPε	impaired granulopoiesis
Stat1	no response to either INFα or INFγ
Stat3	degeneration between 6.5 and 7.5 days of embryo
Stat5β	loss of multiple sexually differentiated responses

Conclusions

The complexity of specific temporal and spatial gene expression is evident in the heterogeneity of regulatory elements effecting these controls. Hepatocyte-specific transcription of genes is regulated, at least in part, by DNA sequences composed of multiple domains which have an additive effect on the rate of transcription of a given gene. Interacting transcription factors are in turn regulated by their own promoter elements, post-translational activation, and the formation of homo- and hetero-dimers.

Varying combinations of these transcription act differentially on a variety of genes, thus limiting the requirement for many different factors. This combinatorial effect results in an increased versatility of the transcription machinery, permitting it to operate with a limited number of different molecules.

The elements which regulate hepatocyte differentiation and proliferation are being elucidated, both from experiments using models of hepatectomy and regeneration as well as animals engineered to be deficient for various transcription factors with roles in hepatocyte development and differentiation. Hepatocytes maintain and control a delicate balance of transcription factors, co-factors, and modulators. Further work with gene targetting techniques may reveal additional mechanisms of gene expression control.

Inducible, tissue-specific and developmentally regulated gene expression is the result of this cascade of regulatory factors, which in the liver is dominated by transcription factors. Continued research is revealing that the cause of some disease states is the alteration of transcription factor expression (94, 95, 96). It is our hope that by accumulating knowledge on the mode of action of transcription factors in we may device novel approachs to control gene expression that may be of significance for gene therapy.

References

1. Michalopoulos, G.K., DeFrances, M.C., 1997, Liver regeneration, *Science* 276:60
2. Jennes, L., Eyigor, O., Janovick, J. A., Conn, P. M., 1997, Brain gonadotropin releasing hormone receptors: localization and regulation, *Recnet. Prog. Horm. Res.* 52: 475
3. Kuiper G. G., Gustafsson, J.A., 1997, The novel estrogen receptor-beta subtype: potential role in the cell- and promoter-specific actions of estrogens and anti-estrogens, *FEBS lett.* 410: 87
4. Pacall, J.C., 1997, Post-transcriptional regualtion of gene expression by androgens: recent observations from the epidermal growth factor gene, *J. Mol. Endocrinol.* 18: 177
5. Stralfors, P., 1997, Insulin second messengers, *Bioessays* 19: 327
6. Rosales, C., O'Brien, V., Kornberg, L., Juliano, R., 1995, Signal transduction by cell adhesion receptors, *Biochim. Biophys. Acta.* 1242:77

7. Peeper, D. S., Bernards, R., 1997, Communication between the exptracellular environment, cytoplasmic, signaling cascades and the nuclear cell-cycle machinery, *FEBS Lett.* 410: 11
8. Bird, AS. P., 1993, Genomic imprinting:imprinting on islands, *Curr. Biol.* 3: 1146
9. Dynan, W. S., 1989, Hepatocarcinogenesis: a dynamic cellular perspective, *Cell* 58: 1
10. Hill, C.S., Treisman, R., 1995, Transcriptional regulation by extracellular signals: mechanisms and specificity, *Cell* 80:199
11. Maniatis, T., Goodbourn, S., Fisher, J.A., 1987, Regulation of inducible and tissue-specific gene expression, *Science* 236:1237
12. Tjian, R., 1994, Transcriptional activation: a complex puzzle with few easy pieces, *Cell* 77:5
13. Bronwell, J.E., Allis, C.D., 1996, Special HATs for special occasions: linking histone acetylation to chromatin assemnbly and gene activation, *Curr. Opin. Genet. Dev.* 6:176
14. Felsenfeld, G., Boyes, J., Chung, J., Clark, D., and Studisky, V., 1996, Chromatin structure and gene expression, *Proc. Natil. Acad. Sci. USA* 93: 9384
15. Kingston, R.E., Bunker, C.A., Imbalzano, A.N., 1996, repression and activation by multiprotein complexes that alter chromatin structure, *Genes. Dev.* 10:905
16. Hershcbach, B. M., and Johnson, S. D., 1993, Transcriptional repression in eukaryocytes, *Annu.REv. Cell Biol.* 9: 479
17. Hertel, K.J., 1997, Common themes in the function of transcrioption andsplicing enhancers, *Curr. Opin. Cell Biol.* 9:350-357
18. Tjian, R., 1996, The biochemistry of transcription in eukaryotes: a paradigm for multisubunit regulatory complexes, *Phil. Trans. R. Soc. Lond.* B. 351:491-499
19. Reines, D., Conaway, J., W., and Conaway, R., C., 1996, The RNA polymerase ll general elongation factors, *Trends. Biochem. Sci.* 21: 351
20. Green, M. R., 1991, Biochemical mechanisms of constitutive and regualted pre-mRNA splicing, *Ann. Rev. Cell Biol.* 7: 559
21. Siomi H., Dreyfuss, G., 1997, RNA-binding proteins as egulators of gene expression, *Curr. Opin. Genet. Dev.* 7:345
22. Kozak, M., 1991, Structral features in eukaryotic mRNAs that modulate the initiation of translation, *J. Biol. Chem.* 266: 19867
23. Curtis, D., Lehmann, R., Zamore, P.D., 1995, Translational regulation in development, *Cell* 81:171
24. St Johnston, D., 1995, The intracellular localization of messenger RNAs, *Cell* 81:161

25. Wickens, M., Kimble, J., Strickland, S., 1996, Translational regulation of developmental decisions, in "Translational Control" Hershey, J.W.B., Mathews, M.B., and Sonenberg., N., eds., Cold Spring Harbor Laboratory Press:411
26. Vandromme, M., Gauthier-Rouvière, C., Lamb, N., and Fernandez, A., 1996, Regulation of transcription factor localization : fine-tuning of gene expression, *Trends Biochem. Sci.* 21: 59
27. Derman, E., Krauter, K., Walling, L., Weinberger, C., Ray, M., and Darnell, J. E. Jr., 1981, Transcriptional control in the production of liver-specific mRNAs, *Cell* 23: 731
28. Sladek, F. M., Zhong, W., Lai, E., and Darnell, J. E. Jr., 1990, Liver-enriched transcription factor HNF-4 is a novel member of the two Zenopous albumin genes, *J. Mol. Biol.* 199: 83
29. Pabo, C. O., and Sauer, R. T., 1992, Transcription factors: structural families and principles of DNA recognition, *Annu. Rev. Biochem.* 61: 1053
30. Cereghini, S., Blumenfeld, M., and Yaniv, M. A., 1988, A liver-specific factor essential for albumin transcription differs between differentiated and dedifferentiated rat hepatoma cells, *Genes Dev.* 2: 957
31. Tenen, D., G., Hromas, R., Licht, J., D., and Zhang, D., E., 1997, Transcription factors, normal myeloid development, and leukemmia, *Blood* 90: 489
32. McEwan, I., J., Wright, A., P., H., and Gustafsson, J., A., Mechanism of gene expression by the glucocorticoid receptor:role of protein-protein interactions, *BioEssays* 19: 153
33. Jackson, S. P., and Tjian, R., 1988, O-glycosylation of eukaryotic transcription factors: implications for mechanisms of transcriptional regualtion, *Cell* 55: 125
34. Hunter, T., and Karin, M., 1992, The regulation of transcription by phosphorylation, *Cell* 70: 375
35. Courtois, G., Morgan, J. G., Campbell, L. A., Fourel, G., and Crabtree, G. R., 1987, Interaction of a liver specific nuclear factor with the fibrinogen and α1-antitrypsin promoters, *Science* 238: 668
36. Monaci, P., Nicosia, A., and Cortese, R., 1988, Two different liver-specific factors stimulate in vitro transcription from the human α1-antitrypsin promotor, *EMBO J.* 7: 2075
37. Schorpp, M., Dobbeling, U., Wagner, U., and Ryffel, G. U., 1988, 5'-flanking and 5'-proxymal exon regions of the two Xenopous albumin genes, *J. Mol. Biol.* 199: 83

38. Xanthopoulos, K. G., Prezioso, V. R., Chen, W., Sladek, F., Cortese, R., and Darnell, J. E. Jr., 1991, The different tissue transcription patterns of genes for HNF-1, C/EBP, HNF-3, and HNF-4, protein factors that govern liver-specific transcription, *Proc. Natl. Acad. Sci USA* 88: 3807

39. Johnson, S., and Andersson, G., 1990, Similar induction of the hepatic EGF receptor in vivo by EGF and partial hepatectomy, *Biochem. Biophys. Res. Comm.* 166: 661

40. Costa, R. H., Grayson, D. R., and Darnell, J. E. Jr., 1989, Multiple hepatocyte-enriched nuclear factors function in the regulation of transthyretin and α1-antitrypsin genes, *Mol. Cell Biol.* 9: 1415

41. Lai, E., Prezioso, V. R., Tao, W., Chen, W. S., and Darnell, J. E. Jr., 1991, Hepatocyte Nuclear Factor 3a belongs to a gene family in mammals that is homologous to the Drosophila homotic gene forkhead, *Gens Dev* 5: 416

42. Ruppert, S., Kelsey, G., Schedl, A., Schmid, E., Thies, E., and Schüta, G., 1992, Deficiency of an enzyme of tyrosine metabolism underlies altered gene expression in newborn liver of lethal albino mice, *Genes Dev.* 6:1430

43. Duncan, S. A., Manova, K., Chen, W. S., Hoodless, P., Weinstein, D. C., Bachvarova, R. F., and Darnell, J. E. Jr., 1994, Expression of transcription factor HNF-4 in the extraembryonic endoderm, gut, and nephrogenic tissue of the developing mouse embryo: HNF-4 is a marker for primary endoderm in the implanting blastcyst, *Proc. Natl. Acad. Sci. USA* 91: 7598

44. Zaret, K. S., 1996, Molecular genetics of early liver development, *Annu. Rev. Physiol.* 58: 231

45. Grange, T., Roux, J., Rigaud, G., and Pictet, R., 1990, Cell-type specific activity of two glucocorticoid responsive units of rat tyrosine aminotransferase gene is associated with multiple bindign sites for C/EBP and a novel liver specific nucear factor, *Nuc. Acid Res.* 19: 131

46. Smit, J. J. M., MOl, C. A. A. M., Deemter, L. V., Wagenaar, E., Schinkel, A. H., and Borst, P., 1995, Characterization of the promoter region of the human MDR3 P-glycoprotein gene, *Biochim. Biophys. Acta* 1261: 44

47. Suchi, M., Sano, H., Mizuno, H., and Wada, Y., 1995, Molecular cloning and structural characterization of the human histidase gene (HAL), *Genomics* 29:98

48. Samadani, U. and Costa, R. H., 1996, The transcriptionla hepatocyte nuclear factor 6 regulates liver gene expression, *Mol. Cell. Biol.* 16: 6273

49. Landschulz, W. H., Johnson, P. F., Adashi, E. Y., Graves, B. J., and McKnight, S. L., 1988, Isolation of a recombinant copy of the gene encoding C/EBP, *Genes Dev* 2: 786
50. Xanthopoulos, K. G., Mirkovitch, J., Decker, T., Kuo, C. F., and Darnell, J. E. Jr., 1989, Cell-specific transcriptional control of the mouse DNA-binding protein mC/EBP, *Proc. Natl. Acad. Sci. USA* 86: 4117
51. Cao, Z., Umek, R. M., and McKnight, S. L., 1991, Regulated expression of three C/EBP isoforms during adipose conversion of 3T3-Li cells, *Genes Dev.* 5: 1538
52. Chang, C.-J., Chen, T.-T., Lei, H.-Y., Chen, D.-S., and Lee, S.-C., 1990, Molecular cloning of a transcription factor, AGP/EBP, that belongs to members of the C/EBP family, *Mol. Cell Biol.* 10: 6642
53. Descombes, P., Chojker, M., Lichtsteiner, S., FAlvey, E., and Schibler, U., 1990, LAP, A novel member of the C/EBP gene family, encodes a liver-enriched transcriptional activator protein, *Genes Dev* 4:1541
54. Metz, R., and Ziff, E., 1991, cAMP stimulates the C/EBP-related transcription factor rNFIL-6 to trans-locate to the nucleus and induce c-fos transcription, *Genes Dev.* 5: 1754
55. Poli, V., Mancini, F. P., and Cortese, R., 1990, IL-6DBP, A nuclear protein involved in Interleukin-6 signal transduction, defines a new family of leucine zipper proteins related to C/EBP, *Cell* 63: 643
56. Williams, S. C., Cantwell, C. A., and Johnson, P. F., 1991, A family of C/EBP-related proteins capable of forming covalently linked leucine zipper dimers in vitro, *Genes Dev.* 5:1553
57. Roman, C., Platero, J. S., Shuman, J., and Calame, K., 1990, Ig/EBP-1: a ubiquitously expressed immunoglobulin enhancer binding protein that is similar to C/EBP and heterodimerizes with C/EBP, *Genes Dev.* 4: 1404
58. Yamanaka, R., Kim, G. D., Radomska, H. S., Himes, J. L., Smith, L. T., Antonson, P., Tenen, D. G., and Xanthopoulos, K. G., 1997, CCAAT/enhancer binding protein ε is preferentially up-regualted during granulocytic differentiation and its functional versality is determined by alternative use of promoters and differential splicing, *Proc. Natl. Acad. Sci. USA*, 94: 6462
59. Costa, R. H., Grayson, D. R., Xanthopoulos, K. G., and Darnell, . E. Jr., 1988, A liver-specific DNA binding protein recognizes multiple nucleotide sites in regulatory regions of transthyretin,

alpha-1-antitrypsin, albumin, and simian virus 40 genes, *Proc. Natl. Acad. Sci. USA* 85: 3840
60. Herrera, R., Ro, H. S., Robinson, G. S., Xanthopoulos,K. G., and Spiegelman, B. M., 1989, A direct role for C/EBP and AP-1 binding site in gene expession linked to adipocyte differentiation, *Mol. Cell Biol.* 9: 533
61. Kuo, C. F., Xanthopoulos, K. G., and Darnell, J. E. Jr., 1990, Fetal and adult localization of C/EBP: evidence for combinational action of transcription factors in cell-specific gene expression, *Development* 109: 473
62. Flodby P., Antonson, P., Barlow, C., Blanck, A., Hällström, I. P., and Xanthopoulos, K. G., 1993, Differentail patterns of expression of three C/EBP isoforms, HNF-1, and HNF-4 after partial hepatectomy in rats, *Exp. Cell Res.* 208: 248
63. Park, E. A., Roseler, W., J., Liu, J., Klemm, D. J., Gurney, A. L., Thatcher, J. D.,Shuman, J., Friedman, A., and Hanson, R. W., 1990, The role of the CCAAT/enhancer-binding protein in the transcriptional regulation of the gene for phosphoenolpyruvate carboxykinase (GTP), *Mol. Cell Biol.* 10: 6264
64. Park, E. A., Gurney, A. L., Nizieski, S. E., Hakimi, P., Cao, Z., Moorman, A., and Hanson, R. W., 1993, Relative roles of CCAAT/enhancer-binding protein β and cAMP reguratory element-binding protein in controling transcription of the gene for phosphoenolpyruvate carboxykinase (GTP), *J. Biol. Chem.* 268: 613
65. Taub, R., 1996, Liver regeneration 4: transcription control of liver, *FASEB letters* 10: 413
66. Bossard, P., McPherson, C. E., and Zaret, K. S., 1997, In vivo footprinting with limiting amounts of embryo tissues: a role for C/EBP beta in early hepatic development, *Methods* 11: 180
67. Cressman D.E., Greenbaum, L. E., Haber, B. A., and Taub R., 1994, Rapid activation of PHF/NFκB in hepatocytes, a primary response in the regenerating liver, *J. Biol. Chem.* 269: 30429
68. Cressman, D. E., and Taub, R., 1994, Physiologic turnover of NF-κB by nuclear proteolysis, *J. Biol. Chem.* 269: 26594
69. Cressman, D. E., Diamond, R. H., and Taub, R., 1995, Rapid activation of Stat3 DNA binding in liver regeneration, *Hepatology* 21: 1443
70. Tewari, M., Doborzanski, P., Mohn, K. L., Cressman, D. E., Hsu, J. -C., Bravo, R., and Taub, R., 1992, Rapid induction in regenerating liver of RL/IF-1, and IκB that inhibits NF-κB, Rel0p50; and PHF, a novel κB site binding complex, *Mol. Cell Biol.* 12: 2898

71. Siebenlist, C., and Franzoso, G., and Brown, K., 1994, Structure, regulation and function of NF-kappa B, *Annu. Rev. Cell Biol.* 10: 405
72. Rice, N., Ernst, M.K., 1993, In vivo control of NF-κB activation by IκBα, *EMBO J.* 12:4685
73. Sun, S.C., Ganchi,P.A., Ballard,D.W., Greene, W.C., 1993, NF-κB controls expression of inhibitor IκBα: evidence for an inducible autoregulatory pathway, *Science* 259:1912
74. Veals, S. A., Schindler, C., Leonard, D., Fu, X. Y., Aebersold, R., Darnell, J. E. Jr., and Levy, D. E., 1992, Subunit of an alpha-interferon-responsive transcription factor is related to interferon regulatory factor and Myb families of DNA-binding proteins, Mol. Cell Biol. 12: 3315
75. Akira, S., Nishio, Y., Inoue, M., Wang, XJ., Wei, S., Matsusaka, T., Yoshida, K., Sudo, T., Naruto, and M., Kishimoto, T.,1994, Molecular cloning of APRF, a novel IFN-stimulated gene factor 3 p91-related transcription factor involved in the gp-130mediated signaling pathway, *Cell* 77:63
76. Azam, M., Erdjument-Bromage, H., Kreider, B. L., Xia, M., Quelle, F., Basu, R., Saris, C., Tempst, P., Ihle, J. N., and Schindler, C., 1995, Interleukin-3 signals through multiple isoforms of Stat5, *EMBO J.* 14: 1402
77. Fu, X-Y., Kessler, D. S., Veals, S. A., Levy, D. E., and Darnell, J. E. Jr., 1990, ISGF3, the transcriptional activator induced by interferon α, consists of multiple interacting polypeptide chains, *Proc. Natl. Acad. Sci. USA*. 87: 8555
78. Fu, X-Y., Schindler, C., Improta, T., Aebersold, R., Darnell, J.E. Jr., 1992, The proteins of ISGF-3, the interferon α-induced transcriptional activator, define a gene family involved in signal transduction, *Proc. Natl. Acad. Sci. USA* 89: 7840
79. Schindler, S. A., Fu, X-Y., Improta, T., Aebersold, R., and Darnell, J. E. Jr., 1992, Proteins of transcription factor ISGF-3: one gene encodes the 91-and 84kDa ISGF-3 proteins that are activated by interferon , *Proc. Natl. Acad. Sci. USA* 89: 7836
80. Wakao, H., Gouilleux, F., and Groner, B., 1994, Mammary gland factor (MGF) is a novel member of the cytokine regulated transcription factor gene family and confers the prolactin response, *EMBO J.* 13: 2182
81. Schindler, C., Darnell, J. E. Jr., 1995, Transcriptional responses to polypeptide ligands: the JAK-STAT pathway, *Annu. Rev. Biochem.* 64:621

82. Cressman, D.E., Greenbaum, L. E., DeAngelis, R. A., Cilibert, G., Furth, E. E., Poli, V., and Taub, R., 1996, Liver failure and defective hepatocyte regeneration in interleukin-6-deficient mice, *Science*, 274: 1379
83. Han, Y., and Brasier, A. R., 1997, Mechanism for biphasic Rel A•NF-κB1 nuclear translocation in Tumor Necrosis Factor α-stimulated hepatocytes, *J. Biol. Chem.* 272: 9825
84. Pontgolio, M., Barra, J., Hadchouel, M., Doyen, A., Kress, C., Bach, J.P., Babinet, C., Yaniv, M., 1996, Hepatocyte nuclear factor 1 inactivation results in hepatic dysfunction, phenylketonuria, and renal Fanconi syndrome, *Cell* 84:575
85. Chen, W.S., Manova, K., Weinstein, D.C., Duncan, S.A., Plump, A.S., Prezioso, V.R., Bachvarova, R.F., Darnell, J.E., Jr., 1994, Disruption of the HNF-4 gene, expressed in visceral endoderm, leads to cell death in embryonic endoderm and impaired gastrulation of mouse embryos, *Genes Dev.* 8:2466
86. Duncan, S.A., Nagy, A., Chan, W., 1997, Murine gastrulation requires HNF-4 regulated gene expression in the visceral endoderm: tetraploid rescue of Hnf-4 -/- embryos, *Development* 124:279
87. Takeda, K., Noguchi, K., Shi, W., Tanaka, T., Matsumoto, M., Yoshida, N., Kishimoto, T., Akira, S., 1997, Targeted disruption of the mouse Stat3 gene leads to early embryonic lethality, *Proc. Natl. Acad. Sci. USA* 94:3801
88. Flodby, P., Barlow, C., Kylefjord, H., Richter, L. Ä., and Xanthopoulos, K. G., 1996, Increased hepatic cell proliferation and lung abnormalities in mice deficient in CCAAT/enhancer binding protein α, *J. Biol. Chem.* 271: 24753
89. Wang, N. D., Finegold, M. J., Bradley A., Ou, C. N., Abdelsayed , S. V., Wilde, M. D., Taylor, L. R., Wilson D. R., and Darlington, G. J., 1995, Impaired energy homeostasis in C/EBPα knockout mice, *Science*, 269: 1108
90. Zhang, D.E., Zhang, P., Wang, N.D., Hetherrington, C.J., Darlington, G.J., Tenen, D.G., 1997, Absence of granulocyte colony-stimulating factor signaling and neutrophil development in CCAAT enhance binding protein α-deficient mice, *Proc. Natl. Acad. Sci. USA* 94:569
91. Meraz, M.A., White, J.M., Sheehan, K.C.F., Bach, E.A., Rodig, S.J., Dighe, A.S., Kaplan, D.H., Rilley, J.K., Greenlund, A.C., Campbell, D., Carver-Moore, K., DuBois, R.N., Clark, R., Aguet, M., Schreiber, R.D., 1996, Targeted disruption of the Stat1 gene in mice reveals unexpected physiologic specificity in the JAK-STAT signaling pathway, *Cell* 84:431

92. Screpanti, I., Romani, L., Musiani, P., Modesti, A., et al., 1995, Lymphoproliferative disorder and imbalanced T-helper response in C/EBPβ-deficient mice, *EMBO J.* 14:1932
93. Tanaka, T., Akira. S., Yoshida, K., Umemoto, M., Yoneda, Y., Shirafuji, N., Fujiwara, H., Suematsu, S., Yoshida, N., and Kishimoto, T., Targeted disruption of the NF-IL6 gene disloses its essential role in bacteria killing and tumor cytotoxicity by macrophages, *Cell* 80, 353
94. Carter, D., A., 1997, Rhythms of cellular immediate-early gene expression: more tahn just an early response, *Exp. Physiol.* 82: 237
95. Latchman, D., S., 1996, Transcription-facotr mutations and disease, *New Engl. J. Med.* 334: 28
96. Miller N., and Whelan, J., 1997, Progress in transcriptionally targeted and regualted vectors for genetic therapy, *hum. gene ther.* 8: 803

Modeling Genetic Diseases in the Mouse

Anthony Wynshaw-Boris, Carrolee Barlow, Amy Chen, Michael Gambello, Lisa Garrett, Theresa Hernandez, Shinji Hirotsune, Wendy Kimber, Denise Larson, Nardos Lijam, Gabriella Ryan and Zoe Weaver

National Human Genome Research Institute, National Institutes of Health, Bethesda, MD 20892

1. Introduction

The use of powerful linkage strategies for the mapping of genetic disease genes has led to the positional cloning of a number of genes associated with human genetic diseases. At the same time, techniques for manipulating the mammalian genome have been refined, so that the modification of the mouse genome via transgenic technology is now routine throughout the world. These two technologies have advanced and proliferated simultaneously to the point that among the first experiments planned upon the cloning of a human disease gene is the creation of an appropriate transgenic or knock-out mouse, with the hope of modeling that disease. A mammalian model is extremely valuable in the understanding of the function of a disease gene in normal animals, as well as its role in the pathophysiology of the disease. A good disease model can be used to test therapeutic options, especially gene therapy vectors. Finally, the genetic and biochemical pathways that a disease gene is part of can be dissected and investigated in animal models.

The mouse is a very good experimental system for the creation of models of human genetic diseases (Wynshaw-Boris, 1996). The genome size and number of genes are similar between human and mouse, as are patterns of development. It is relatively inexpensive to house mice, in comparison to other mammals. Mice have a relatively short gestation period, brief time of maturation to sexual maturity, and large litter sizes. Most importantly, it is the best mammalian species for genetic manipulation, and the only one in which it is currently possible to perform germline knock-outs in embryonic stem cells.

We will outline technical aspects of the generation of transgenic and knock-out mice. For further information about these techniques, the reader is referred to the excellent laboratory manual from Brigid Hogan, Rosa Beddington, Frank Costantini and Elizabeth Lacy (Hogan *et al.*, 1994). Then, we will briefly describe examples of mouse models of genetic diseases that have been created in our laboratory. Our research is described on the World Wide Web on our home page:

http://www.nhgri.nih.gov/Intramural-research/People/boris.html

2. Production of Transgenic Animals

Transgenic mice can be produced in one of two ways: direct injection of cloned DNA into the male pronucleus of a fertilized mouse egg; or the more complicated procedure of targeted introduction of transgenes by homologous recombination using ES cells. These techniques have allowed for the stable transfer of normal, altered, or chimeric genes into the mouse

NATO ASI Series, Vol. H 105
Gene Therapy
Edited by Kleanthis G. Xanthopoulos

germ line, thus providing powerful tools with which to study mechanisms underlying gene expression within a physiologic context.

2.1 Direct Injection

Transgenic mice are produced by microinjection of DNA directly into pronuclei of fertilized mouse eggs, and results in the random integration of foreign DNA into the mouse genome. Therefore, in order for the gene to be expressed, promoter elements as well as signals important for RNA processing are included in the transgene construct. Additionally, it has been shown that plasmid sequences in commonly used cloning vectors interfere with the expression of the gene of interest and that linearized DNA integrates more readily. This requires that the cloned transgene is purified and then cleaved so that vector sequences are removed to prepare the DNA for injection.

After fertilization, the male and female pronuclei do not fuse immediately. The male pronucleus is large and, using micromanipulators to stabilize the egg, 100-200 copies of the purified DNA can be injected into the male pronucleus. The DNA generally integrates at the one cell stage and therefore foreign DNA is present in every cell of the "transgenic" animal. The manipulated embryo can be implanted at this time, or maintained in culture until the two cell stage, and is then transferred into the oviduct of a pseudopregnant female where the embryo continues to develop. Offspring are analyzed and transgenic animals which have integrated the gene into their DNA are referred to as founders. Several founders can be created during each day of injection. These founders, in which the DNA has integrated into their germ cells, are able to transmit the transgene as a heritable trait and hence a transgenic line can be established.

In general, the transgene integrates as 1-50 copies arranged in a head-to-tail orientation (tandem repeats). Founders must then be screened for expression of the transgene as sequences surrounding the integration site may result in undesired patterns of expression or may even prevent the gene from being expressed. In general, the expression pattern from integrated transgenes remains stable over many generations.

2.2 Homologous Recombination

In contrast to the random integration of DNA which occurs due to non-homologous recombination during pronuclear injection of DNA, DNA can also integrate in a site-specific manner by homologous recombination. Foreign DNA introduced into mammalian cells generally integrates into the host genome via non-homologous recombination, with a frequency of homologous to non-homologous recombination of 1 in 1000. A major technical hurdle was overcome with the ability to grow cells in culture and select for those which have undergone homologous recombination. The development of cell culture methods for growing, transfecting, and selecting for homologous recombination in ES cells while maintaining the ES cells in an undifferentiated state (pluripotent) has allowed for the extensive manipulation of the mouse genome that is occurring in several laboratories throughout the world. Meticulous culture procedures are required to maintain ES cells in culture so that they retain a normal karyotype and do not differentiate, to insure that the genetically manipulated cells can contribute to the germline after injection into the host blastocyst.

ES cells are grown in culture and manipulated DNA of interest is transferred by electroporation. The manipulated DNA contains a marker which allows the selection of cells which have "taken up" the DNA, whether it be *via* homologous or non-homologous recombination. Following electroporation, cells are subcloned so that DNA can be extracted. Genomic DNA is screened using standard molecular biology procedures to identify the clones which have undergone homologous recombination at one allele. Once ES cell clones have been identified they are injected into blastocysts (host blastocysts) obtained by flushing the uterus of a three and one half day pregnant female. The injected blastocysts are then transplanted into the uterus of a pseudopregnant female where the embryo continues to develop. Offspring are screened for the presence of the altered allele. Offspring are referred to as a chimeras, as some of the animal is derived from the host blastocyst and some from the injected ES cells which contain the altered allele or transgene. If the manipulated ES cells contribute to the germline, then the chimera can be bred and some of its offspring will be heterozygous for the altered allele. This is referred to as germline transmission. Germline transmission is essential for establishing a transgenic line.

Several factors have been defined which facilitate the growth of ES cells, enhance the frequency of homologous recombination and simplify screening procedures. For example, use of isogenic DNA (DNA from the identical strain of mice as the ES cells), use of linearized DNA, and the use of two selectable markers improve the ratio of homologous to non-homologous clones surviving selection to the range of 1 in 15-300. In general, rates of homologous to nonhomologous recombination using optimized systems are 1 in 25-100.

The use of specific strains for the derivation of ES cells which have dominant coat color, in combination with the isolation of blastocysts from strains with a recessive coat color allow for the simplified screening procedure and identification of offspring in which the ES cells have contributed to the mouse. The most commonly used mouse strain for derivation of ES cells is one of the 129 strains. This strain carries the agouti coat color locus which is dominant over the recessively inherited black coat color locus of the C57BL/6J mouse strain from which the host blastocysts are obtained. Offspring in which the 129 ES cells have contributed to the embryo will have agouti and black fur due to the difference in origin of the cells, agouti from the ES cells or black from the cells of the host blastocyst. This is referred to as coat color chimerism. If ES cells have also contributed to the germline, then mating of the chimera to a female with a recessive coat color results in offspring with the coat color of the ES cell derived strain. ES cell derived offspring can then be screened for the presence of the mutated allele. Once animals heterozygous for the recombined allele have been identified, they can be interbred to generate animals homozygous for the mutated allele. In this way, null mutations or "knock outs", of a specific gene can be obtained.

3. Examples of Mouse Models of Mammalian Genetic Diseases

3.1 *Atm*-deficient Mice: Understanding Pleiotropic Functions of Atm by Modeling in the Mouse.

Ataxia-telangiectasia (AT) is an autosomal recessive disorder characterized by cerebellar degeneration and oculocutaneous telangiectasia, accompanied by immunodeficiency, infertility, small size, sensitivity to the effects of ionizing radiation and increased cancer predisposition. The AT gene product appears to participate in cell cycle checkpoints, and cell lines from AT patients show profound sensitivity to the cytotoxic and clastogenic effects of ionizing radiation and radiomimetic chemicals. Heterozygous carriers appear to have an increased predisposition to cancer, particularly for breast cancer. To address the complex relationship between gene function and the pleiotropic AT phenotype, a murine model of ataxia-telangiectasia was created by disrupting the *Atm* locus via gene targeting (Barlow *et at,* 1996). Mice homozygous for the disrupted *Atm* allele displayed growth retardation, neurologic dysfunction, male and female infertility secondary to the absence of mature gametes, defects in T lymphocyte maturation, and extreme sensitivity to γ-irradiation. The majority of animals developed malignant thymic lymphomas between two and four months of age. Several chromosomal anomalies were detected in these tumors. Fibroblasts from these mice grew slowly, and exhibited abnormal radiation-induced G1 checkpoint function. Therefore, the *Atm*-disrupted mice recapitulate the AT phenotype in humans, providing a mammalian model in which to study the pathophysiology of this disorder. We are analyzing the phenotype of these mice in great detail in order to understand the mechanisms by which a single gene disruption can result in such a pleiotropic phenotype, and to understand pathways which are regulated by Atm. In particular, we have used these mice to further study the molecular basis of infertility and radiation sensitivity.

Atm-deficient mice are completely infertile. To gain further insight into the role of Atm in meiosis, we have examined meiosis in *Atm*-deficient mice, and in mice doubly deficient for Atm and p53 or Atm and p2l. Male and female gametogenesis are severely disrupted in *Atm*-deficient mice as early as leptotene of prophase I, resulting in apoptotic degeneration (Barlow *et al,* 1997a). Rad5l foci are not assembled properly on unpaired axial elements in leptotene of mutant spermatocytes, and p53, p2l and Bax are elevated in these mutant testes. In *Atmlp53 or Atmlp2l* double mutants, spermatogenesis progresses further into pachytene stages, but not to diplotene. Assembly of Rad5l foci on axial elements remains defective, and p53, p2l and Bax remain elevated unless genetically eliminated. Thus, it appears that *Atm* is absolutely required for Rad5l assembly onto the axial elements, as well as for suppressing p53, p2l and Bax levels, suggesting that *Atm* participates in the regulation or surveillance of meiotic recombination and progression. The mislocalization of Rad5l in the *Atm -/-* mice is of particular interest, since Rad5l is important for double strand break repair and meiotic recombination, and associates with both BRCA1 and BRCA2.

In response to ionizing radiation (IR), Atm is part of a DNA damage-response pathway that involves p53. p53 is a multifunctional protein that simultaneously regulates distinct downstream pathways controlling cell cycle progression and apoptosis. However, the mechanisms by which p53 differentially activates downstream pathways are unknown. We have investigated this problem in *Atm*-deficient mice (Barlow *et al,* 1997b). Our data demonstrate that after IR, Atm acts via p53 to activate cell cycle checkpoint pathways and induce p2l, but p53-dependent or p53independent apoptotic responses and Bax induction are independent of Atm function. IRinduced thymic apoptosis was suppressed in *Atmlp53* double mutant mice, but not in *Atmlp2l* double mutants, demonstrating that these IR-mediated apoptotic responses are p53-dependent. Our results support a model in which upstream effectors such as Atm selectively activate p53 to regulate specific downstream pathways, providing a mechanism for controlling distinct cell cycle and apoptotic responses.

Finally, in collaboration with the laboratories of Jean Wang (UCSD), David Baltimore (MIT) and Michael Kastan, (Johns Hopkins University School of Medicine), we have demonstrated that c-abl is directly phosphorylated by Atm (Baskaran *et al*, 1997). We are currently investigating the biological significance resulting from this phosphorylation event.

3.2 *Dishevelled* Mutant Mice: Models for Human Neuropsychiatric Disorders and Neural Tube Defects.

Dishevelled is one of several segment polarity genes required for the wingless signal transduction pathway in Drosophila. This developmental pathway appears to be conserved in the mouse, where mutations in murine homologs of this pathway result in a variety of specific abnormalities. Three murine *Dishevelled genes, Dvl1, Dvl2 and Dvl3,* have been isolated by the laboratory of Daniel Sussman (University of Maryland School of Medicine), and we are collaborating to study the function of these genes *in vivo.* All three Dvl genes are expressed in a similar pattern in embryos and adults. To address the function of these genes during mammalian development, we have generated mice with targeted disruptions of each of the Dvl genes. Mice completely deficient for *Dvl1,* are viable, fertile and structurally normal (Lijam *et al.,* 1997). Surprisingly, these mice exhibited reduced social interaction, including differences in whiskertrimming, deficits in nest-building, less huddling contact during home cage sleeping, and subordinate responses in a social dominance test. Sensorimotor gating was abnormal, as measured by deficits in prepulse inhibition of acoustic and tactile startle. These experiments were done in collaboration with Richard Paylor and Jacqueline Crawley at the National Institutes of Mental Health. These results are consistent with an interpretation that common genetic mechanisms underlie abnormal social behavior and sensorimotor gating deficits and implicate *Dvl1* in processes underlying complex behaviors. Mice homozygous for null alleles of *Dvl2* and *Dvl3* survive to adulthood and are fertile, but are born in reduced numbers from heterozygous crosses. However, all *Dvl1/Dvl2* double homozygotes display a completely open neural tube and exencephaly, demonstrating an essential role for Dvl genes in neural tube closure. Thus, Dvl mutants provide genetic models for aspects of several human psychiatric disorders including autism, schizophrenia, and Tourette syndrome, as well as for neural tube defects.

3.3 Miller-Dieker Syndrome and Isolated Lissencephaly

Isolated lissencephaly (ILS) is a human brain developmental disorder in which the brain has a smooth cerebral surface and disorganized cortical layers due to abnormal neuronal migration. ILS is often associated with haploinsufficiency at chromosome 17pl3, or with heterozygous mutations in *LIS1,* a subunit of platelet activating factor acetylhydrolase. Miller-Dieker syndrome (MDS) is associated with larger deletions and consists of classical lissencephaly and more complex phenotypes including facial abnormalities. To determine the contribution of various genes on 17p13.3 to MDS, we have determined the structural organization of the syntenic region in mouse (Hirotsune *et al,* 1997). Our results demonstrate that the human and mouse MDS regions are conserved, suggesting that the mouse can be used to model ILS and MDS.

We have disrupted various genes in the syntenic region of the mouse (chromosome 11B2) and we are creating mice with deletions spanning the MDS critical region in mouse using the Cre-*lox*P system. We have created mice with full inactivation and conditional disruption of three genes in this region: *Lis*1, a gene mutated in ILS; *Mnt,* a Max-interacting bHLH protein; and *14-3-3ε,* a signaling molecule in the Ras/Raf pathway that binds cdc25 and p53. At the same time, *lox*P sites were introduced into these genes to allow deletions to be made with Cre. In a *Lis l* targeted ES clone, a second *lox*P site was introduced at the *Mnt* locus. Mice with each of these disruptions are currently being analyzed to determine the phenotypes caused by these disruptions. These experiments have been done in collaboration with David Ledbetter (University of Chicago School of Medicine) and Robert Eisenman (Fred Hutchinson Cancer Center).

In the course of these experiments, we created a mouse model of lissencephaly by targeted disruption of *Lisl* in the mouse. Heterozygous Lisl mutant mice were viable and fertile. However, as in ILS human patients, heterozygous Lisl mutants displayed neuronal migration abnormalities manifested by disorganized and trabeculated neurons at cerebral cortical layers and large ventricles. In particular, the disorganization of hippocampal pyramidal cells was quite pronounced. The brains of chimeric mice derived form *Lis*l targeted ES cells also displayed a mosaic pattern of disorganization suggesting that the abnormal neuronal cell migration resulting from hemizygous loss of *Lisl* function is cell autonomous. Homozygous *Lis*l disruption resulted in embryonic lethality, demonstrating an essential role for *Lisl* in early embryonic developmental events. Thus, *Lis*l mutant mice provide an excellent model for human lissencephaly, and will provide a tool to elucidate the molecular mechanisms responsible for the neuronal migration defects seen in ILS patients. In addition, *Lisl* is essential for murine embryonic development.

3.4 DiGeorge Syndrome and Related 22 q11 Deletion Syndromes

A number of human syndromes including DiGeorge (DGS), and Velocardiofacial syndromes are considered together as the 22ql1 microdeletion syndromes based on the observation that the majority of cases arise from deletion of a common region of chromosome 22ql1. Phenotypically these syndromes share one or more common features. However, the phenotypic severity varies tremendously even within syndromes, with no correlation evident with the extent of the deletion. At the present time the underlying genetic basis of these disorders remains unclear, and so to further a better understanding we are conducting a genetic dissection of these syndromes using the mouse as a model system, in collaboration with Peter Scambler (Institute of Child Health, University College London School of Medicine). As for Miller-Dieker lissencephaly, we plan to recreate these deletions in the mouse using the Cre-*lox*P recombinase system. A comparison of different deletions and their transgenic complementation should allow the phenotypic contribution of subregions and genes within the DGS critical region to be elucidated.

The contribution of individual genes such as *Hira* and *Idd,* which lie within the minimal deleted region in the human, are being determined by targeted gene disruption. This has been achieved in ES cells, and mice bearing these mutations are being generated for analysis. In addition, the region syntenic to a balanced translocation breakpoint in a human DGS family is being studied by the creation of a relatively small deletion of 140 kb in the mouse. Mice bearing this deletion have been generated and their phenotype is under analysis. Finally, deletions of the same magnitude as those observed in the human 22ql1 microdeletion syndromes, i.e. 2 Mb, are being created through the use of the *lox*P and Cre gene targeting strategy. *Lox*P sites have been targeted to various points in the mouse syntenic region, and exposure to the Cre protein should result in precise deletions of the desired size. Through comparison of the phenotypes expressed by these models, it may be possible to identify regions within 22ql1 which have significance for the disease phenotype, and perhaps lead to the elucidation of the underlying genetic defect.

3.5 Limb Deformity: a Model of Renal Agenesis and Incomplete Penetrance

We are studying the mechanisms responsible for incomplete penetrance of the renal agenesis phenotype of the mouse mutant limb deformity (ld). Ld was originally identified in mice as an autosomal recessive syndrome consisting of completely penetrant reduction and fusion anomalies of the long bones and digits of all four limbs, and incompletely penetrant renal agenesis/dysgenesis. Several alleles of ld mutations have arisen on different genetic backgrounds, three of which are ld^{Bri}, ld^{In2}, and ld^{U}. The ld^{Bri} and ld^{In2} mutants have a 20-30% penetrance of renal abnormalities, whereas the ld^{U} mutants have a 98% penetrance. The ld gene encodes four alternatively-spliced isoforms (I-IV). Isoform IV null mice (ld^{G}) have been generated by gene targeting. The limb and kidney phenotypes were distinguished in the null mice. ld^{G}/ld^{G} mice have a 6% incidence of renal agenesis, yet have perfectly normal limbs (Wynshaw-Boris *et al*, 1997). We are attempting to understand the incomplete penetrance of the renal phenotype of the various ld mutants is allele-specific, background-specific or a combination of the two. We have generated various combinations of compound ld heterozygotes that have been mated. The F2 offspring have been scored for limb and renal abnormalities. Our results indicate that the renal abnormalities correlate with the ld allele, thus suggesting that it is the mutation itself, not the genetic background, that governs the penetrance of renal agenesis at this locus. To further investigate the nature of these ld mutations, we plan to remove the entire ld locus by gene targeting in ES cells. The phenotype of this null allele will be compared with the phenotypes of other ld alleles. These mice will serve as the recipient for various modified ld alleles, introduced via YAC transgenesis.

3.6 Brca1-deficient Mice: Understanding the Function of a Familial Tumor Suppressor

The human breast and ovarian cancer susceptibility gene, *BRCA1,* has recently been cloned. BRCA1 encodes a protein of unknown function. The mutant alleles are predicted to be null alleles or loss-of-function mutations. To study the role of *BRCA1* in tumorigenesis, we have created a mouse with a null allele at the murine *Brca*1 locus in exon 11. This targeted disruption results in an early embryonic lethal phenotype, between 5.5 and 6.5 days post-conception. Thus, the mice have not provided specific information on the role of *Brca1* in tumorigenesis. Therefore, in collaboration with Chu-Xia Deng and Lothar Hennighausen (NIDDK), we are attempting to create mice with conditional knock-outs of *Brca1,* so that we can disrupt the gene only in the breast.

4. Conclusions

It is now feasible to manipulate the mouse genome to make precise genetic alterations, and to create gain-of-function and loss-of-function alleles of specific genes. Thus, the mouse is an excellent organism to model specific mammalian genetic diseases. We have briefly reviewed the techniques required to create genetically manipulated mice using pronuclear injection or homologous recombination in ES cells, and highlighted the use of these techniques by presenting examples of mouse models studied in our laboratory. Such mouse models have already provided us with additional insights into the pathophysiologic mechanisms responsible for several of the phenotypes in these models. We hope that we will now be able to use these models to test specific therapies for these genetic diseases, including gene therapies.

5. References

Barlow, C., Hirotsune, S. Paylor, R., Liyanage, M. Eckhaus, M., Collins, F., Shiloh, Y., Crawley, J.N., Ried, T., Tagle, D. and Wynshaw-Boris, A. (1996) *Atm*-deficient mice: a paradigm of ataxiatelangiectasia. Cell. 86:159-171.
Barlow, C., Liyanage, M., Moens, P.B., Deng, C.-X., Ried, T. and Wynshaw-Boris, A. (1997a) Partial rescue of the severe prophase I defects of *Atm*-deficient mice by p53 and p2l mutant alleles. Submitted.
Barlow, C., Brown, K.D., Deng, C.-X., Tagle, D. and Wynshaw-Boris, A. (1997b) Atm selectively regulates distinct p53-dependent cell cycle checkpoint and apoptotic pathways. Submitted.
Baskaran, R., Wood, L.D., Whitaker, L.L., Canman, C.E., Morgan, S.E., Xu, Y., Barlow, C., Baltimore, D., Wynshaw-Boris, A., Kastan, M.B. and Wang, J.Y.J. (1997) Ataxia telangiectasia mutated gene product activates c-Abl tyrosine kinase in response to ionizing radiation. Nature. 387:516-519.
Hirotsune, S., Chong, S.S., Pack, S.P., Robbins, C., Pavan, W., Ledbetter, D. and WynshawBoris, A. (1997) Genomic organization of the murine Miller-Dieker/Lissencephaly region: conservation of linkage with the human region. Genome Res. 7:625-634.
Hogan, B., Beddington, R., Costantini, F. and Lacy E. (1994) Manipulating the mouse embryo: a laboratory manual. Cold Spring Harbor Laboratory Press, Cold Spring Harbor.
Lijam, N., Paylor, R., McDonald, M.P., Crawley, J.N., Deng, C., Herrup, K. Stevens, K., Macaferri. G., McBain, C.J., Sussman, D.J. and Wynshaw-Boris, A. (1997) Social interaction and sensorimotor gating abnormalities in mice lacking Dvl-1. Cell. 90:895-905.
Wynshaw-Boris, A. (1996) Model mice and human disease. NatureGenet. 13:259-260.
Wynshaw-Boris, A., Ryan, G., Deng, C., Chan, D., Jackson-Grusby, L., Larson, D., Dunmore, J. and Leder, P. (1997) The role of a single formin isoform in the limb and renal phenotypes of *limb deformity*. Mol Med. 3:372-384.

The Use of Genetic Switches for the Generation of Conditional Mutants at the Level of Cells and Animals

Hermann Bujard, Zentrum für Molekulare Biologie Heidelberg (ZMBH), Universität Heidelberg, Im Neuenheimer Feld 282, 69120 Heidelberg, Germany

Introduction

Most of the fundamental insights into gene function have been gained from the study of organisms susceptible to both efficient genetic dissection and biochemical analysis. Accordingly, bacterial systems - particularly *E.coli* - as well as "low" eukaryotes such as yeast but also Drosophila have been instrumental in establishing our present days' understanding of biological processes. Despite of their continuing importance in biological and medical research, these model systems are limited with respect to questions directly concerning development, gene function and genetic disorders of mammals, questions most relevant to human medicine. As a result, the mouse became the principle species to investigate specifically mammalian problems. However, the complexity of the genome, embryonic development and the technical difficulties of studying large numbers make the classical genetic approach in this model difficult and in many respects impossible. The technique of transgenesis (1) and of gene targeting in mice (2) has been an important breakthrough but the irreversibility of the mutational alterations which may lead to compensatory developments, developmental defects and even embryonic mortality limit this approach. One way to partially overcome such limitations is gene targeting with the site-specific Cre/lox recombination system (3) as pioneered by Byrne & Ruddle (4), by Westphal et al. (5) and particularly by K. Rajewsky and coworkers (6, 7). In this strategy, the Cre recombinase controlled by an appropriate promoter is used to activate, inactivate or alter a gene during a defined differentiated state of cells in the developping organism. Again, however, the genetic changes are irreversible and follow a program that cannot be influenced after its onset.

"Genetic switches" that could be operated at will and that would permit the control of individual gene activities quantitatively and reversibly in a temporal and spatial manner would thus be of great advantage. Several lines of research were devoted towards the development of such genetic switches (8 - 13) which all are aiming at specifically controlling the activity of individual genes at the level of transcription. The different developments should they turn out to be successful may allow to develop different experimental strategies since they will have different specific strengths. Moreover, they may provide means to regulate several genes independently of each other and thus to analyze more complex phenotypes which arise through the combined action of more than one gene.

In the following, I will concentrate on a regulatory system (14) which is based on control elements of the tetracycline resistance operon of bacteria. It is the most advanced system, so far, and has been widely applied at the level of cells including yeast, plant and mammalian cells as well as in transgenic organisms such as plants, mice and Drosophila. The progress achieved to date in various systems via tet regulation allows to envisage the impact of genetic switches on future research strategies, particularly concerning conditonal mouse mutants as models of human diseases. Moreover, they have possibly opened up new approaches for cell and gene therapy in humans.

NATO ASI Series, Vol. H 105
Gene Therapy
Edited by Kleanthis G. Xanthopoulos

The principle of the tet regulatory systems.

By using transcription control elements of an evolutionarily distant species like the bacterium *E.coli*, it was intended to develop a regulatory circuit that would not interfere with the cellular metabolism when superimposed onto the complex regulatory network of a higher eukaryotic cell. Consequently, the elements originating from *E.coli* were expected to confer high specificity to this artificial control system. The bacterial elements chosen are part of a tetracycline resistance operon and thus the controlling effector molecule is tetracycline (Tc) or one of its many derivatives. The broad knowledge in the pharmacology of tetracyclines of which some are widely used as antibiotics in human medicine is of great advantage, particularly when tet regulation is established in animals.

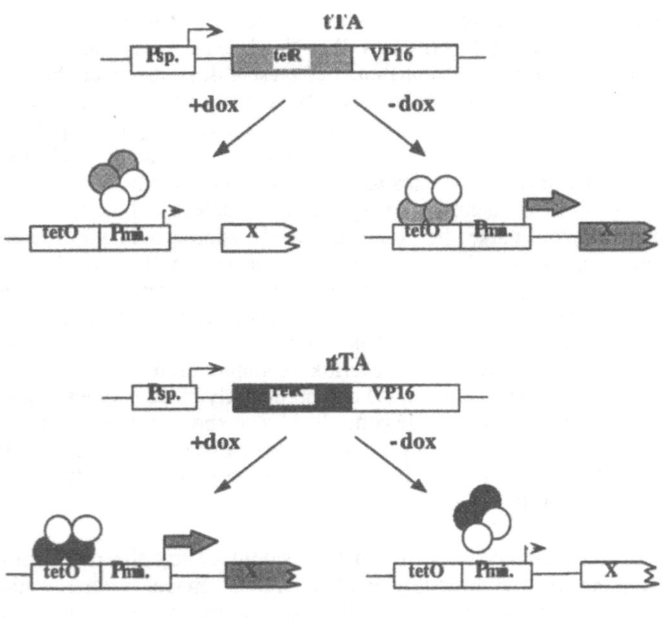

Fig. 1. Schematic outline of the tet regulatory systems.
Upper part: mechanism of action of the Tc controlled transactivator (tTA). The transactivator is composed of the repressor (TetR) of the Tn10 Tc-resistance operon of Escherichia coli and a C-terminal portion of VP16 that functions as a strong transcription activator. tTA binds in absence of Dox (but not in its presence) to an array of 7 cognate operator sequences (tetO) and activates transcription from an adjacent minimal promoter (14). The minimal promoter-tet operator fusion is referred to as $P_{hCMV^{*}-1}$.
The lower part shows the rtTA system which is identical to the tTA system with the exception of 4 amino acid exchanges in the TetR moiety. These changes convey a reverse phenotype to the Tet repressor (rTetR). The resulting rtTA requires Dox for binding to tetO and thus for transcription activation (14). Tissue-specificity of these systems is achieved by placing the tTA or rtTA gene under the control of a tissue-specific promoter (P_{sp}*). Thus, in mouse lines where tTA and rtTA synthesis is controlled by* P_{hCMV} *(*TA^{CMV} *and* rTA^{CMV} *line, see Fig. 2), Dox-regulated expression of indicator genes is found - as expected - in a number of tissues (17). By contrast,* P_{LAP} *leads to mouse lines (*TA^{LAP}*, see Fig. 3) producing tTA exclusively in hepatocytes (17).*

Upon stable transfer of the tTA/rtTA system into cell lines, regulation factors up to 10^5 fold were measured provided that the gene of interest controlled by a tTA/rtTA responsive promoter is integrated into an appropriate locus of the chromosome (14).
Tc-controlled expression of a variety of genes has been reported over the last few years which also shows that the Tet systems function in many different cell lines and under various conditions (16).

By fusing in frame the gene of a tetracycline repressor protein (TetR) to the coding region of the transcriptional activating domain of Herpes simplex virus protein 16 (VP16), a transcription activation factor was generated that has retained the DNA-binding specificity and the induction properties of the Tet repressor. Activation of transcription, thus, depends on the binding of the "tetracycline controlled trransactivator" (tTA) to *tet* operator sequences (*tetO*) fused to a minimal RNA polymerase II promoter whose activity totally depends on the presence of an activation domain as provided by the VP16 moiety. Binding of the activator to its target sequence is, however, prevented by the effector molecule tetracycline or some of its derivatives, and thus, transcription activation is abolished.
A second transactivator protein was generated that differs dramatically in its response to the effector substance. The amino acid sequence of this protein deviates from tTA by only 4 amino acids, yet its DNA-binding behaviour is reversed (15). Thus, in contrast to tTA, rtTA (the reverse tetracycline controlled transactivator) will only bind to *tetO* in the presence of certain tetracycline derivatives such as doxycycline (Dox) with subsequent activation of transcription (Fig. 1).

Tet-dependent regulation in transgenic mice.

To examine the regulatory potential of the Tet systems in transgenic animals, two classes of mouse lines were generated: (a) mice which produce tTA or rtTA under the control of the human cytomegalovirus IE promoter (P_{hCMV}); (b) mice which contain a gene encoding an indicator function - luciferase or ß-galactosidase - under the control of the tTA/rtTA responsive promoter (17). Among the "indicator mouse lines", some showed no or extremely little background activity demonstrating that the luciferase expression unit was integrated in a chromosomal locus where no or very little outside activation of the tTA/rtTA responsive promoter would occur.

When such animals were crossed with individuals producing tTA or rtTA, high expression of indicator functions was observed in those tissues where P_{hCMV} is known to be active e. g. in muscle or pancreas. This expression was completely dependent on the presence or absence of Dox as shown in Fig. 2. The expression levels could be modulated via the concentration of Dox supplied in the drinking water of the animals and induction kinetics, particularly with the rtTA system were found to be rapid (17).

High cell type-specific regulation could be shown with animals expressing tTA exclusively in hepatocytes. Here, the homogeneous expression in all cells and a regulatory range spanning more than 5 orders of magnitude (Fig. 3) underlines the potential of the Tc-controlled genetic switches in transgenic animals.

Application of Tc-controlled expression to specific questions.

Since its first description in 1992, Tet regulation has been successfully used to address numerous questions at the cellular as well as at the organismal level. More than 20 different cell lines stably expressing tTA or rtTA were established so far and used to control the expression of various genes. Many of these genes were for the first time stably maintained in cell lines since their expression could be controlled. They include for example genes which play a role in cell cycle control, programmed cell death or viral genes involved in pathogenic pathways (ref. 18 - 27). Here, it should suffice to point out just a few examples which hallmark novel experimental strategies now feasable.

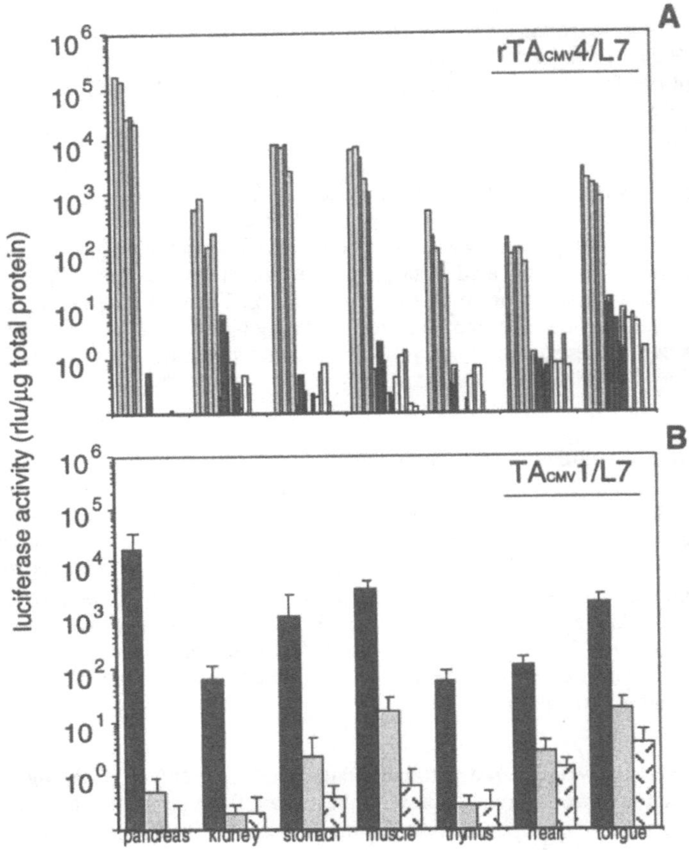

Fig. 2. Luciferase induction in double transgenic rtTA^{CMV}/L7 and tTA^{CMV}/L7 mice. Mice producing tTA or rtTA under the control of P_{hCMV} were crossed with animals of the L7 indicator mouse line (17) which contains the luciferase gene under the control of a tTA/rtTA-responsive promoter (see Fig. 1). Luciferase activity was measured in various organs of double transgenic offsprings that were kept in presence and absence of Dox.

(A) Pattern of luciferase activity in various organs of individual rtTA^{CMV}/L7 animals. Bars: darkly shaded, luciferase activities in absence; lightly shaded, luciferase activities in presence of Dox (2 mg/ml, 1 week); open, background activities of L7 animals (note the logarythmic scale).

(B) Pattern of luciferase activity in various organs of TA^{CMV}/L7 animals. Bars darkly shaded, luciferase activities in absence of Dox; lightly shaded, luciferase activities in presence of Dox (2 mg/ml, 1 week); stippled, luciferase background of the L7 mouse line. Error bars show standard deviations.

Thus, the group of Nienhuis developed a packaging cell line which produces high titers of amphotropic retroviruses for gene therapeutic experiments (28). These viruses gain their broad host range and their stability by the capsid protein of the Vesicular stomatitis virus. When constitutively expressed in amounts required for efficient virus production, this protein is not tolerated by cells. By contrast, when placed under Tc-control, the VSG gene can be stably maintained and activated to proper extend whenever the packaging of viruses is initiated. Virus production is thus increased 100 to 1000 fold.

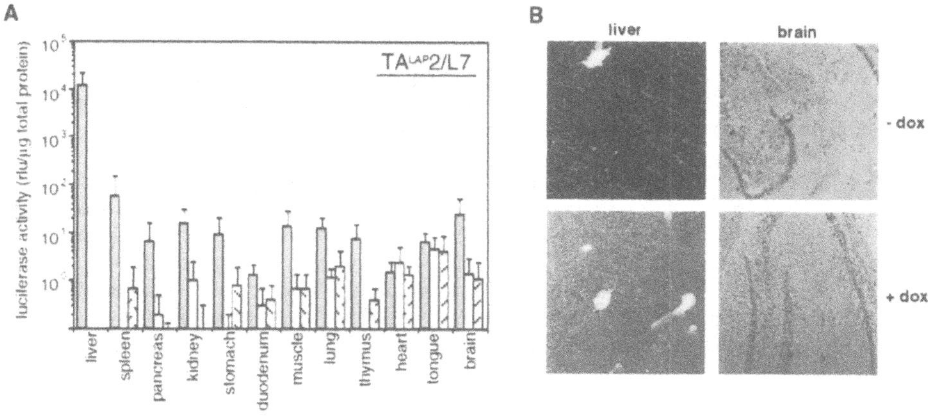

Fig. 3. Liver-specific regulation of indicator functions.
(A) Mice synthesizing tTA highly specifically in hepatocytes (TA^{LAP2}) were crossed with L7 indicator animals (17). Double transgenic offsprings were kept in presence or absence of Dox and analyzed for luciferase activity in the organs indicated. Darkly shaded bars, animals kept in absence; lightly shaded bars, animals kept in presence of Dox. Missing bars indicate that luciferase activity was not measurable. Stippled bars give the L7 background and error bars indicate standard deviations.
(B) Mice producing tTA in hepatocytes as in (A) were crossed with indicator mice containing the β-galactosidase gene under the control of the tTA-responsive promoter and kept in presence or absence of doxycycline. Liver and brain (hypocampus area) cryosections of animals treated with or without Dox were subjected to 4-bromo-3-chloro-2-indolyl-β-D-galactoside staining followed by counterstaining with nuclear fast red. The in situ analysis demonstrates a highly cell type-restricted, homogeneous regulation of β-galactosidase via tTA activation.

F. Gage and coworkers succeeded in constructing a retrovirus which expressed the *v-myc* gene under Tc control (29). Using this virus, they generated conditionally proliferating hypocampal progenitor cells which would proliferate as long as Tc is absent from the culture. Addition of the antibiotic, however, leads to growth arrest and differentiation of the progenitors into neuronal cells. With this and similar approaches, a new type of primary cell lines may become accessible.

Finally, Sodroski and collaborators have used Tc controlled expression to identify molecular determinants responsible for acute single cell lysis by Human immunodeficiency virus type 1 (27).

In a good number of projects, cell lines were developed which may become useful for high throughput screening systems for the identification of drugs directed against specific targets.

Remarkable results were also obtained with transgenic mice. The group of Efrat succeeded in generating two most interesting mouse lines (30). The first one expresses the tTA gene specifically in ß-cells of the pancreas, the second one contains the SV40 Tag under the control of the tTA/rtTA responsive promoter. When individuals of the two lines were crossed, adult double transgenic animals developed very specifically ß-cell tumors. These proliferating ß-cells could be propagated *in vitro*. Addition of Tc, however, induced growth arrest. When such cells were implanted into diabetic mice and grown to a proper population size, the diabetic phenotype was cured provided that Tc was supplied to prevent further growth. These results are highly suggestive not only with respect to certain strategies of cell therapy but also for creating new types of conditionally proliferating primary cells.

By crossing SV40 Tag individuals with animals expressing tTA under the control of the MMTV promoter, Efrat's and Hennighausen's groups generated animals which develop after several months tumors of the mammary gland as long as no tetracycline was supplied in the drinking water (31). Together with the work of Efrat mentioned above, these are the first demonstrations of a controlled *in vivo* induction of tumors by a single gene activity.

Finally, Yu et al. (32) have described mouse lines which specifically synthesize tTA in cardiac myocytes. An approach which may have direct implication for gene or cell therapy, has been followed by M. Heard and collaborators (33). Using retroviruses, they introduced the erythropoietin gene controlled by tTA or rtTA into mouse primary myoblasts. Upon transplantation of such myoblasts into mice, they were capable of modulating the hèmatocrite of the animals in a doxycycline-dependent manner. Interestingly, they did not find any immune response directed against the components of the tet system within the six-month-duration of the experiment (34).

More than 150 reports have appeared over the last 4 years in which tet regulation has been used to introduce conditionality of various gene functions in higher eukaryotic systems. Many of these gene functions could not be studied in a satisfying way by expressing the corresponding genes constitutively or in a transient situation. There can be little doubt that the Tc-controlled genetic switches which are operated by non toxic substances with excellent tissue penetration properties have opened up a broad spectrum of approaches for the study of gene function *in vivo*, for the generation of novel high throughput screening systems for pharmaca and for the development of human disease models in transgenic animals.

Acknowledgements

This work was supported by the Deutsche Forschungsgemeinschaft (SFB 229) and by the Fonds der Chemischen Industrie Deutschlands.

References

1. JAENISCH, R. and MINTZ, B. (1974), Simian virus 40 DNA sequences in DNA of healthy adult mice derived from preimplantation blastocysts injected with viral DNA. *Proc. Natl. Acad. Sci. USA* 71, 1250-1254

2. THOMAS, K. R. and CAPECCHI, M. R. (1987), Site-directed mutagenesis by gene targeting in mouse embryo-derived stem cells. *Cell* **51**, 503-512.

3. STEMBERG, N. and HAMILTON, D. (1981), *J. Mol. Biol.* **150**, 467-486.

4. BYRNE, G. and RUDDLE, F. H. (1989), Multiplex gene regulation: a two-tiered approach to transgene regulation in transgenic mice. *Proc. Natl. Acad. Sci. USA* **86**, 5473-5477.

5. LAKSO, M., SAUER, B., MOSINGER, B. JR., LEE, E. J., MANNING, R. W., YU, S. H., MULDER, K. L. and WESTPHAL, H. (1992), Targeted oncogene activation by site-specific recombination in transgenic mice. *Proc. Natl. Acad. Sci. USA* **89**, 6232-6236.

6. GU, H., ZOU, Y. R. and RAJEWSKY, K. (1993), Independent control of immunoglobulin switch recombination at individual switch regions evidenced through Cre-loxP-mediated gene targeting. *Cell* **73**, 1155-1164.

7. GU, H., MARTH, I. D., ORBAN, P. C., MOSSANN, H. and RAJEWSKY, K. (1994), Deletion of a DNA polymerase beta gene segment in T cells using cell type specific gene targeting. *Science* **265**, 103-106.

8. YARRANTON, G. T. (1992), Inducible vectors for expression in mammalian cells. *Curr. Op. Biotech.* 3, 506-511

9. EILERS, M., PICARD, D., YAMAMOTO, K. and BISHOP, J.M. (1989), Chimeras between the MYC oncoprotein and steroid receptors cause hormone-dependent transformation of cells. *Nature* 340, 66-68

10. PICARD, D., SALSER, S.J. and YAMAMOTO, K.R. (1988), A movable and regulatable inactivation function within the steroid binding domain of the glucocorticoid receptor. *Cell* 54, 1073-1080

11. HO, D. Y., McLAUGHLIN, J.R. and SAPOLSKY, R.M. (1996), Inducible gene expression from defective herpes simplex virus vectors using the tetracycline responsive promoter system. *Mol. Brain Res.* 41, 200-209

12. CHRISTOPHERSON, K.S., MARK, M.R., BAJAJ, V. and GODOWSKI, P.J. (1992) *Proc. Natl. Acad. Sci. USA* 89, 6314-6318

13. WANG, Y., O'MALLEY, B.W.JR., TSAI, S.Y. and O'MALLEY, B.W. (1994), A regulatory system for use in gene transfer. *Proc. Natl. Acad. Sci. USA* 91, 8180-8184

14. GOSSEN, M. and BUJARD, H. (1992), Tight control of gene expression in mammalian cells by tetracycline responsive promoters. *Proc. Natl. Acad. Sci. USA* 89, 5547-5551

15. GOSSEN, M., FREUNDLIEB, S., BENDER, G., MÜLLER, G., HILLEN, W. and BUJARD, H. (1995), Transcriptional activation by tetracyclines in mammalian cells. *Science* 268, 1766-1769

16. FREUNDLIEB, S., BARON, U., BONIN, A.L., GOSSEN, M. and BUJARD, H. (1997), Use of tetracycline controlled gene expression systems to study the mammalian cell cycle. *Methods Enzymol.*, in press

17. KISTNER, A., GOSSEN, M., ZIMMERMANN, F., JERECIC, J., ULLMER, C., LÜBBERT, H. and BUJARD, H. (1996), Doxycycline-mediated, quantitative and tissue-specific control of gene expression in transgenic mice. *Proc. Natl. Acad. Sci. USA* 93, 10933-10938

18. RESNITZKY, D., GOSSEN, M., BUJARD, H. and REED, S.I. (1994), Acceleration of the G1/S phase transition by expression of cyclins D1 and E using an inducible system. *Mol. Cell. Biol.* 14, 1669-1679

19. AGARWAL, M.L., AGARWAL, A., TAYLOR, W.R. and STARK, G.R. (1995), p53 controls both the G2/M and the G1 cell cycle checkpoints and mediates reversible growth arrest in human fibroblasts. *Proc. Natl. Acad. Sci. USA* 92, 8493-8497

20. CHEN, Y.Q., CIPRIANO, S.C., ARENKIEL, J.M. and MILLER, F.R. (1995), Tumor suppression by p21WAF1. *Cancer Res.* 55, 4536-4539

21. GALAKTIONOV, K., CHEN, X. and BEACH, D. (1996), Cdc25 cell cycle phosphatase as a target of c-myc. *Nature* 382, 511-517

22. ELDREDGE, E.R., CHIAO, P.J. and LU, K.P. (1995), Use of tetracycline operator system to regulate oncogene expression in mammalian cells. *Methods Enzymol.* 254, 481-491

23. MAHESWARAN, S., ENGLERT, C., BENNETT, P., HEINRICH, G. and HABER, D.A. (1995), The WT1 gene product stabilizes p53 and inhibits p53-mediated apoptosis. *Genes Dev.* 9, 2143-2156

24. ENGLERT, C., HOU, X., MAHESWARAN, S., BENNETT, P., NGWU, C., RE, G.G., GARVIN, A.J., ROSNER, M.R. and HABER, D.A. (1995), WT1 suppresses synthesis of the epidermal growth factor receptor and induces apoptosis. *EMBO J.* 14, 4662-4675

25. BOLDIN, M.P., VARFOLOMEEV, E.E., PANCER, Z., METT, I.L., CAMONIS, J.H. and WALLACH, D. (1995), A novel protein that interacts with the death domain of Fas/APO1 contains a sequence motif related to the death domain.*J. Biol. Chem.* **270**, 7795-7798.

26. BOLDIN, M.P., METT, I.L., VARFOLOMEEV, E.E., CHUMAKOW, I., SHEMER-AVNI, Y., CAMONIS, J.H. and WALLACH, D. (1995), Self-association of the "death domains" of the p55 tumor necrosis factor (TNF) receptor and Fas/APO1 prompts signaling for TNF and Fas/APO1 effects. *J. Biol. Chem.* **270**, 387-391.

27. CAO, J., PARK, I., COOPER, A. and SODROSKI, J. (1996), Molecular determinants of acute single-cell lysis by Human Immunodeficiency Virus Type 1. *J. Virol.* 70, 1340-1354

28. YANG, Y., VANIN, E.F., WHITT, M.A., FORNEROD, M., ZWART, R., SCHNEIDERMAN, R.D., GROSVELD, G. and NIENHUIS, A.W. (1995), Inducible, high-level production of infectious murine leukemia retroviral vector particles pseudotyped with vesicular stomatitis virus G envelope protein. *Hum. Gene Ther.* **6**, 1203-1213.

29. HOSHIMARU, M., RAY, J., SAH, D.W.Y. and GAGE, F.H. (1996), Differentiation of the immortalized adult neuronal progenitor cell line HC2S2 into neurons by regulatable suppression of the v-myc oncogene. *Proc. Natl. Acad. Sci. USA* 93, 1518-1523

30. EFRAT, S., FUSCO-DEMANE, D., LEMBERG, H., EMRAN, O.A. and WANG, X. (1995), Conditional transformation of a pancreatic ß-cell line derived from transgenic mice expressing a tetracycline-regulated oncogene. *Proc. Natl. Acad. Sci. USA* 92, 3576-3580

31. EWALD, D., LI, M., EFRAT, S., AUER, G., WALL, R., FURTH, P. and HENNIGHAUSEN, L. (1996), Time-sensitive reversal of hyperplasia in transgenic mice expressing SV40 T antigen. *Science* 273, 1384-1386

32. YU, Z., REDFERN, C.S. and FISHMAN, G.I. (1996), Conditional transgene expression in the heart. *Circulation Res.* 79, 691-697

33. BOHL, D. and HEARD, J.M. (1997), Modulation of erythropoietin delivery from engineered muscles in mice. *Human Gene Ther.* 8, 195-204

34. BOHL, D., NAFFAKH, N. and HEARD, J.M. (1997), Long term control of hematocrit by doxycycline in mice transplanted with engineered primary myoblasts. *Nature Med.*, in press

Application of Tetracycline Regulatable Systems for Gene Therapy

Delphine Bohl and Jean-Michel Heard, Laboratoire Rétrovirus et Transfert Génétique, Institut Pasteur, 28 rue du Docteur Roux, 75724 Paris cedex 15. Phone: 33 (0) 1 45 68 82 46, Fax: 33 (0) 1 45 68 89.40, e-mail: jmheard@pasteur.fr.

Abstract

Many diseases candidates for gene therapy require that the therapeutic gene expression level is controlled in order to ensure biological efficacy and to prevent toxic effects. Various systems have been described which allow transcriptional regulation by artificial chimeric transactivators in mammalian cells. This paper describes the tetracycline regulatable systems and discuss their potential application for gene therapy.

Introdution

Many diseases candidates for gene therapy require that the expression level of therapeutic protein is adjusted to prevent toxicity and to ensure biological efficacy after in vivo gene transfer. Although tight regulation in response to endogenous physiological stimuli is ultimately desirable, this cannot be presently achieved in most considered diseases. Indeed, endogenous response elements to physiological stimuli are often poorly characterized or too large to be accomodated in gene transfer vehicles. Alternatively, inducible or repressible systems responsive to exogenous stimuli have been reported, which can provide a satisfactory control of foreign gene expression in mammalian cells.

Physiological regulatory controls have been transposed for modulating foreign gene expression (review in Yarranton et al.). The transactivator element is a natural protein present in mammalian tissue, which controls gene expression in response to a physiological stimulus. Control of foreign genes is obtained by inserting DNA sequences responsive to the natural transactivating element in the promoter-enhancer regions located upstream of the foreign protein coding sequences in the gene transfer vehicle. Inducible eukaryotic promoters used in that goal include: the metallothionein promoter or systems responsive to heavy metal ions (Hu and Davidson, 1990; Mayo et al., 1982; Searl et al., 1985), the heat shock promoter (Schweinfest et al., 1988; Wurm et al., 1986), interferon-inducible promoters (Staeheli et al., 1986; Totzke et al., 1992) and promoters responsive to steroid hormone (Israel and Kaufman, 1989; Ko et al., 1989; Mader and White, 1993. These systems have some advantages and a major drawback. Advantages include that: i) there is no need to introduce a gene coding for the transactivating protein, since it is already present in the cell; ii) there is no risk of an immune response ellicited by the transactivating factor since it is an endogenous protein. The major drawback is the absence of selectivity. Upon administration of the stimulus, all promoters containing responsive elements recognized by the transactivator, and not only that one introduced in the gene transfer vector, are activated, leading to pleiotropic effects which may unmask the biological consequences of modulating the expression of the foreign protein. Moreover those systems generally suffer from leakiness of the inactive state (e.g., the metallothionein promoter) and from modest (10-20 fold) induction factors. Practically, regulatory systems based on physiological stimuli are inapproriate for gene therapy applications.

The limitations of natural regulatory systems stressed for the search of non-natural systems allowing a selective control of the expression of foreign sequences. To reach that goal, artificial transactivating proteins are required. The non-natural compound is expressed in target cells only. Responsive elements controlling foreign gene expression are not present

NATO ASI Series, Vol. H 105
Gene Therapy
Edited by Kleanthis G. Xanthopoulos

in mammalian cell DNA. The stimulus required for transactivtion is usually a chemical compound which, hopefully, has little impact on normal cell and physiology. As a non-natural protein, the transactivator has to be encoded by a foreign gene, implying that the regulatory system consists of two separated transcriptional units. Of course, the artificial transactivator is potentially immunogenic, thus possibly exposing genetically modified cells to cytotoxic T cell attack. Investigating the feasibility of introducing two transcriptional units in target cells and the existence of an immune response against modified tissues largely determines whether these systems are appropriate for gene therapy applications.

We are now going to describe two systems based on artificial transactivators dereived from the tetracycline operon of E. coli. Research in that field is quite intense and new systems are actively investigated, which are expected to provide more powerful stimulation, to be less immunogenic and in which the stimulating compound is devoid of any biological activity on non-modified cells. However, general principles conducting the conception of the future regulatory systems will very likely be close to that of the existing ones.

I - The tetracycyline-repressible system.

- Description

A system based on elements derived from the tetracycline resistance operon of the Escherichia coli transposon Tnl0 was developed in 1992 by the group of H. Bujard (fig. lA) (Gossen and Bujard, 1992). A chimeric tetracycline-controlled transactivator (tTA) was constructed by fusing the C-terminal 129 amino acids of the acidic domain of the VP16 protein of the herpes simplex virus (HSV) (Triezenberg et al., 1988) to the carboxyl terminus of the bacterial tetracycline repressor protein (tetR). In the absence of tetracycline, the tetR moiety of tTA binds a regulatory region comprising 7 repeats of the 19-bp tetO bacterial operator with high affinity and specificity. This element was placed immediately upstream of a minimal human cytomegalovirus promoter (tetO-CMV promoter), which by itself is devoid of any transcriptional activity. When bound on the promoter, the VP16 moiety of tTA induces the assembly of a transcriptional iniation complex (Ingles et al., 1991; Lin et al., 1991). In the presence of tetracycline or tetracycline derivatives such as anhydrotetracycline (Gossen and Bujard, 1993), tTA preferentially binds the antibiotic, leading to conformational modifications decreasing the affinity for the tet operator. Bujard et al reported that in a clone of human HeLa cell line stably expressing tTA, the expression of a luciferase gene could be modulated over a range of 5 orders of magnitude depending on the absence or the presence of 1 mM tetracycline in the culture medium (Gossen and Bujard, 1992). The kinetics of the regulation was fast: gene expression decreased 10 fold within 8 hours and 50 fold within 12 hours after antibiotic addition.

The tet-suppressible system offers advantages that makes it particularly attractive for regulation of gene expression in vivo and for gene therapy applications. One characteristic feature of the regulatory system is that both the tetr moiety of tTA and its cognate binding site tetO are of prokaryotic origin, thus ensuring that transactivation is restricted to transgenic sequences and avoid side effects. Moreover the low toxicity of tetracycline in mammalian cells and the high affinity of tetracycline for the tetR protein, $10^9 M^{-1}$ (Degenkolb et al., 1991), enables the use of the antibiotic at concentrations that cause little adverse effects on mammalian cells and transgenic animals (Furth et al., 1994).

A possible drawback consists in the potentially immunogenicity of the chimeric transactivator protein. To overcome this problem new transactivation domains of proteins have been proposed. For example the KRAB repressor domain of the human zinc finger protein Koxl was fused to tetR. The binding of this chimeric protein to tetO sequences of the tetO-CMV promoter blocked transcription efficiently in vitro (Deuschle et al., 1995). The activation domain located in the carboxy terminal region of the human transcription factor NF-kB p65 protein has also been proposed as an alternative to the HSV VP16 sequences (Rivera et al., 1996).

- Applications for research.

Very rapidly after the initial description, the use of the tetracycline suppressible system has been extensively investigated. Reports concerned a variety of eukaryotic organisms including yeasts (Dingermann et al., 1992), parasites (Wirtz and Clayton, 1995), viruses (Cao et al., 1996; Kim et al., 1995), plantes (Gatz et al., 1991; Röder et al., 1994), ameba (Dingermann et al., 1992) and cultured mammalian cells (Agarwal et al., 1995; Cayrol and Flemington, 1995; Gossen and Bujard, 1992; Howe et al., 1995; Resnitzky et al., 1995; Shockett et al., 1995; Wimmel et al., 1994). It has proved useful in a broad range of investigations, among which: i) studies of the role of cellular gene products during cell differentiation and developmental processes (Rizzino and Miller, 1995); ii) establishment of stable cell lines for constitutive expression of deleterious genes with potentially cytotoxic effects (Howe et al., 1995; Resnitzky et al., 1995; Schaack et al., 1995; Shan and Lee, 1994; Wimmel et al., 1994); iii) studies of functions of mammalian gene products in development and oncogenesis in transgenic animals (Efrat et al., 1995; Furth et al., 1994; Passman and Fishman, 1994; Shockett et al., 1995).

- Application for gene therapy.

Gene therapy applications have special requirements. They imply i) the efficient transduction of the two components of the tetracycline regulation system into a large proportion of target primary cells; ii) negligeable gene expression in the absence of inducer; iii) intense, modulable and reversible inducibility; iv) sustained regulation over long periods of time in animals.

The simultaneous introduction of both components of the system in poorly transfectable cells requires the use of viral vectors. The retrovirus-mediated transfer of both elements of the tet-suppressible system was obtained by different strategies, using either a single recombinant genome with several internal promoters or internal ribosome entry sites (IRES) (Hofman et al., 1996; Hoshimaru et al., 1996; Hwang et al., 1996; Iida et al., 1996; Paulus et al., 1996) or two separated retrovirueses vectors simultaneously introduced into the target cells (Bohl and Heard, 1997). Toxic squelching effects of the VP16 domain (Gossen and Bujard, 1992) often complicated the isolation of packaging cells producing high vector titers.

Immortalized cell lines (Fishman et al., 1994; Hoshimaru et al., 1996; Hwang et al., 1996; Iida et al., 1996; Paulus et al., 1996; Schultze et al., 1996) or primary muscle cell cultures (Hoffman et al., 1996) were transduced with retrovirus vectors and a regulation of gene expression in the range of 300 fold in population of NIH-3T3 cells (Hwang et al., 1996; Paulus et al., 1996) and 600 fold in stably transduced clones of primary myogenic cells (Hoffman et al., 1996) was obtained. The background level of expression of the minimal tetO-CMV promoter was influenced by cell type (Ackland-Berglund and Leib, 1995) and the presence of several promoters in the same vector (Iida et al., 1996; Paulus et al., 1996). Although this strategy removes the need for transducing two different vectors, it exposes to possible drawbacks such as suboptimal gene transfer efficiency, generation of recombinant vectors genomes, difficulty to maintain stable vector-producing cell lines, and promoter interferences (Emerman and Temin, 1984).

To eliminate promoter interferences Hoffman et al. designed a self-inactivating 5' LTR retrovirus vector and the expression of both tTA and the reporter gene were placed under the control of the tetO-CMV promoter. The background level of expression of the reporter gene expressed from the tetO-CMV promoter in this vector was relatively high and the kinetics of induction by tetracycline were lower than in other systems, because of the vector loop principle itself.

Whereas the system is extensively used in cell cultures, its in vivo peformances were assessed in one situation relevant to gene therapy (Bohl and Heard, 1997). Levels of erythropoietin secretion were measured after gene transfer into mouse skeletal muscle in animals receiving or not tetracycline in the drinking water. Hematocrit levels significantly differed between treated mice and controls. However, the high background expression level of the particular retrovirus vectors used in this study impaired accurate regulation,

II - The tetracycline-inducible system.

A second system also constructed by the group of H. Bujard allows the induction of gene expression in response to tetracycline-derivatives (Gossen et al., 1995). The transactivator protein, referred to as reverse transactivator (rtTA), requires tetracycline derivatives, such as doxycycline, for binding the tetO (fig. 1B). In HeLa cells constitutively synthetizing rtTA, the expression of an integrated luciferase reporter gene inserted downstream of a tetO-CMV promoter was stimulated more than a thousandfold by the addition of doxycycline in the culture medium. Maximum induction occured after less than 24 hours. In double transgenic mice a doxycycline-regulation was obtained which vary between 1 and 7000 fold induction depending on the tissue (Gossen et al., 1995).

The tetracycline-inducible system has been much less investigated than the suppressible system. However, a definitive demonstration of its usefullness for gene therapy applications has been recently provided (Bohl et al., 1997). Mouse primary myogenic cells were transduced ex vivo with two retroviruses, encoding the rtTA and the murine Epo under the control of a op-CMV promoter, respectively. After transplantation in recipients mice, Epo secretion could be iteratively switched on and off over a five-month period depending on the presence or the absence of doxycycline in the drinking water. The absence of an immune response directed against muscle fibers expressing the rtTA chimera was one of the important observation of this study, even if its does not allow definitive conclusion about the antigenicity of the protein in others biological contexts.

Conclusion.

The tetracycline regulatable systems are potentially suitable for the transcriptional regulation of therapeutic genes expression in response to the administration of drugs. Others regulatable systems based on the use of modified steroid hormone receptors (No et al., 1996; Mahfoudi et al., 1995; Feil et al., 1996; Wang et al., 1994; Delort and Capecchi, 1996) or the heterodimerization of transcriptional activator in the presence of rapamycin (Rivera et al., 1996; Blau et al., 1997; Freiberg et al., 1997; Freiberg et al., 1996; Spencer et al., 1996) have also been proposed.

The feasibility of controlling gene expression in engineered cells suggests that foreign genes could also be controlled by physiological stimuli in ectopic tissues, provided that the required transactivating factors are present. For example, it is be much preferrable to regulate Epo secretion in response to hypoxia than to tetracycline. Response elements to hypoxia located 3' to the Epo human gene (Blanchard et al., 1992; Madan and Curtin, 1993) were identified as the DNA binding site for the hypoxia-inducible factor (HIF-1)(Jiang et al., 1996; Wang and Semenza, 1993). These sequences for hypoxic regulation could be inserted in gene transfer vectors. Modulation of gene expression was in the range of 20 to 50 fold, i.e. significantly less than with artificial systems. It is expectable that modulation will be even lower in a polyclonal cell population in which foreign genes are integrated randomly. In deed, chromatin environment is crucial for regulation, and a majority of randomly integrated sequences will be located in a suboptimal environment for expression and regulation (Iida et al., 1996; Xu et al., 1989). Therefore, stable chromosomal integration may not be the preferrable situation for regulation of foreign gene expression in response to physiological stimuli.

Regulation at the level of mRNA translation is an alternative approach for the control of protein expression. In most cases regulation takes place at the level of initiation of translation, which is often attributable to the interaction of proteins with structural secondary structures of the 5'-untranslated region of the mRNA. For example, iron-responsive elements have been identified in mRNAs coding for proteins implicated in iron homeostasis (Cox and Adrian, 1993). A iron-dependent translation to heterologous genes has also been reported (Goossen et al., 1990) and was constructed. The insertion of an iron response element into the LTR of a retroviral vector allowed translational regulation in vitro in response to iron concentration in the culture medium (Davis et al., 1997).

New strategies are being developed with aim to design high affinity ligands by combinatirial selection. This process relies on the selection of a subset of a random population of potential ligands for binding a specific target. The target molecule can be an enzyme, an antibody, a receptor, or any regulatory molecule (Kenan et al., 1994). The capacity to produce ligand agaisnt any potential target will likely prove useful for the future development of regulatable vectors for gene therapy.

References.

Ackland-Berglund, C. E., and Leib, D. A. (1995). Efficacy of Tetracycline-Controlled Gene Expression is Influenced by Cell Type. BioTechniques 18, 196-200.

Agarwal, M. L., Agarwal, A., Taylor, W. R., and Stark, G. R. (1995). p53 controls both the G2/M and the G1 cell cycle checkpoint and mediates reversible growth arrest in human fibroblasts. Proc. Natl. Acad. Sci. USA 92, 8493-8497.

Blanchard, K. L., Acquaviva, A. M., Galson, D. L., and Bunn, H. F. (1992). Hypoxic induction of the human erythropoietin gene: cooperation between the promoter and enhancer, each of which contains steroid receptor response elements. Mol. Cell. Biol. 12, 5373-5385.

Blau, C. A., Peterson, K. R., Drachman, J. G., and Spencer, D. M. (1997). A proliferation switch for genetically modified cells. Proc. Natl. Acad. Sci. 94, 3076-3081.

Bohl, D., and Heard, J. (1997). Modulation of erythropoietin delivery from engineered muscles in mice. Human Gene Ther. 8, 195-204.

Bohl, D., Naffakh, N., and Heard, J. M. (1997). Long term control of erythropoietin secretion levels by tetracycline in mice transplanted with engineered primary myoblasts. Nature Med. 3, 299-312.

Cao, J., Park, I., Cooper, A., and Sodroski, J. (1996). Molecular determinants of acute single-cell lysis by human immunodeficiency virus type 1. J. Virol. 70, 1340-1354.

Cayrot, C., and Flemington, E. K. (1995). Identification of cellular target genes of the Epstein-Barr virus transactivator Zta activation of transforming growth factor beta igh3 (TGF-beta igh3) and TGF-beta1. J. Virol. 69, 206-212.

Cox, L. A., and Adrian, G. S. (1993). Posttranscriptional regulation of chimeric human transferrin genes by iron. Biochemistry 32, 4738-4745.

Davis, J. L., Gross, P. R., and Danos, O. (1997). Development and characterization of a translationally regulated retroviral vector. Abstract, Keystone Meeting, April 13-19, 21.

Degenkolb, J., Takahashi, M., Ellestad, G. A., and Hillen, W. (1991). Structural requirements of tetracycline-tet repressor interaction: determination of equilibrim binding constants for tetracycline analogs with the tet repressor. Antimicrob. Agents Chemother. 35, 1591-1595.

Delort, J. P., and Capecchi, M. R. (1996). TAXI/UAS: a molecular switch to control expression of gens m vivo. Hum. Gene Ther. 7, 809-820.
Deuschle, U., Meyer, W. K.-H., and Thiesen, H.-J. (1995). Tetracycline-Reversible Silencing of Eukaryotic Promoters. Mol. Cell. Biol. 15, 1907-1914.
Dingermann, T., Frank-Stoll, U., Werner, H., Wissman, A., Hillen, W., Jacquet, M., and Marschalek, R. (1992). RNA polymerase III catalysed transcription can be regulated in Saccharomyces cerevisiae by the bacterial tetracycline repressor-operator system. EMBO J. 11, 1487-1492.
Dingermann, T., Werner, H., Schütz, A., Zündorf, I., Nerke, K., Knecht, D., and Marschalek, R. (1992). Establishment of a system for conditional gene expression using an inducible tRNA suppressor gene. Mol. Cell. Biol. 12, 4038-4045.
Efrat, S., Fusco-DeMane, D., Lemberg, H., Al Emran, O., and Wang, X. (1995). Conditional transformation of a pancreatic b-cell line derived from transgenic mice expressing a tetracycline-regulated oncogene. Proc. Natl. Acad. Sci. USA 92, 3576-3580.
Emerman, M., and Temin, H. M. (1984). Genes with promoters in retrovirus vectors can be independently suppressed by an epigenetic mechanism. Cell 39, 459-467.
Feil, R., Brocard, J., Mascrez, B., LeMeur, M., Metzger, D., and Chambon, P. (1996). Ligand-activated site-specific recombinasion in mice. Proc. Natl. Acad. Sci. 93, 10887-10890.
Fishman, G. I., Kaplan, M. L., and Buttrick, P. M. (1994). Tetracycline-regulated cardiac gene expression in vivo. J. Clin. Invest. 93, 1864-1868.
Freiberg, R. A., Ho, S. N., and Khavari, P. A. (1997). Transcriptional control in keratinocytes and fibroblasts using synthetic ligands. J. Clin. Invest. 99, 2610-2615.
Freiberg, R. A., Spencer, D. M., Choate, K. A., Peng, P. D., Schreiber, S. L., Crabtree, G. R., and Khavari, P. A. (1996). Specific triggering of the fas signal transduction pathway in normal human keratinocytes. J. Biol. Chem. 271, 31666-31669.
Furth, P. A., St. Onge, L., Böger, H., Gruss, P., Gossens, M., Kistner, A., Bujard, H., and Hennighausen, L. (1994). Temporal control of gene expression in transgenic mice by a tetracycline-responsive promoter. Proc. Natl. Acad. Sci. USA 91, 9302-9306.
Gatz, C., Kaiser, A., and Wendenburg, R. (1991). Regulation of a modified CaMV 35S promoter by the Tn10-encoded Tet repressor in transgenic tobacco. Mol. Gen. Genet. 227, 229-237.
Goossen, B., Wright Caughman, S., Harford, J. B., Klausner, R. D., and Hentze, M. W. (1990). Translational repression by a complex between the iron-responsive element of ferritin mRNA and its specific cytoplasmic binding protein is position dependent in vivo. EMBO J. 9, 4127-4133.
Gossen, M., and Bujard, H. (1993). Anhydrotetracyclin, a novel effector for tetracycline controlled gene expression in eukaryotic cells. Nucl. Acid Res. 21, 4411-4412.
Gossen, M., and Bujard, H. (1992). Tight control of gene expression in mammalian cells by tetracycline-responsive promoters. Proc. Natl. Acad. Sci. USA 89, 5547-5551.
Gossen, M., Freundlieb, S., Bender, G., Müller, G., Hillen, W., and Bujard, H. (1995). Transcriptional activation by tetracyclines in mammalian cells. Science 268, 1766-1769.
Ho, D. Y., McLaughlin, J. R., and Sapolsky, R. M. (1996). Inducible gene expression from defective herpes simplex virus vectors using tetracycline-responsive promoter system. Mol. Brain Res. 41, 200-209.
Hoffman, A., Nolan, G., and Blau, H. B. (1996). Rapid retroviral delivery of tetracycline-inducible genes in a single autoregulatory cassette. Proc. Natl. Acad. Sci. USA 93, 5185-5190.
Hoshimaru, M., Ray, J., Sah, D. W. Y., and Gage, F. H. (1996). Differentiation of the immortalized adult neuronal progenitor cell line HC2S2 into neurons by regulatable suppression of the v-myc oncogene. Proc. Natl. Acad. Sci. USA 93, 1518-1523.
Howe, J. R., Skryabin, B. V., Belcher, S. M., Zerillo, C. A., and Schmauss, C. (1995). The responsiveness of a tetracycline-sensitive expression system differs in different cell lines. J. Biol. Chem. 270, 14168-14178.
Hu, M. C. T., and Davidson, N. (1990). A combination of derepression of the lac operator-repressor system with positive induction by glucocorticoid and metal ions provides a high-level-inducible gene expression system based on the human metallothionein-II_A promoter. Mol. Cell. Biol. 10, 6141-6451.
Hwang, J. J., Scuric, Z., and Anderson, W. F. (1996). Novel retroviral vector transferring a suicide gene and a selectable marker gene with enhanced gene expression by using a tetracyclineresponsive expression system. J. Virol. 70, 8138-8141.
Iida, A., Chen, S. T., Friedmann, T., and Yee, J. K. (1996). Inducible gene expression by retrovirus-mediated transfer of a modified tetracycline-regulated system. J. Virol. 70, 6054-6059.
Ingles, C. J., Shales, M., Cress, W. D., Triezenberg, S. J., and Greenblatt, J. (1991). Reduced Binding of TFID to Transcriptionally compromised Mutants of VP16. Nature 351, 588-590.
Israel, D., and Kaufman, R. J. (1989). Highly Inducible Expression From Vectors Containing Multiple GRE's in CHO Cells Overexpressing the Glucocorticoid Receptor. Nucleic. Acids. Res. 17, 4589-4604.
Jiang, B., Rue, E., Wang, G. L., Roe, R., and Semenza, G. L. (1996). Dimerization, DNA binding and transactivation properties of hypoxia-inducible factor 1. J. Bio. Chem. 271, 17771-17778.
Kenan, D. J., Tasi, D. E., and Keene, J. D. (1994). Exploring molecular diversity with combinatorial shape libraries. TIBS 19, 57-64.
Kim, H. J., Gatz, C., Hillen, W., and Jones, T. R. (1995). Tetracycline repressor-regulated gene repression in recombinant human cytomegalovirus. J. Virol. 69, 2565-2573.

Lin, Y.-S., Ha, I., Maldonado, E., Reinberg, D., and Green, M. R. (1991). Binding of General Transcription Factors TFIIB to an Acidic Activating Region. Nature 353, 569-571.
Madan, A., and Curtin, P. T. (1993). A 24-base-pair sequence 3' to the human erythropoietin gene contains a hypoxia-responsive transcriptional enhancer. Proc. Natl. Acad. Sci. USA 90, 3928-3932.
Mader, S., and White, J. S. (1993). A Steroid-Inducible Promoter for the Controlled Overexpression of Cloned Genes in Eukaryotic Cells. Proc. Natl. Acad. Sci. USA 90, 5603-5607.
Mahfoudi, A., Roulet, E., Dauvois, S., Parker, M. G., and Whali, W. (1995). Specific mutations in the estrogen receptor change the properties of antiestrogens to full agonists. Proc. Natl. Acad. Sci. 92, 4206-4210.
Mayo, K. E., Warren, R., and Palmiter, R. D. (1982). The Mouse Metallothionein-I Gene Is Transcriptionally Regulated by Cadmium Following Transfaction into Human or Mouse Cells. Cell 29, 99-108.
No, D., Yao, T., and Evans, R. M. (1996). Ecdysone-inducible gene expression in mammalian cells and transgenic mice. Proc. Natl. Acad. Sci. 93, 3346-3351.
Passman, R. S., and Fishman, G. I. (1994). Regulated expression of foreign genes in vivo after germline transfer. J. Clin. Invest. 94, 2421-2425.
Paulus, W., Baur, I., Boyce, F. M., Breakefield, X. O., and Reeves, S. A. (1996). Self-contained, tetracycline-regulated retroviral system for gene delivery to mammalian cells. J. Virol. 70, 62-67.
Resnitzky, D., Hengst, L., and Reed, S. I. (1995). Cyclin A-associated kinase activity is rate limiting for entrance into S phase and is negatively regulated in G1 by p27Kip1. Mol. Cell. Biol. 15, 4347-4352.
Rivera, V. M., Clackson, T., Natesan, S., Pollock, R., Amara, J. F., Keenan, T., Magari, S. R., Phillips, T., Courage, N. L., Cerasoli, F., Dennis, A. H., and Gilman, M. (1996). A humanized system for pharmacologic control of gene expression. Nat. Med. 2, 1028-1032.
Rizzino, A., and Miller, K. (1995). The function of inducible promoter systems in F9 embryonal carcinoma cells. Exp. Cell Res. 218, 144-150.
Röder, F. T., Schmülling, T., and Gatz, C. (1994). Efficiency of the tetracycline-dependent gene expression system: complete suppression and efficient induction of the rolB phenotype in transgenic plants. Mol. Gen. Genet. 243, 32-38.
Schaack, J., Guo, X., Ho, Y., Karlok, M., Chen, C., and Oernelles, D. (1995). Adenovirus type 5 precursor terminal protein-expressing 293 and HeLa cell lines. J. Virol. 69, 4079-4085.
Schultze, N., Burki, Y., Lang, Y., Certa, U., and Bluethmann, H. (1996). Efficient control of gene expression by single step integration of the tetracycline system in transgenic mice. Nature Biotech. 14, 499-505.
Schweinfest, C. W., Jorcyk, C. L., Fujiwara, S., and Papas, T. S. (1988). A heat-shock inductible eukaryotic expression vector. Gene 71, 207-210.
Searl, P. F., Stuart, G. W., and Palmiter, R. D. (1985). Building a Metal-Responsive Promoter With Synthetic Regulatory Elements. Mol. Cell. Biol. 5, 1480-1489.
Shan, B., and Lee, W. (1994). Deregulated expression of E2F-1 induces S-phase entry and leads to apoptosis. Mol. Cell Biol. 14, 8166-8173.
Shockett, P., Difilippantonio, M., Hellman, N., and Schatz, D. (1995). A modified tetracycline-regulated system provides autoregulatory, inducible gene expression in cultured cells and transgenic mice. Proc. Natl. Acad. Sci. USA 92, 6522-6526.
Spencer, D. M., Belshaw, P. J., Chen, L., Ho, S. N., Randazzo, F., Crabtree, G. R., and Schreiber, S. L. (1996). Functional analysis of fas signaling in vivo using synthetic inducers of dimerization. Current Biol. 6, 839-847.
Staeheli, P., Danielson, P., Haller, O., and Sutcliffe, J. G. (1986). Transcriptional activation of the mouse Mx gene by type 1 interferon. Mol. Cel. Biol. 6, 4770-4774.
Totzke, F., Marmé, D., and Hug, R. (1992). Inducible expression of human phospholipase C-g2 and its activation by platelet-derived growth factor B-chain homodimer and platelet-derived growth factor A-chain homodimer in transfected NIH 3T3. Eur. J. Biochem. 203, 633-639.
Triezenberg, S. J., Kingsbury, R. C., and McKnight, S. L. (1988). Functional Dissection of VP16, the Trans-Activator of Herpes Simplex Virus Immediate Early Gene Expression. Genes and development 2, 718-729.
Wang, G. L., and Semenza, G. L. (1993). General involvment of hypoxia-inducible factor 1 in transcriptional response to hypoxia. Proc. Natl. Acad. Sci. USA 90, 4304-4308.
Wang, Y., O'Malley Jr., B. W., Tasai, S. Y., and O'Malley, B. W. (1994). A regulatory system for use in gene transfer. Proc Natl. Acad. Sci. USA 91, 8180-8184.
Watsuji, T., Okamoto, Y., Emi, N., Katsuoka, Y., and Hagiwara, M. (1997). Controlled gene expression with a reverse tetracycline-regulated retroviral vector system. Bioche. Biophys. Res. Comm. 234, 769-773.
Wimmel, A., Lucibello, F. C., Sewing, A., Adolph, S., and Muller, R. (1994). Inducible acceleration of G1 progression through tetracycline-regulated expression of human cyclin E. Oncogene 9, 995-997.
Wirtz, E., and Clayton, C. (1995). Inducible gene expression in trypanosomes mediated by a prokaryotic repressor. Science 268, 1179-1183.
Wurm, F. M., Gwinn, K. A., and Kingston, R. E. (1986). Inducible overexpression of a mouse c-myc protein in mammalian cells. Proc. Natl. Sci. 83, 5414-5418.
Xu, L., Yee, J. K., Wolff, J. A., and Friedmann, T. (1989). Factors influencing long-term stability of Moloney leukemia virus-based vectors. Virology 171, 331-341.
Yarranton, G. T. (1992). Inducible vectors for expression in mammalian cells. Cur. Op. in Biotech. 3, 506-511.

Aknowledgements

This work was supported by grants from the Pasteur Institute, Vaincre les Maladies Lysosomales, the Association Française contre les Myopathies.

Alphavirus-Retrovirus Vectors

Henrik Garoff and Kejun Li
Karolinska Institutet, Dept. of Biosciences at Novum
141 57 HUDDINGE, Sweden

Introduction

During recent years the concept of gene therapy has established itself as a potentially very interesting possibility to cure cancer, genetic and infectious diseases. However, as clearly stated in a recent N.I.H. report by S. Orkin and A. Motulsky (see commentary in Science 1995, vol 270, p.1751) the huge potential of gene therapy for medicine can only be released if the present technology of gene transduction is significantly improved.

The retrovirus vector system

The retrovirus vectors are perhaps the most important of all vector systems that are used in gene therapy. These are, in practice, the only ones that can efficiently integrate a foreign gene into a chromosome of a cell and thereby allow for long-lasting gene expression. Below, we will shortly discuss the retroviral life cycle, the existing retroviral vector systems and their short commings. We will also introduce a novel system for the packaging of retroviral vectors that we recently introduced (Li and Garoff, 1996) and describe its advantages.

Retroviruses are enveloped viruses that have the unique property of converting their RNA genome into double-stranded DNA and further to integrate that DNA into a host chromosome. The virus encodes three precursor proteins, gag, gag-pol and env. The gag precursor protein directs encapsidation of the RNA genome and the particle budding and release at the plasma membrane (PM) of an infected cell (Hunter, 1994). The env precursor is the membrane protein of the retrovirus. The membrane proteins are incorporated into the viral membrane during the budding process and are responsible for targeting of the virus to new host cells and also for a virus-host membrane fusion event. The membrane fusion process results in release of the viral capsid, which contains the genome, into the cytoplasm of the new cell. The gag-pol precursor contains three protein units in addition to gag. These are the viral protease, which cleaves the viral precursors during and shortly after budding, the reverse transcriptase, which converts the viral RNA to DNA, and the integrase, which mediates integration of the viral DNA into the chromosome. New viral RNA genomes and mRNAs are then transcribed from the integrated viral DNA (also called provirus).

The original retroviral vector system consists in principal of two provirus variants: one in which the region encoding the RNA encapsidation signal (Ψ) has been removed and another one in which the structural protein region has been replaced with that of a foreign gene (Mann et al., 1983). Coexpression of both vectors results in precursor protein production from the first vector and encapsidation competent RNA production from the second. Consequently, particles will be assembled that contain only foreign genes (in their RNA form). If the foreign gene represents a therapeutic gene then the particles can be used for gene therapy by infecting somatic cells. For practical reasons the retroviral vectors are produced in so called packaging cell lines (Miller and Rosman, 1989). These are transformed cells that produce the retroviral precursor proteins. For safety reasons the latter proteins are usually encoded by two separate genes (Fig. 1). When the packaging cells are transformed a second time with the provirus containing the therapeutic gene a so called producer cell line is generated. This continuously releases retroviral particles containing the therapeutic gene.

NATO ASI Series, Vol. H 105
Gene Therapy
Edited by Kleanthis G. Xanthopoulos

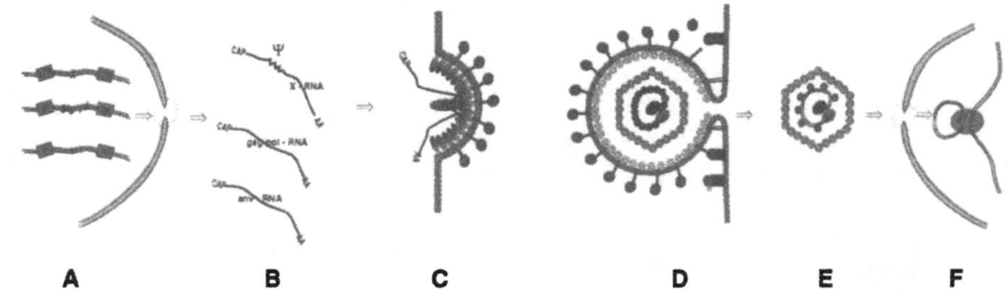

Fig. 1. Production and function of a retrovirus vector. The therapeutic gene X, the gag-pol and the env genes are parts of separate expression units in the producer cell (A). The unit for the X gene contains functional *cis* acting elements of the provirus. Transcription of the units results in the production of corresponding mRNA molecules (B). The X mRNA represents a recombinant retroviral genome with an encapsidation signal Ψ. The other ones serve as mRNAs for gag, gag-pol and env proteins. The latter will package X gene into particles (C), that via receptor binding and membrane fusion (D), will transport the X gene into new cells (E). Inside the new cell the X RNA is converted into dsDNA that finally is integrated into the chromosal DNA (F).

Problems with the retrovirus vector system

The mostly commonly used retroviral vectors are those that are derived from the Moloney Murine leukemia virus (MLV). The major problems with the MLV system are the following:

i) Only low amounts of transduction competent particles are released from the producer (Miller and Rosman, 1989). Usually 2-3 x 10^6 cells release only about 10^6 (titre 10^5/ml) infectious particles during a three day incubation period. This is less than one infectious particle per cell! The reasons for this are unclear. One possibility is that producer cell lines generate large amounts of incomplete retrovirus-like particles that compete with the complete ones for entry into cells.

ii) The particles only mediate a low level of gene expression (Huang and Gorman, 1990). Furthermore, the expression level tends to decrease with time.

iii) The particles lack specific targeting properties. Most packaging cell lines that are used today, express the env of the amphotropic MLV. This membrane protein binds to phosphate transporters that are present in varying amounts on the PM of cells of most tissues and in many different species including human.

vi) There are many safety risks involved. One inherent risk with the retroviral vector system is that replication proficient particles can arise due to RNA recombination during vector production in the producer cell line. Another risk is combined with the integration process, which can be regarded as an event of insertional mutagenesis of the chromosomal DNA.

iv) The system is very tedious and inflexible to use. It takes months to establish a good producer cell line. Afterwards one is completely fixed to the viral functions, e.g. targeting properties, that were originally introduced into the packaging cell line.

The principle of the alpha-retrovirus vector system

We have recently introduced a new vector system called alphavirus-retrovirus vectors (Li and Garoff, 1996). In this, alphaviral vectors are used to produce retroviral vectors. In order to explain the principle of this vector system we will first describe the alphaviral vector system (Liljeström and Garoff, 1991). It is based on the RNA replicon of Semliki Forest virus, SFV, another enveloped RNA virus. However, in contrast to the retroviral RNA the SFV RNA is replicated in the cytoplasm of the infected cells. The replication is directed by a viral polymerase which is translated from the 5' end of the viral genome. The 3' end of the genome is transcribed by a modified form of the polymerase into an mRNA encoding the structural viral proteins, a capsid and two membrane proteins. The SFV expression is about 100 times more efficient than that of a retrovirus. As a consequence SFV infected cells are lysed after about 2 days. The SFV vectors are composed of plasmids that contain the viral polymerase region and all *cis* -acting replication signals. The structural genes of SFV are replaced by a polylinker region into which a foreign gene can be inserted. The whole region is under the control of an SP6 promotor so that recombinant SFV RNA can be transcribed *in vitro*. When this is introduced into unifected cells it will produce SFV polymerase, start replication of recombinant RNA and finally transcription of mRNA that will be used for translation of the foreign protein.

The principle of the alphavirus-retrovirus vector system is shown in Fig. 2. The system is composed of three SFV plasmid vectors that can be used for *in vitro* transcription of SFV RNAs encoding the MLV env and gag-pol and a recombinant retrovirus gene (LTR-X-LTR), respectively. Cotransfection of all three SFV RNAs into BHK-21 cells results in an efficient production of the corresponding retroviral precursor proteins and the recombinant genome. These are assembled into retroviral particles that are released into the medium. The particles produced can transduce the X-gene into mouse cells. The most unique feature of this system is that it directs retroviral RNA genome expression in the cytoplasm and not in the nucleus.

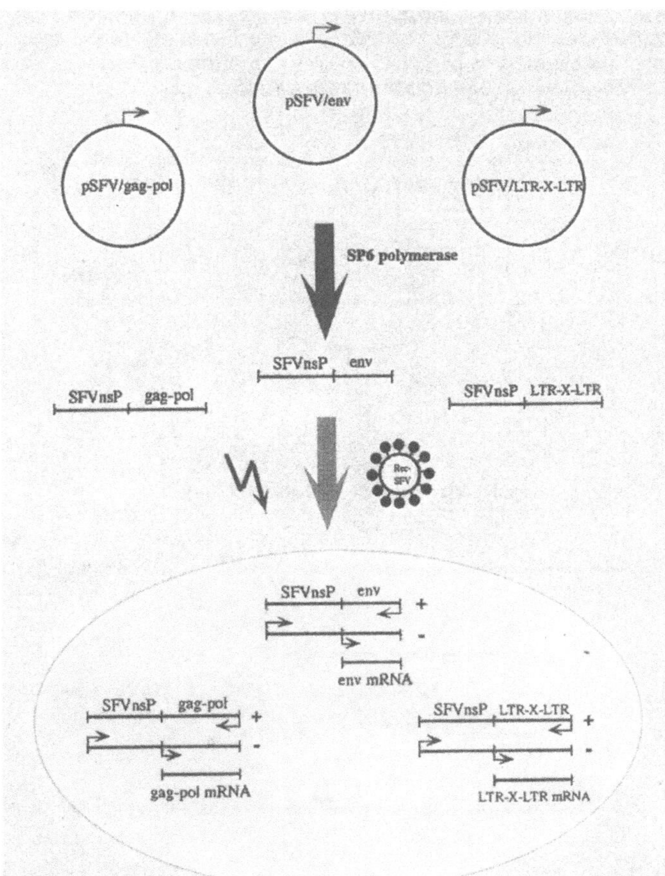

Fig. 2. The alphavirus-retrovirus vector system. Three SFV plasmid vectors (top) are used for *in vitro* transcription of recombinant SFV RNAs (middle). These are transfected into BHK-21 cells by electroporation. Alternatively the RNAs can be packaged into recombinant SFV particles and transferred by SFV

infection. In the BHK-21 cells the nsP gene regions of the recombinant SFV RNAs are translated into SFV polymerase compexes. These start replication of SFV RNAs in the cell cytoplasm and transcription of their 3' region into mRNAs (bottom part of figure). The env and gag-pol mRNAs will be translated into MLV precursor proteins that will encapsidate the recombinant retrovirus genomes (LTR-X-LTR) into retroviral vectors. These bud-off from the PM into the medium.

Potential problems in developing the alphavirus-retrovirus vector system

We could foresee two major difficulties in our strategy to produce functional alphavirus retrovirus vectors. Firstly, in all retrovirus packaging systems described so far the retrovirus genes are produced in nucleus and not in cytoplasm as is the case when using alphavirus expression vectors. If a nuclear localisation of the retrovirus genome is required for its efficient packaging the alphavirus driven expression system will most likely be inappropriate. Secondly, it is not possible to produce an authentic retrovirus genome in the form of an alphavirus subgenome as the latter requires some alphavirus specific sequences at its 5' and 3' ends. These are a 5' end sequence, which constitutes both the 3' region of the alphavirus subgenome-promoter and the coding sequence of nonstructural protein 4 (fig. 3), and a 3' end sequence that constitutes a viral RNA replication signal (Strauss and Strauss, 1994). Thus, the addition of these sequences to 5' and 3' ends of the retrovirus genome is necessary for its expression by the alphavirus vector. It was not clear to us to what extent such sequence addition influences retrovirus genome packaging into particles, reverse transcription, polymerization into double-stranded DNA, chromosome integration and expression (Fig. 3).

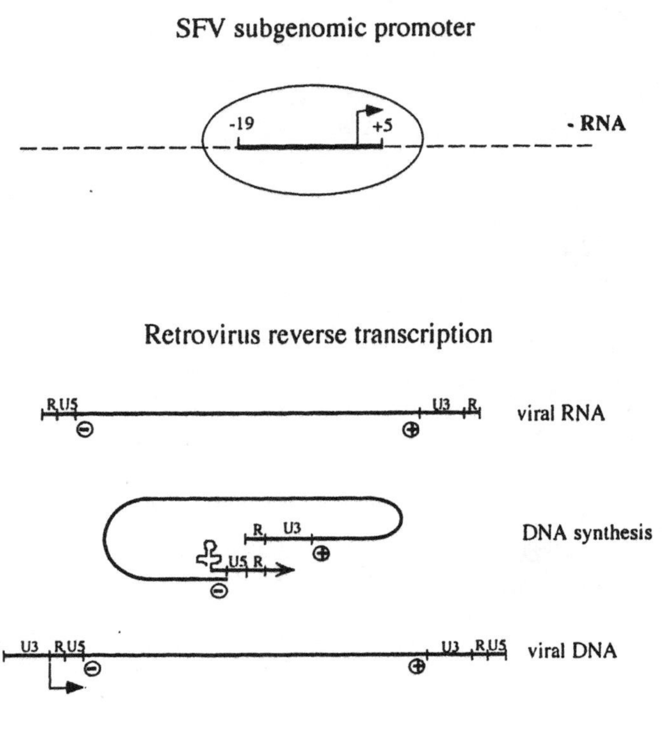

Fig. 3. Upper panel shows schematically how SFV promotor region overlaps with the 5' end of the SFV transcript. Lower panel shows the initial steps of DNA synthesis in a retrovirus.

Construction of SFV vectors for the expression of functional retrovirus recombinant genes

We made two kinds of SFV vector constructs to express the recombinant retrovirus genome, R-U5-Ψ^{+}-neo^{R}-U3-R from pLN, (Miller and Rosman, 1989). In the first one, pSFV/LN3i, a 38 pairs long sequence from the 5' end of the SFV subgenome was left attached to the 5' end of the recombinant retroviral genome. In order to facilitate DNA synthesis from the corresponding subgenomic RNA the same 38 base sequence was added between the U3 and R regions at the 5' end of the recombinant retroviral genome in pSFV1/LN3i (Fig. 4, upper panel). Reverse transcription of the encoded RNA subgenome and subsequent (+)DNA strand synthesis should generate DNA provirus as depicted in Fig. 4, upper panel (Luciw and Leung, 1992). In this provirus the SFV sequence will be inserted at the initiation site of the U3 directed transcription process. As it was not clear to what extent the expression of a downstream gene will be inhibited by the presence of the 38 basis long SFV sequence at the 5' end of the transcript we constructed also another plasmid, pSFV1/LN-U3insert. In RNA transcribed from this one the necessary SFV sequence at the 5' end was followed by an extra U3 sequence (Fig. 4, lower panel). At the 3' end, the same SFV sequence was inserted after the right integrase attachment site (att_r) of the 3' U3 region. When this RNA is converted to a provirus the SFV sequence will be found at its extreme 5' end leaving an intact U3-R-U5 region for transcription.

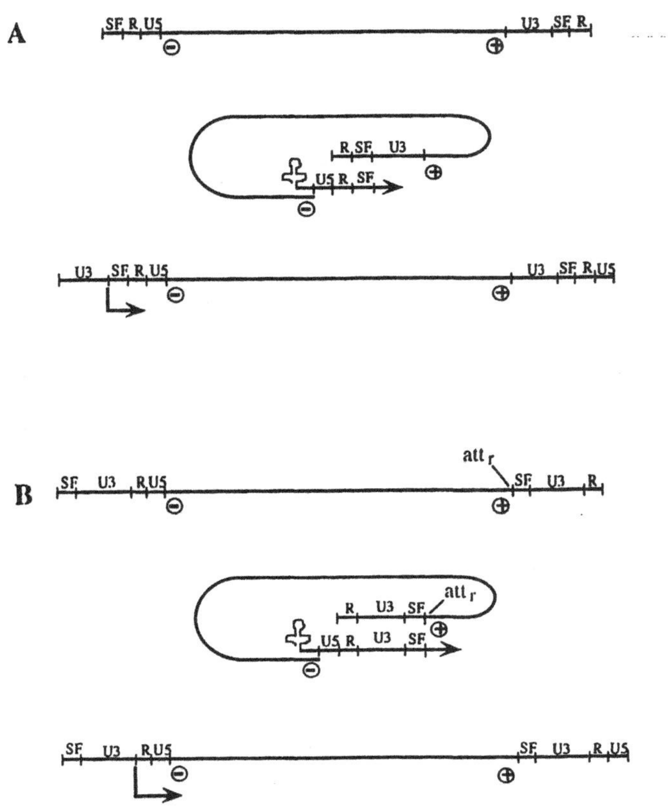

Fig. 4. Reverse transcription in alphavirus-retrovirus vectors. (A) This shows the initial stages of reverse transcription of recombinant retroviral RNA from SFV-1/LN3i. At top the SFV produced retroviral RNA

genome is shown to contain the necessary 5' SFV sequence (SF), the R and U5 elements of the MLV LTR, the site of (-) DNA strand synthesis, a coding region, the site for (+) DNA strand synthesis, the U3 elements of MLV LTR, a repeated SFV sequence and the R element of the MLV LTR. In the middle it is shown how the different RNA elements are copied into the (-) DNA strand. At the bottom the resulting proviral DNA is depicted. Initiation of transcription from the provirus is indicated with a red arrow. (B) This shows the structure of the recombinant retroviral RNA from pSFV-1/LN3i-U3insert (top), how it is reversed transcribed (middle) and the expected structure and function of the provirus (bottom). att_r, denotes the right integrate attachment site.

Production of retrovirus vectors

In order to produce retrovirus vectors that contain the neo^R gene we transcribed SFV vector RNA from pSFV1/LN3i *in vitro* and cotransfected this RNA and two other SFV-RNA vectors that were expressing the retrovirus structural proteins and enzymes. The latter RNAs were transcribed form the plasmids pSFV-C/gag-pol (or pSFV-1/gag-pol) and pSFV-1/AMenv (or pSFV-1/Pr80env). The pSFV-1/gag-pol plasmid contains the coding-region of the gag-pol of MLV (Suomalainen and Garoff, 1994). This is also present in pSFV-C/gag-pol. The latter plasmid contains in addition the SFV capsid coding-region in front of that of gag-pol. In the transcribed C-gag-pol RNA the C-region specifies increased translation efficiency as compared to gag-pol mRNA (Sjöberg et al., 1994).The pSFV-1/Pr80 env contains the coding region of the homologous (ecotropic) MLV env precursor protein and the SFV-1/AM env that of the heterologous env precursor protein of the Amphotropic Murine Leukemia virus. The latter envelope protein has the capacity to target the recombinant retrovirus particle to a broad range of animal host cells including several human cells, whereas the ecotropic env only recognizes mouse cells. The protein synthesis in cells cotransfected with SFV-1/LN3i, SFV-C/gag-pol and SFV1AMenv RNAs was followed by metabolic labelling with $[^{35}S]$methionine. This showed that all retrovirus structural proteins were expressed in the cells. The formation of virus particles was followed by analysis of the media from the cotransfected cells. Particles with correct protein compositions were found. The infectivity of the particles was studied by using the media containing the particles to infect NIH 3T3 cells and then selecting for neo^R transformants with the drug G418. The results (Table 1) showed that infectious particles were formed by our production procedure. This was a significant finding since it indicates that (1) a recombinant retrovirus genome that has been produced in cell cytoplasm, and not in the nucleus as during wild-type retrovirus infection, can be packaged into infectious retrovirus particles; and (2) that insertion of foreign SFV-derived RNA sequences in the subgenomic recombinant retrovirus RNA molecule is compatible with for efficient recombinant retrovirus RNA packaging into recombinant retrovirus particles, reverse transcription, dsDNA synthesis, integration of retrovirus DNA into host chromosomes and expression of the integrated gene. The time course of recombinant particle production (Table 1) shows of that, when using RNA combinations including SFV-C/gag-pol RNA, not more than 10 h inbucation is required for the generation of a vector preparation with more than 10^6 particles/ml of culture medium. This is true for recombinant retrovirus particles with both ecotropic and amphoptropic env proteins. The analyses of the media from the three subsequent 5 h incubations showed that 2-4 x 10^6 particles were released during each of the incubation intervals. These recombinant retrovirus particles-concentrations are very high and corresponds to the highest ones reported for recombinant retrovirus vectors that have been produced by other stable or transient producer cell systems (Finer et al., 1994) (Landau and Littman, 1992) (Miller and Rosman, 1989) (Pear et al., 1993) (Soneoka et al., 1995)

Table 1. Release of infectious recombinant retrovirus particles from transfected BHK-21 cells*

experiment	RNA †	G418^R CFU‡/ml				
		0 - 5 hr	5 - 10 hr	10 - 15 hr	15 - 20 hr	20 -25 hr
1	SFV1/LN3i + SFV-C/gag-pol	—§	0	—	—	—
2	SFV1/LN3i + SFV1/Pr80env	—	0	—	—	—
3	SFV1/LN3i + SFV1/AMenv	—	0	—	—	—
4	SFV1/LN3i + SFV1/gag-pol + SFV1/Pr80env	3.7×10^4¶	8.0×10^5	1.1×10^6	8.5×10^5	6.5×10^5
5	SFV1/LN3i + SFV-C/gag-pol + SFV1/Pr80env	7.3×10^4	4.0×10^6	4.0×10^6	2.1×10^6	4.0×10^6
6	SFV1/LN3i + SFV-C/gag-pol + SFV1/AMenv	1.0×10	2.2×10^6	2.3×10^6	2.0×10^6	3.4×10^6

* In each experiment about 4×10^6 transfected BHK-21 cells were plated into a 60mm culture dish and incubated at 37°C. The medium was collected and replaced at 5 hour intervals. Media samples were passed through 0.45µm filter and stored at -130 C before being used for titration.
† RNA used for transfection of BHK-21 cells
‡ CFU, colony forming units.
§ —, not analysed.
¶ 3T3 cells were incubated with diluted medium of transfected BHK-21 cells and then subjected to G418 selection. The numbers refer to resistant colonies formed after 12 days incubation.

Similar studies were performed with RNA made from pSFV1/LN-U3insert. The titer of the corresponding recombinant retrovirus preparation was approximately the same as obtained with RNA from pSFV/LN3i.

Production of replication competent particles

When considering to use recombinant retrovirus vectors for gene therapy in humans it is important to assess the safety risks that might be associated with such an undertaken. The major risk with recombinant retrovirus vector preparations is contamination by replication competent retrovirus particles. A replication competent particle has acquired all retrovirus structural protein genes and hence it has the capacity to spread from cell to cell. Such particles can be generated in the producer cell through the process of RNA recombination. The possible generation of replication-competent particles in our production system was tested using a marker rescue assay (van Beusechem et al., 1990). No replication-competent particles were found in a sample containing 2.6×10^6 infectious recombinant particles.

Concluding remarks

To this end we have shown that our expression system can be used for production of neo^R-transduction competent retroviral particles with both eco (mouse cell specific) and amphotropic (broad host-cell range) env. The system can be used for the production of stocks of retrovirus vector with high titer. The production efficiency is about 10^7 infectious particles (titre 4×10^6/ml) in 15 h using 3×10^6 transfected cells. We have not detected any replication proficient particles in our preparations. Because our system is based on three separate RNA transcription plasmids and a short-time particle production, it is both flexible and fast to use. Therefore, the system should, among others, be very useful for testing long series of recombinant retrovirus variants that are, for instance, designed for targeted gene delivery. Furthermore, the transcient but efficient nature of the system should allow for production of recombinant retroviruses containing toxic gene products. The most interesting feature with the new system is, however, the cytoplamic production of the retroviral genome. This should, in theory, enable the packaging of intron containing genes into retroviruses

Future potential of the alphavirus-retrovirus vector system

Normally eukaryotic genes contain introns. These are, among others, important for correct 3' processing of mRNA molecules and for mRNA transport from the nucleus to cytoplasm. Comparisons of the gene expression levels in cells that have been transformed with the same gene with or without an intron have clearly demonstrated the importance of the intron for efficient gene expression. In these experiments, which were done both in cell cultures and in animals, the intron increased the gene expression up to 100 fold (Huang and Gorman, 1990) (Choi et al., 1991). Therefore, it appears reasonable to include intron sequences also into therapeutic genes that are to be transduced into somatic cells by retroviral vectors. Most likely this should result in an increased expression level of the therapeutic gene and also in a more durable expression. However, the mode of retroviral gene expression excludes the possibility of expressing genes with introns. If such genes were engineered into a proviral DNA, and used for transformation of a packaging cell line, the intron sequences would be removed during nuclear processing of the RNA that is transcribed from the recombinant provirus. Thus, only intronless recombinant RNA genomes will reach the cytoplasm and be packaged into retroviral particles. The alphavirus-retrovirus vector system that we have introduced circumvents this problem. In this case the recombinant retroviral RNA genome is produced in the cytoplasm, not in the nucleus where the splicing machinery is located. The system therefore provides, in theory, an unique possibility to package genes with introns into retroviral vectors. Such vectors should be very useful for efficient transduction of intron containing therapeutic genes into somatic cells. A series of experiments is planned to test this possibility.

References

Choi, T., Huang, M., Gorman, C., and Jaenisch, R. (1991). A generic intron increases gene expression in transgenic mice. Mol. & Cell. Biol. *11*, 3070-3074.

Finer, M. H., Dull, T. J., Qin, L., Farson, D., and Roberts, M. R. (1994). A high-efficiency retrivirual transducton system for primary human T lymphocytes. Blood *83*, 43-50.

Huang, M. T. F., and Gorman, C. M. (1990). Intervening sequences increase efficiency of RNA 3' processing and accumulation of cytoplasmic RNA. Nucleic Acids Res. *18*, 937-947.

Hunter, E. (1994). Macromolecular interactions in the assembly of HIV and other retroviruses. Virology *5*, 71-83.

Landau, N. R., and Littman, D. R. (1992). Packaging system for rapid production of murine leukemia virus vectors with variable tropism. J. Virol. *66*, 5110-5113.

Li, K., and Garoff, H. (1996). Production of infectious recombinant Moloney Murine leukemia virus particles in BHK cells using Semliki Forest virus-derived RNA expression vectors. Proc. Natl. Acad. Sci. USA *93*, 11658-11663.

Liljeström, P., and Garoff, H. (1991). A new generation of animal cell expression vectors based on the Semliki Forest virus replicon. BioTechnology *9*, 1356-1361.

Luciw, P. A., and Leung, N. J. (1992). Model for RT. In The Retroviridae, J. A. Levy, ed. (New York: Plenum), pp. 194-198.

Mann, R., Mulligan, R. C., and Baltimore, D. (1983). Construction of a retrovirus packaging mutant and its use to produce helper-free defective retrovirus. Cell *33*, 153-159.

Miller, A. D., and Rosman, G. J. (1989). Improved retroviral vectors for gene transfer and expression. BioTechniques *7*, 980-990.

Pear, W. S., Nolan, G. P., Scott, M. L., and Baltimore, D. (1993). Production of high-titer helper-free retroviruses by transient transfection. Proc. Natl. Acad. Sci. USA *90*.

Sjöberg, M., Suomalainen, M., and Garoff, H. (1994). A significantly improved Semliki Forest virus expression system based on translation enhancer segments from the viral capsid gene. Bio/Technology *12*, 1127-1131.

Soneoka, Y., Cannon, P. M., Ramsdale, E. E., Griffiths, J. C., Romano, G., Kingsman, S. M., and Kingsman, A. J. (1995). A transient three-plasmid expression system for the production of high titer retroviral vectors. Nucleic Acids Res. *23*, 628-633.

Strauss, J. H., and Strauss, E. G. (1994). The alphaviruses: Gene Expression, Replication and Evolution. Microbiol. Rev. *58*, 491-562.

Suomalainen, M., and Garoff, H. (1994). Incorporation of homologous and heterologous proteins into the envelope of Moloney murine leukemia virus. J. Virol. *68*, 4879-4889.

van Beusechem, V. W., Kukler, A., Einerhan, M. P. W., Bakx, T. A., van der Eb, E. J., Bekkum, D. W., and Valerio, D. (1990). Expression of human adenosine deaminase in mice transplanted with hemopoietic stem cells infected with amphotropic retroviruses. J. Exp. Med. *172*, 729-736.

Integration of Retroviruses into a Predetermined Site

Kelvin Davies & Ganjam Kalpana

Dept. of Molecular Genetics, Ullman 823, 1300 Morris Park Ave, Bronx, NY 10461.

a. Introduction

Retroviruses are currently the preferred vehicle for transduction of mammalian cells to be used in gene therapy. However, safety concerns limit their use at present to *ex vivo* procedures. One of the ultimate challenges for gene therapy will be to develop a vehicle for gene delivery that can be used *in vivo*. The vector of choice is likely to remain a retrovirus; one of its major advantages being that it can link therapeutic sequences to the host chromosome with stable covalent bonds. However, this positive attribute of retroviral vectors is potentially one of its major drawbacks. The integration of the viral genome, catalyzed by the enzyme integrase (IN), shows no discernible sequence specificity. There is the danger that such unpredictable integration of retroviral sequences could activate proto oncogenes potentially leading to carcinogenisis, or conversely, could lead to gene inactivation. However, integration is not completely random. Several studies have indicated that *in vivo*, many mouse retroviruses tend to integrate into DNAseI hypersensitive or transcriptionally active sites suggesting that there may be a bias towards open chromatin structures[1, 2, 3, 4, 5]. An even more pronounced bias in target site selection has been observed for retrotransposable elements such as Ty1 and Ty3 of yeast. Emerging evidence suggests that target site selection may be governed by host factors that associate with retroviral/retrotransposable element integration machinery[6]. By understanding the mechanism of target site selection by retroviruses, one may be able to modulate the process in such a way that retroviral integration can be directed to predetermined safe sites on the chromosome.

b. Integration

Integration into the host cell genome is an essential step in the life cycle of all retroviruses and is catalyzed by the virally encoded enzyme integrase (IN). Soon after the entry of retrovirus into the host cell, the RNA genome is converted into double stranded cDNA, which resides within a huge (~160S) nucleoprotein complex called the pre-integration complex or PIC[7]. This complex contains both viral proteins, that include IN, and host cell factors, such as non-histone protein, HMG I(Y)[8]. Even though the structural and functional organization of the PIC is not completely understood, it is believed that the viral cDNA ends are tightly protected by binding of proteins, probably IN, while the rest of the viral cDNA is accessible to nucleases such as restriction enzymes and DNAseI. The PIC that is formed in the cytoplasm of an infected cell, traverses the nucleus and then the viral cDNA integrates into the host DNA.

NATO ASI Series, Vol. H 105
Gene Therapy
Edited by Kleanthis G. Xanthopoulos

The integrase reaction has three distinct steps: i) removal of 2 terminal bases from each 3' end of the viral DNA called "3' processing" or "clipping" that results in the formation of 3' -OH at the end of conserved CA dinucleotides[9] , ii) joining of 3' end to the 5' phosphate of the host strand called the strand transfer" or "joining" reaction. In this reaction, each strand of the viral DNA joins one strand of the host DNA separated by 4-6 base pairs[10]; and iii) repair of gaps generated at the target site and removal of 5' overhangs of the viral unjoined strand. The first two steps of integration can be carried out by IN alone in vitro. The third repair step has been thought to be carried out by the host enzymes.

c. Target site selection during Integration

Retroviruses integrate at many sites on the host chromosome without any sequence specificity. However, several reports have implicated a bias in target site selection for many retroviruses in vivo. For example, MLV seems to preferentially integrate at DNAseI hypersensitive sites or transcriptionally active regions [1, 2, 3, 4, 5]. A higher degree of bias has been documented for the target site selection by retrotransposons such as Ty elements of yeast. For example, Ty 5 elements are all located at telomeres or regions that have telomeric chromatin. Ty1 through 4 show a preference for integration upstream of PolIII transcribed units[11]. Ty1-3 show a gradation in the specificity for target site selection--while Ty1 inserts few hundred bases upstream of tRNA genes, Ty3 integrates within 5 bases of transcriptional initiation site[12].

The mechanism for target site selection is poorly understood. The bias in target site selection could be governed by the modification or structural alteration in the DNA. Several reports indicate that CpG methylation of DNA may be preferred target sites for in vitro integration[13]. Experiments using chromatin DNA as a target *in vitro* suggests that sequences that are exposed or have kinks in their DNA due to the packaging of the nucleosomes may be preferred target sites in vitro[14,15,16]. IN itself might play a role for target site selection, at least *in vitro*. For example, studies show that IN from different retroviruses have different integration patterns[16] which appear to be determined by the central core domain of IN[17]. However, it is unknown whether any of these factors would be major determinants of bias in target site specificity *in vivo*.

There is gathering evidence that cellular host factors, in particular transcription factors, play a role in determining the sites of retroviral and retrotransposon integration. The best studied example is the integration by Ty3 element, which integrates within few bases around the transcription initiation sites. Level of transcription does not seem to influence the integration into this site. However, in vitro integration studies suggests that binding of transcription factors TFIIIB and TFIIIC to the promoter sites seems to be essential for integration. Similarly, retroviral integration may be influenced by the host transcription factors. In an attempt to identify host factors that are necessary for integration, HIV-1 IN was used as bait in the yeast two hybrid system[6]. A single gene Ini1 was identified as a IN binding protein in this assay. IniI has homology to SNF5, a component of the yeast

SWI/SNF complex[18]. Mammalian equivalents of SWI/SNF complex have been purified and these complexes has been shown to contain Ini1 and other human homologues [19,20,21]. Like the yeast SWI/SNF complex, the mammalian SWI/SNF complex has the ability to remodel chromatin and allow the access of sequence specific DNA binding proteins [22,23] *in vitro*. Based on these observation we hypothesize that binding of IN to Ini1 present in the SWI/SNF complex may target the retroviral integration to open chromatin (Fig. 1.). Since transcriptionally active regions also have high rate of transcription coupled repair, this not only would ensure the transcription of integrated provirus but also facilitates repair of the integration intermediate.

d. Using Host Factors to Direct Integration

In our laboratory we are testing the hypothesis shown in Fig. 1, that Ini1 can target retroviral integration. We are hoping to demonstrate that Ini1 fused to a specific DNA binding protein can target retroviral integration into predetermined sequences preferred by the DNA binding element in mammalian cells. The sites recognized by the DNA binding protein will either be present transiently on a plasmid, or integrated into the host genome. Upon infection of the cells expressing the Ini1 DNA binding fusion protein we hope to see integration into the vicinity of the target site. Once we are successful in the intial targeting efforts, we hope to utilize this strategy for targeting retroviral vectors into specific sites on the host chromosome.

One approach for developing a targeted integration has been to fuse DNA binding domains to IN itself. Initial experiments have demonstrated the feasibility of this idea in vitro. LexA [24, 25], lambda repressor[26] or Zif[27] have been fused to the C-terminus of IN and the ability of the integrases to target integration was investigated. However, the retroviral virus particles carrying Zif-IN fusion was non infectious[27]. Since IN plays a pivotal role in integration, it is likely that altering the three dimensional structure of IN by the addition of DNA binding elements lowered the efficiency of the integration reaction. Altering the host protein may overcome the need to alter IN protein itself and may provide more efficient method of targeting integration.

e. Conclusion and Perspectives

In order to fulfill the great potential offered by gene therapy it will be necessary to design a system by which genes are safely and stably integrated into target cells. At present the vector with most promise for achieving these goals is a retrovirus. Random insertion of foreign DNA into a patients genome is unlikely to receive widespread approval for clinical use. Therefore, an understanding of the factors that determine retroviral integration and the development of methods to

Fig.1 Targeting of Retroviral Integration by Ini1

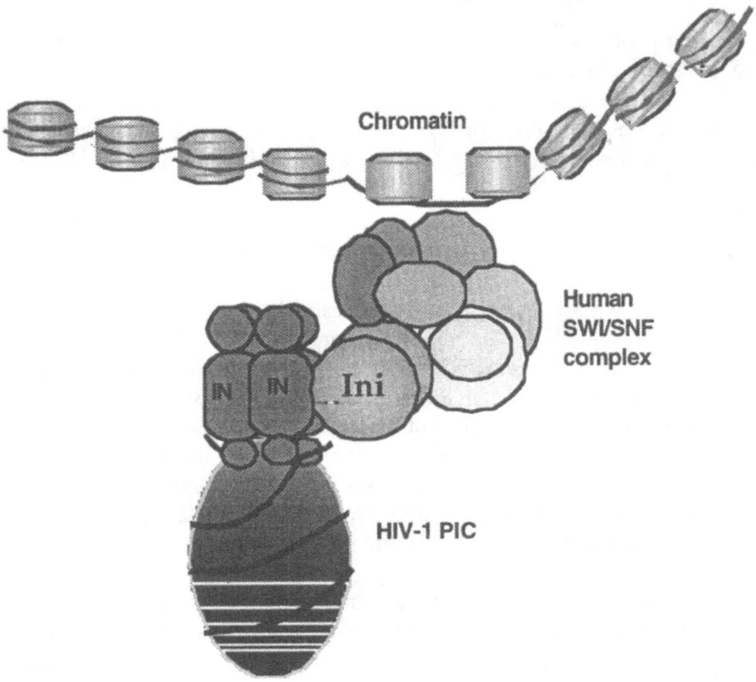

exploit these factors should be a concern in the strategic development of gene therapy. One of the approaches may be to target the integration to predetermined sites by using a host factor that interacts with IN.

References

1. Vijaya S, Steffen DL, Robinson HL, 1986. Acceptor sites for retroviral integrations map near DNAse I-hypersensitive sites in chromatin. *J. Virol. 60*: 683-692.

2. Rohdewohld H, Weiher H, Reik W, Jaenisch R, Breindl M, 1987. Retrovirus integration and chromatin structure: Moloney murine leukemia proviral integration sites map near DNAse I-hypersensitive sites. *J. Virol. 61*: 336-343.

3. Scherdin U, Rhodes K, Breidl M, 1990. Transcriptionally active genome regions are preferred targets for retrovirus integration. *J. Virol. 64*: 907-912.

4. Shih C-C, Stoye JP, Coffin JM, 1988. Highly preferred targets for retroviral integration. *Cell 53*: 531-537.

5. Withers-Ward ES, Kitamura Y, Barnes JP, Coffin JM, 1994. Distribution of targets for avian retrovirus DNA integration in vivo. *Genes and Development 8*: 1473-1487.

6. Kalpana GV, Marmon S, Wang W, Crabtree GR, Goff SP, 1994. Binding and stimulation of HIV-1 integrase by a human homolog of yeast transcription factor SNF5. *Science 266*: 2002-2006.

7. Bowerman B, Brown PO, Bishop JM, Varmus HE, 1989. A nucleoprotein complex mediates the integration of retroviral DNA. *Genes Dev. 3*: 469-478.

8. Farnet CM, Bushman F.D., 1997. HIV-1 cDNA integration: Requirement of HMG I(Y) protein for function of preintegration complexes in vitro. *Cell 88*: 483-492.

9. Katzman M, Katz RA, Skalka AM, Leis J, 1989. The avian retroviral integration protein cleaves the terminal sequences of linear viral DNA at the in vivo sites of integration. *J. Virol. 63*: 5319-5327.

10. Craigie R, Fujiwara T, Bushman F, 1990. The IN protein of Moloney murine leukemia virus processes the viral DNA ends and accomplishes their integration in vitro. *Cell 62*: 829-837.

11. Ji H, Moore DP, Blomberg MA, Braiterman LT, Voytas DF, Natsoulis G, Boeke JD, 1993. Hotspots for unselected Ty1 transposition events on yeast chromosome III are near tRNA genes and LTR sequences. *Cell 73*: 1007-1018.

12. Chalker DL, Sandmeyer SB, 1992. Ty3 integrates within the region of RNA polymerase III transcription initiation. *Genes Dev. 6*: 117-128.

13. Kitamura Y, Ha Lee YM, Coffin JM, 1992. Nonrandom integration of retroviral DNA *in vitro*: Effect of CpG methylation. *Proc. Natl. Acad. Sci. 89*: 5532-5536.

14. Wolffe AP, 1994. Switched-on chromatin. *Current Biology 4*: 525-528.

15. Pryciak PM, Varmus, H.E., 1992. Nucleosomes, DNA-binding proteins, and DNA sequence modulate retroviral integration target site selection. *Cell 69*: 769-780.

16. Pryciak PM, Sil A, Varmus HA, 1992. Retroviral integration into minichromosomes in vitro. *EMBO J. 11*: 769-780.

17. Shibagaki Y., Chow, SA., 1997. Central Core Domain of Retroviral Integrase is Responsible for Target Site Selection. *J. Biol. Chem. 272*: 8361-8369.

18. Laurent BC, Treitel MA, Carlson M, 1990. The SNF5 protein of Saccharomyces cerevisiae is a glutamine- and proline-rich transcriptional activator that affects expression of a broad spectrum of genes. *Mol. Cell. Biol. 10*: 5616-5625.

19. Muchardt C, Reyes, J.C., Bourachot, B., Leguoy, E. and Yaniv, M., 1996. The hbrm and BRG-1 proteins components of the SNF/SWI complex, are phosphorylated and excluded from the condensed chromosomes during mitosis. *EMBO J 15*: 3394-3402.

20. Wang W, Cote J, Zhou S, Muchardt C, Khavari P, Bigger SR, Xue Y, Kalpana GV, Goff SP, Yaniv M, Tjian R, Workman J, Crabtree G, 1996. Identification of multiple forms of SWI/SNF complexes in mammalian cells: implication for its diverse function in development and tissue-specific gene expression. *Submitted.*

21. Wang W, Xue, Y., Zhou, S. Kuo, A., Cairns, B.R. and Crabtree, G.R., 1996. Diversity and specialization of mammalian SWI/SNF complexes. *Genes and development 10*: 2117-2130.

22. Imbalzano AN, Know H, Green MR, Kingston RE, 1994. Facilitated binding of TATA-binding protein to nucleosomal DNA. *Nature 370*: 481-485.

23. Kwon H, Imbalzano A, Khavari PA, Kingston RE, Green MR, 1994. Nucleosome disruption and enhancement of activator binding by a human SWI/SNF complex. *Nature 370*: 477-481.

24. Goulaouic H, Chow SA, 1996. Directed Integration of Viral DNA mediated by Fusion Proteins Consisting of Human Immunodeficiency Virus Type 1 Integrase and *Escherichia coli* LexA Protein. *J. Virology 70*: 37-46.

25. Katz RA, Merkel G, Skalka AM, 1996. Targeting of Retroviral Integrase by Fusion to a Heterologous DNA Binding domain: *In vitro* Activities and Incorporation of a Fusion Protein into Viral Particles. *Virology 217*: 178-190.

26. Bushman FD, 1994. Tethering human immunodeficiency virus 1 integrase to a DNA site directs integration to nearby sequences. *Pro. Natl. Acad. Sci. U.S.A. 91*: 9233-9237.

27. Bushman FD, Miller MD, 1997. Tethering Human Immunodeficiency Virus Type 1 Preintegration Complexes to Target DNA Promotes Integration at Nearby Sites. *J. Virology 71*: 458-464.

Amphotropic Envelope/Receptor Interactions

Pierre Rodrigues and Jean-Michel Heard, Laboratoire Rétrovirus et Transfert, Génétique, Institut Pasteur, 28 rue du Docteur Roux, 75724 Paris cedex 15., Phone: 33 (0) 1 45 68 82 46, Fax: 33 (0) 1 45 68 89 40, e-mail: jmheard@pasteur.fr.

Abstract

The amino-terminal domain of murine leukemia virus (MuLV) envelope glycoproteins (SU) is sufficient for binding cell surface receptors and mediating entry into cells. This domain is an anti-parallel ß sandwich with two helical subdomains forming loops adjacent to the ß-sandwich. The loops contains the determinants involved in receptor recognition. A purified 208 aminoacid fragment containing the amphotropic receptor binding domain competes with the binding of amphotropic particles and inhibits the entry of amphotropic retrovirus vectors. Concentrations inhibiting entry appeared much lower than that required to abolish binding. This suggested that only a fraction of the receptors are competent for processing retrovirus entry. Requirement for cell factors or for association with specific cell structure may account for this restriction. Alternatively, the association of several receptor molecules may be required for processing entry. Scatchard analysis performed with ^{125}I-labeled AS208 showed curvilinear plots with downward concavity, indicating that receptor cooperativity participates in binding efficiency.

The amphotropic receptor is the Pit-2 molecule encoded by the *ram*-1 gene. It is a multiple transmembrane protein which functions as a phosphate/Na symporter. Sequence analysis predict 10 transmembrane domains, 5 extracellular loops and intracellular N- and C-terminal extremities. We inserted a 11 amino acid epitope of the VSV G protein at various locations including the extracellular loop 5 and the C-terminal extremity. The tagged receptors were expressed in CHO cells which do not express the amphotropic receptor naturally. Virus particle binding and infection mediated by tagged or wild type receptors were equivalent. Axiti-VSV-G mAbs immunoprecipitated a 7OkDa glycosylated receptor molecules in transfected cells. Flowcytometry and immunofluorescence analysis revealed that, in contrast with the predicted topology, the C-terminal epitope is extracellular. In naive cells, the signal was homogeneously spread over the plasma membrane. After one hour incubation with virus particles, the signal condensed as large granulations. This observation was consistent with the hypothesis of a clustering of the amphotropic receptor in response to particle binding. Spatial reorganization of the receptors was observed in response to phosphate. High phosphate concentration induced spreading of the receptor and inhibited virus entry. In contrast, phosphate starvation induced receptor aggregation, induced pictures evoking stress cable formation and membrane ruffling, and increased virus entry. Actin staining confirmed colocalization of cell surface receptors with intracellular actin filaments. Colocalization was still observed after Cytochalasine D treatment, which disrupt actin network. Virus entry was completely abolished in the presence of cytochalasine. These data show that association of the amphotropic receptor with cytoskeleton structures plays a crucial role in virus entry. We are currently investigating the signalisation pathways stimulated by virus particle binding which induce receptor reorganisation and allow for virus entry into cells.

NATO ASI Series, Vol. H 105
Gene Therapy
Edited by Kleanthis G. Xanthopoulos

Introduction

Retrovirus infection is initiated by the attachment of viral particles to specific receptor proteins present at the target cell surface. Receptor recognition is mediated by the SU (surface) subunits of viral envelope glycoprotein oligomers, which are bound at the virions surface through interaction with TM (transmembrane) subunits. Five Murine Leukemia Virus (MLV) subgroups have been identified, which bind different cell surface receptors (Weiss, 1993). The ecotropic and amphotropic MLV subgroups interact with multiple membrane-spanning transporters for cationic amino-acids (Rec-1)(Kim, 1991; Wang, 1991a) and inorganic phosphate (Pit-2)(Kavanaugh, 1994), respectively. The receptor binding domain of MLV SUs has been located in the first half of the SU (Heard, 1991) and two hypervariable regions, VRA and VRB, have been shown to contribute to receptor recognition (Battini, 1992; Ott, 1992; Morgan, 1993). Fusion between the viral and the cytoplasmic lipid bilayers is likely to be triggered by conformational changes of the SU-TM heterodimers, which follow receptor binding. A fusogenic peptide most probably located at the N-terminal extremity of the TM subunit (Jones, 1993) and the C-terminal half of the SU are involved in the fusion process (Pinter, 1986; Nussbaum, 1993). The N-terminal receptor binding domain and the C-terminal fusogenic domain of the SU are separated by a proline-rich hinge.

I - Receptor binding determinants of the amphotropic SU are located in a 14 amino-acid hydrophilic loop.

The map of disulphide bridges is available for the ecotropic (Linder, 1992) and polytropic (Linder, 1994) MLV SUs and the crystal structure of the ecotropic N-terminal domain has recently been established by Berger and Cunningham (unpublished). Sequence alignment of SU N-terminal halfs indicates that most cystein residues engaged in disulphide bridge formation are conserved between MLV subgroups (Battini, 1992), suggesting that the maps of amphotropic, xenotropic and 10Al N-terminal disulphide bridges must be closely related. According to these findings, the formation of hydrophilic loops can be predicted in three different locations of the MLV SU receptor binding site: the N-terminal half of VRA, the C-terminal half of VRA and the VRB domain. An additional hydrophilic loop may exist at the N-terminal edge of the amphotropic VRB. These structures are candidates for mediating interaction with cell surface receptors. Point mutations introduced in the ecotropic SU revealed that the loop-forming structure located in the N-terminal half of VRA may participate in the recognition of the ecotropic receptor (McKrell, 1996).

We have substituted segments of 6 to 15 aminoacids located in the amphotropic SU receptor binding site with an 11 aminoacid linear epitope of the VSV-G protein. This peptide, which does not interact with any cell surface receptor, was used as a tagging epitope for studying synthesis, incorporation into virus particles and receptor interaction of the modified envelope glycoproteins. Substitutions were performed at two locations of VRA and two locations of VRB.

Substitutions in predicted loop structures were compatible with envelope processing, transport and incorporation into virus particles. However, these processes were much less efficient for substituted than for wild type molecules at 37°C. Significant improvement was observed when temperature was shifted to 32°C. Temperature sensitive mutants of the N-terminal extremity (Szurek, 1990) or in the proline-rich region (Gray, 1993) of the ecotropic SU have been reported. The ts phenotype of substituted SUs suggests that altered processing at 37°C resulted from unstable folding. Folding unstability probably also accounted for spontaneous shedding.

Despite structural unstability, loop-substituted mutant gave rise to mature envelope glycoproteins which were stably expressed at the cell surface and incorporated into virus particles, as shown by flowcytometry and western blot analysis performed with the anti-SU antibodies. The detection of a cell surface signal indicates that substituted segments were exposed on SU oligomers, and, consistently with the prediction of hydrophilic loop-forming structures, easily accessible to antibody molecules.

Interference assays, binding studies and infection of target cells indicated that substitution in potential loop forming segments of the receptor binding site of the amphotropic SU does not impair receptor recognition except for a loop located at the N-terminal edge of VRA. Therefore, whereas epitopes located in other potential loop forming are dispensable for receptor interaction, a sequence of 14 amino acids of VRA was identified as the main determinant involved in the recognition of the amphotropic receptor by the amphotropic SU.

II - Binding of the amphotropic envelope to the cell receptor.

Ecotropic envelope-receptor interactions have been previously analyzed using radiolabeled purified SUs (De Larco, 1976; Fowler, 1977; Kalyanaraman, 1978; Choppin, 1981; Johnson, 1986; Eiden, 1993) or by cytofluorimetric analysis of cells incubated with ecotropic virus particles (McGrath, 1978; McGrath, 1979; Kadan, 1992). Affinity constants were determined in the range of $10^9 M^{-1}$. Amphotropic envelope-receptor interactions have also been investigated using similar approaches. However, since purified preparations of the amphotropic envelope were not available, the binding parameters have not been determined (Kadan, 1992; Kozak, 1995).

We have purified to homogeneity a fragment of the amphotropic SU glycoprotein (referred to as AS208) containing the amphotropic receptor recognition domain, and its binding properties to different cell types have been characterized. The purified ^{125}I-labeled fragment (AS208) specifically bound cells susceptible to amphotropic MuLV infection. The number of AS208 binding sites was in the range of 7 to 17×10^4 per cell. Binding of the purified protein competed with that of amphotropic virus particles and led to complete resistance to entry of amphotropic virus pseudotypes into cells.

Interestingly, several lines of evidence suggested that, in living cells, the interactions between amphotropic envelopes and cell surface receptors, which mediate virus entry, are more complex than a bimolecular, two orders, reversible reaction. A first indication came from AS208 binding studies on NIH 3T3 cells. Scatchard plot of AS208 interactions with binding sites appeared curvilinear with a downward concavity. This aspect suggests that binding sites are heterogeneous with several classes of order differing in binding affinity. It is consistent with the hypothesis of a positive cooperativity between receptors. A second indication came from competition experiments. The concentrations of AS208 required to inhibit the binding or the entry of amphotropic virus particles into cells were determined. A reduction of roughly 50% of the number of bound virus particles required 1.8 to 3 nM of AS208, whereas a reduction of roughly 50% of cell infection was observed with 0.5 to 1 nM. Therefore, the inhibition of cell infection appeared for lower AS208 concentrations than the inhibition of receptor binding. For example, in the presence of 0.6 nM of AS208, infection was reduced 5 fold, whereas particle binding was not affected, indicating that the number of available binding sites was sufficient to bind significant amounts of virus particles, whereas the number of functional receptors was drastically reduced. This suggests that only a fraction of the envelope binding sites participated to the internalization of virus particles and actually functioned as virus receptors. A third indication for receptor heterogeneity was suggested by virus binding kinetics studies. Whereas AS208 binding progressively increased over time until a plateau was reached, virus particle binding showed a peak followed by an early decrease of the number of attached particles, then a plateau. Since this experiment was performed with a large excess of unbound virus, one possible interpretation is that a fraction of the receptors may have internalized bound particles rapidly whereas others did not. Another indication of receptor heterogeneity came from CHO cell clones transduced with various copy numbers of the *ram-1* cDNA. In these cells, the susceptibility to infection with an amphotropic *lacZ* retrovirus pseudotype appeared independant of the number of AS208 binding sites. Similar observations have previously been reported by Wang et al. on cells transduced with *rec*-1, which suggested the existence of an accessory factor or process limiting infection with ecotropic viruses (Wang, 1991b).

It is conceivable that only a fraction of the amphotropic envelope binding sites present at the cell surface are located in a suitable subcellular environment, where the post-binding events required for infection can be processed efficiently. This would result in a functional heterogeneity regarding the function of Pit-2 proteins as retrovirus receptors. Efficient binding and internalization of retrovirus particles may also require a cooperation between several Pit-2 molecules, which may not be necessary for the natural function of this molecules as phosphate transporters.

Binding and penetration are dynamic processes which seems to require very specific features of both the envelope glycoprotein and the cell surface receptor. The motility and/or the capacity of receptor molecules to agregate may be relevant to this process. Such features may be difficult to maintain in modified envelope molecules, while it may be presumed that only specific classes of cell surface molecules have the capacity to manage retrovirus entry into cells. This would account for the usual low efficiency of infection observed after retargeting the binding of retrovirus envelope glycoproteins to cell surface molecules that have not been naturally selected as retrovirus receptors (Roux, 1989; Etienne-Julan, 1992; Kasahara, 1994; Valsesia-Wittmann, 1994; Cosset, 1995; Kabat, 1995; Somia, 1995; Tearina, 1995).

III - Dynamic of the amphotropic receptor at the cell surface and role of the cytoskeleton in amphotropic virus entry.

The Pit-2 protein is a multiple transmembrane molecule which functions as a phosphate/Na symporter and as the receptor for amphotropic MLVs (Kavanaugh, 1994). Sequence analysis predict 10 transmembrane domains, 5 extracellular loops and intracellular N- and C-terminal extremities.

With the aim to investigate the motility and/or the capacity of receptor molecules to agregate in response to virus binding, we have studied the dynamic of Pit-2 at the cell surface. An 11 amino acid epitope of the VSV G protein was inserted at various locations including the extracellular loop 5 and at the C-terminal extremity. DNA encoding the tagged receptors was inserted in a CMV expression vector and transfected in CHO cells, which do not express Pit-2. Virus particle binding and infection mediated by tagged or wild type receptors were equivalent. The anti-VSV-G epitope mAb P5D4 immunoprecipitated 70kDa glycosylated receptor molecules in transfected cells. Flowcytometry and immunoflurescence analysis were performed on permeabilized and non-permeabilized cells. Loop 5 epitope was detected at the cell surface, as expected.

Surprisingly, in contrast with the predicted topology, the C-terminal epitope was also detected at the surface of non-permeabilized cells, indicating that the C-terminal extremity of the molecule is extracellular.

The cell surface localization of the C-terminal tagged receptor was studied by immunofluorecsence and confocal microscopy. In naive cells, the signal was homogeneously spread over the plasma membrane. After one hour incubation with ultrafiltrated virus particles, the signal condensed as large granulations, suggesting a clustering of receptor molecules. This observation was in agrement with the hypothesis of a receptor aggregation in response to virus binding. With the aim to examine whether receptor oligomerization participates in the virus entry process, expression vectors encoding C-terminal truncated receptors were cotransfected with wild type or C-terminal tagged receptors in CHO cells. Coexpression of non-functional receptors abolished virus binding and infection, suggesting that these molecules exhibited a dominant negative effect and that receptor oligomerization may be required for processing MLV-A entry.

Spatial reorganization of receptors was observed in response to varying phosphate concentration in the culture medium. In the presence of high phosphate concentration (50 mM, 5 fold higher than normal), receptors were homogeneously spread at the cell surface. Amphotropic pseudotype particle binding was not affected, however, entry was poorly efficient. Control experiments with VSV-G pseudotypes showed no modification. Phosphate starvation induced receptor aggregation, pictures evoking stress cable formation and membrane ruffling. The binding of amphotropic pseudotypes was not affected. However virus entry was much more efficient than in the presence of 10 mM phosphate. Since immunofluorecence pictures evoked cytoskeleton images, confocal analysis was performed to examine possible colocalization with intracellular microfilaments. Colocalization of cell surface receptors with intracellular actin filaments was observed. Colocalization was still visible after disruption of the actin network by cytochalasine D treatment. However, whereas cytochalasine D treatment did not affect virus binding, it completely abolished virus entry. These data indicate that interaction of Pit-2 with the actin network is crucial for its function as amphotropic virus receptor. They are consistent with the hypothesis that all receptor molecules may not be equivalent for porcessing virus entry, depending on the local subcellular environment and on the interactions they establish with the cytoskeleton. It is very likely that these interactions can be modulated overtime, for example in response to modified phosphate concentrations.

References.

Battini, J. L., Heard, J. M. and Danos, O. (1992). Receptor choice determinants in the envelope glycoproteins of amphotropic, xenotropic and polytropic murine leukemia viruses. J. Virol. 66, 1468-1475.

Choppin, J., Schaffar-Deshayes, L., Debré, P. and Lévy, J. P. (1981). Lymphoïd cell surface receptor for moloney leukemia virus envelope glycoprotein gp7l. J. Immunol. 126, 2347-2351.

Cosset, F. L., Morling, F. J., Takeuchi, Y., Weiss, R. A., Collins, M. K. L. and Russel, S. J. (1995). Retroviral retargeting by envelopes expressing N-terminal binding domain. J. Virol. 69, 6314-6322.

De Larco, J. and Todaro, G. J. (1976). Membrane receptors for Murine Leukemia Viruses: characterization using the purified viral envelope glycoprotein, gp7l. Cell 8, 365-371.

Eiden, M. V., Farrell, K., Warsowe, J., Mahan, L. C. and Wilson, C. A. (1993). Characterization of a naturally occurring ecotropic receptor that does not facilitate entry of all ecrotropic murine retroviruses. J. Virol. 67, 4056-4061.

Etienne-Julan, M., Roux, P., Carillo, S., Jeanteur, P. and Piechaczyk, M. (1992). The efficiency of cell targeting by recombinant retroviruses depends on the nature of the receptor and the composition of the artificial cell-virus linker. J. Gen. Virol. 73, 3251-3255.

Fowler, A. K., Twardzik, D. R., Reed, C. D., Weislow, O. S. and Hellman, A. (1977). Binding characteristics of Rauscher leukemia virus envelope glycoprotein gp7l to murine lymphoid cells. J. Virol. 24, 729-735.

Gray, K. D. and Roth, M. (1993). Mutational analysis of the envelope gene of Moloney murine leukemia virus. J. Virol. 67, 3489-3496.

Heard, J. M. and Danos, O. (1991). An amino-terminal fragment of the Friend Murine Leukemia Virus envelope glycoprotein binds the ecotropic receptor. J. Virol 65, 4026-4032.

Johnson, P. A. and Rosner, M. R. (1986). Characterization of murine-specific leukemia virus receptor from L cells. J. Virol. 58, 900-908.

Jones, J. S. and Risser, R. (1993). Cell fusion induced by the murine leukemia virus envelope glycoprotein. J Virol 67, 67-74.

Kabat, D. (1995). Targeting retroviral vectors to specific cells. Science 269, 417.

Kadan, M. J., Sturm, S., Anderson, W. F. and Eglitis, M. A. (1992). Detection of receptorspecific murine leukemia virus binding to cells by immunofluorescence analysis. J. Virol. 66, 2281-2287.

Kalyanaraman, V. S., Sarngadharan, M. G. and Gallo, R. C. (1978). Characterization of Rauscher murine leukemia virus envelope glycoprotein receptor in membranes from murine fibroblasts. J. Virol 28, 686-696.

Kasahara, N., Dozy, A. M. and Kan, Y. W. (1994). Tissue-specific targeting of retroviral vectors through ligand-receptor interactions. Science 266, 1373-1376.

Kavanaugh, M. P., Miller, D. G., Zhang, W., Law, W., Kozak, S. L., Kabat, D. and Miller, A. D. (1994). Cell-surface receptors for gibbon ape leukemia virus and amphotropic murine retrovirus are inducible sodium-dependent phosphate symporters. Proc Natl Acad Sci USA 91, 7071-7075.

Kim, J. W., Closs, E. I., Albritton, L. M. and Cunningham, J. M. (1991). Transport of cationic amino acids by the mouse ecotropic retrovirus receptor. Nature 352, 725-728.

Kozak, S. L., Siess, D. C., Kavanaugh, P., Miller, A. D. and Kabat, D. (1995). The envelope glycoprotein of an amphotropic murine retrovirus binds specifically to the cellular receptor/phosphate transporter of susceptible species. J. Virol. 69, 3433-3440.

Linder, M., Linder, D., Hahnen, J., Schott, H. H. and Stirm, S. (1992). Localization of the intrachain disulfide bonds of the envelope glycoprotein 71 from Friend murine leukemia virus. Eur. J. Biochem 203, 65-73.

Linder, M., Wenzel, V., Linder, D. and Stinn, S. (1994). Structural elements in glycoprotein 70 from polytropic Friend mink cell focus-inducing virus and glycoprotein 71 from ecotropic Friend murine leukemia virus, as defined by disulfide-bonding pattern and limited proteolysis. J. Virol. 68, 5133-5141.

McGrath, M. S., Declève, A., Lieberman, M., Kaplan, H. S. and Weissman, L. (1978). Specificity of cell surface virus receptors on radiation leukemia virus and radiation-induced thymic lymphomas. J. Virol. 28, 819-827.

McGrath, M. S. and Weissman, I. L. (1979). AKR leukemogenesis: Identification and biological significance of thymic lymphoma receptors for AKR retroviruses. Cell 17, 65-75.

McKrell, A. J., Soong, N. W., Curtis, C. M. and Anderson, F. (1996). Identification of a subdomain in the Moloney murine leukemia virus envelope protein involved in receptor binding. J. Virol. 70,1768-1774.

Morgan, R. A., Nussbaum, O., Muenchau, D. D., Shu, L., Couture, L. and Anderson, W. F. (1993). Analysis of the functional and host range-determining regions of the murine ecotropic and amphotropic retrovirus envelope proteins. J. Virol. 67, 4712-4721.

Nussbaum, O., Roop, A. and Anderson, W. F. (1993). Sequences determining the pH dependence of viral entry are distinct from the host range-determining region of the murine ecotropic and amphotropic retrovirus envelope proteins. J Virol 67, 7402-7405.

Ott, D. and Rein, A. (1992). Basis for receptor specificity of nonecotropic murine leukemia virus surface glycoprotein $gp70^{su}$. J. Virol. 66, 4632-4638.

Pinter, A., Chen, T. E., Lowy, A., Cortez, N. G. and Siligari, S. (1986). Ecotropic murine leukemia virus-induced fusion of murine cells. J. Virol. 57, 1048-1054.

Roux, P., Jeanteur, P. and Piechaczyk, M. (1989). A versatile and potentially general approach to the targeting of specific cell types by retroviruses: Application to the infection of human cells by means of histocompatibility complex class I and class II antigens by ecotropic murine leukemia virus-derived viruses. Proc. Natl. Acad. Sci. USA 86, 9079-9083.

Somia, N. V., Zoppé, M. and Verma, I. M. (1995). Generation of targeted retroviral vectors by using single-chain variable fragment: An approach to *in vivo* gene delivery. Proc. Natl. Acad. Sci. USA 92, 7570-7574.

Szurek, P. F., Yuen, P. H., Ball, J. K. and Wong, P. K. Y. (1990). A val-25-to-Ile substitution in the envelope precursor polyprotein gPr80env is responsible for the temperature sensitivity, inefficient processing of gPr80 and neurovirulence of tsl, a mutant of Moloney murine leukemia virus TB. J. Virol. 64, 467-475.

Tearina, T. H. and Dornburg, R. (1995). Retroviral vectors particles displaying the antigen-binding site of an anitbody enable cell-type gene transfer. J. Virol. 69, 2659-2663.

Valsesia-Wittmann, S., Drynda, A., Deléage, G., Aumailley, M., Heard, J. M., Danos, O., Verdier, G. and Cosset, F. L. (1994). Modifications in the binding domain of avian retrovirus envelope protein to redirect the host range of retroviral vectors. J. Virol. 68, 4609-4619.

Wang, H., Kavanaugh, M. P., North, R. A. and Kabat, D. (1991a). Cell-surface receptor for ecotropic murine retroviruses is a basic amino-acid transporter. Nature 352, 729-731.

Wang, H., Paul, R., Burgeson, R. E., Keene, D. R. and Kabat, D. (1991b). Plasma membrane receptors for ecotropic murine retroviruses require a limiting accessory factor. J.Virol. 65, 6468-6477.

Weiss, R. A. (1993). *Cellular receptors and viral glycoproteins involved in retrovirus entry.* New York and London: Plenum Press, 1-108.

Transduction of CD34+ PBSC Mobilized in Cancer Patients Using a Novel Fr-MuLV Retrovirus Vector Derived from FB29 Highly Titrating Strain

Luz Marina Restrepo[1], Michel Masset[1], Jan Bayer[1], Liliane dal Cortivo[2], Jean Pierre Marolleau[2], Marc Benbunan[2], Michel Marty[1,3], Odile Cohen-Haguenauer[1,3]

[1] Laboratoire Transfert Génétique et Oncologie Moléculaire, Institut d'Hématologie
[2] Centre de Transfusion sanguine Paris-Est
[3] Departement d'Oncologie Médicale, Hôpital Saint-Louis 75 475 Paris Cedex 10-France

1. Introduction

In addition to potential therapeutic applications in genetic disorders and malignancies, gene transfer into haematopoietic stem cells can provide insights into the mechanism of growth and differentiation of haematopoietic progenitors[1]. Mechanisms of disease relapse can be studied[2,3]. Peripheral blood stem cells (PBSC), mobilized from the bone marrow compartment by treatment with colony stimulating factors, are considered a favourable target for therapeutic gene transfer. Full in vivo haematopoietic reconstitution has been demonstrated ; access to this cell-source proves non-invasive and innocuous[4]. Samples included in this study were obtained from middle aged cancer patients who had received multiple cycles of chemotherapy. Patients material inherently displays variations according to the age, case history and chemotherapy regimen. Efforts towards transduction of haematopoietic progenitors from such patients is relevant to : 1°/ gene marking of autografts in order to trace the origin of relapse and investigate improved purging procedures and 2°/ gene therapy of disorders where the haematopoietic stem pool is already compromised such as inherited disease like Fanconi Anemia or infectious disease like AIDS.

Retrovirus mediated gene transfer is the best developed technique to introduce foreign genes in eukaryotic cells in a stable manner. Retroviral gene transfer has been extensively evidenced *in vitro* in human haematopoietic progenitors, and in peripheral blood leukocytes (PBLs)[5-8], although until recently transduction efficiencies have remained quite low. In large animal models, retroviral gene transfer proves less efficient than in small laboratory animals, nevertheless long-term expression of transferred genes has been evidenced[9]. In ongoing clinical trials, the presence of the foreing gene has been demonstrated over 18 months following transduction. Sustained and significant expression of the transgene in multiple haematopoietic lineages is yet to be achieved[7,8,10,11]. In order to achieve these goals technological improvements are required along two lines : 1°/ qualitative increase in transduction of haematopoietic progenitors and stem cells in particular (which require evaluation assays : from LTC-IC to immunocompromized mice repopulating cells and ultimately *in vivo* reconstitution in human) ; 2°/ quantitative increase in the number of progenitors available for engraftment ; by means of procedures likely to spare both functional characteristics and cell-numbers.

NATO ASI Series, Vol. H 105
Gene Therapy
Edited by Kleanthis G. Xanthopoulos

The retroviral vector FOCH29 was derived from the FB29 strain of Friend Murine Leukemia Virus. The FB29 strain was selected according to high infectivity extending to all haematopoietic lineages in mice and in particular to potential of U3 regulatory sequences within the LTR[12]. The aim of this study consists in achieving efficient transduction and long-term expression in CD34+ PBSC from cancer patients (solid tumours, in particular breast and ovarian cancer and lymphomas) making use of this novel vector.
PBSC mobilized from patients following chemotherapy might already be triggered into cell cycle and thus represent an activated subset of haematopoietic progenitors. We investigated the feasibility of simple procedures towards transduction of PBSC as target for retroviral gene transfer ; taking into account high levels of transduction achieved on various sources of CD34+ human haematopoietic progenitors, i.e. : cord-blood, bone-marrow and PBMSC [13]. In some cases, poor reconstitutive potential of the cell sample[14,15] has been recorded. Transduction was performed making use of viral supernatant and in no instance applying cocultivation with virus producing cells.

2. Fr-MuLV vector from FB29 highly titrating strain

The FB29 strain of Friend Murine Leukemia virus has been shown to harbour properties which make it an interesting candidate for the backbone of gene expression vectors. The retroviral vector FOCH29 was designed in order to investigate a viral construct which could achieve efficient transduction based on high titers virus production and result in interesting expression of genes of interest. We have demonstrated that high titers producer clones could be generated (over 10^7 cfu/ml), in the absence of replication competent particles, harbouring a log range increase when compared to a reference vector derived from Mo-MuLV. This holds true on two independent target cell-lines tested. These data were confirmed when titers were compared on pools and of a series of producer clones generated in parallel. The use of two different amphotropic packaging cell-lines resulted in similar data (Cohen-Haguenauer et al, submitted).

3. Patient samples

A total of 14 clinical samples was subjected to retroviral gene transfer. These samples were derived from cancer patients with various malignancies following several courses of chemotherapy. CD34+ enriched fractions were obtained. Purity and viability were determined in each sample (Table 1). Mean purity was 85% and mean viability 91%. Initial clonogenic capacity of the cells was established by seeding 1,000 viable cells in CFU-GM assay. Results were variable from patient to patient; the number of colonies ranging from 7 to 135. In order to assess for potential endogenous neoR CFU-GM, duplicate wells were seeded in the presence of G418 (1 mg/ml). Endogenous resistant colonies were detected in 7 samples (Table 2). A reduction was observed in CFU-GM number following *in vitro* culture and transduction. Following transduction 5,000 cells/well were seeded. Unselected colonies ranged from 8 to 880. Again some variations were observed from patient to patient (Table 2).

SAMPLE	PURITY (%)	VIABILITY (%)	CFU-GM COLONIES Nb Pre-transduction 1000 /well	CFU-GM COLONIES Nb Post-transduction 5000 /well
1	-	-	-	184
2	90	77	-	34
3	93	94	-	880
4	95	-	-	43
5	78	92	76	47
6	87	100	0	28
7	97	97	74	114
8	97	97	73	201
9	93	97	81	46
10	93	83	135	18
11	46	92	7	111
12	97	93	39	50
13	93	84	38	45
14	51	92	134	275
Mean	85	91	73	158

Table 1 : Viability and clonogenic capacity of CD34+ PBSC from cancer patients After CD34+ selection, cell viability was determined by trypan blue dye exclusion. 1000 viable cells/well were plated in CFU-GM assay to determine clonogenic capacity. Cultures were initiated in the presence or absence of G418, in order to detect endogenous resistant colonies. After transduction 5000 viable cells/well were plated. CFU-Nb : number of CFU-GM colonies counted in non-selective medium.

SAMPLE NUMBER	GM-HAEMATOPOIETIC COLONIES Pre-G418R(%)	Post-G418R (%)	Net G418R (%)	% PCR
1		8		
2		50		80
3		54		37
4		70		75
5	6	47	41	42
6	0	14	14	100
7	13	24	11	40
8	5	16	11	30
9	11	26	15	70
10	3	44	41	
11	0	1	1	
12	0	100	100	90
13	5	9	4	
14	0	49	49	50
Mean	4	37	29	61

Table 2: Transduction efficiency of chemotherapy and G-CSF mobilized CD34+ PBSC isolated from cancer patients. 12 hours after the fourth cycle of infection, cells were spun down and counted. 5000 viable cells/well were plated in methylcellulose medium with or without G418 (1 mg/ml). The percentage of pre-transduction (endogenous) neo-R CFU-GM was subtracted from the percentage of post-tranduction neo-R CFU-GM in order to obtain net percentages (Net G418R %).

4. Post-infection analyses

CD34-enriched samples were subjected to prestimulation for 36 - 48 hours, using a minimal growth factor cocktail (MC). Subsequently, CD34+ cells were exposed to cell-free virus supernatant during 48 hours, adding fresh supernatant four times (MOI 10, each time). Sixteen hours after last supernatant addition, cells were washed and viability was determined. Post transduction cell viability was 55% on average (starting from 93%). Transduction efficiency was evaluated on the basis of, both the presence of the transgene in individual CFU-GM (PCR) as well as its expression (growth in the presence of G418). Results are shown in Table 2. Colonies resistant to G418 ranged from 1% to 100% (mean 29%). PCR proved positive in 30% to 100% (mean 61%) of CFU-GM colonies tested. The presence of neoR-gene sequences in the retroviral context was detected by PCR in at least twice more colonies on average with respect to G418 resistance. Similar observations have been reported making use of Moloney-based retroviral vector transduction of haematopoietic stem cells (24) (Dubé, personal communication). This might relate to the use of too high doses of selective drug. Transduction might result in poor gene expression in case viral LTRs prove poor insulators in some instances, according to the site of integration. Translation of RNAs generated off the construct can also remain mediocre. Alternatively, expression of the neomycine-resistance gene product might impair the metabolism and the functionality of transduced cells as suggested by Valera et al (25).

5. Prestimulation conditions

The effect of prestimulation growth factor cocktail composition was compared in four randomly chosen patient samples. In each case, one half of the sample was prestimulated with a "minimal cocktail" (MC) and the other half of the sample with "Highly concentrated cocktail"(HCC). Post-transduction viability proved on average higher with HCC. Loss of clonogenic progenitors during the transduction procedure was 61% (MC) vs 70% (HCC). Net-percentage of G418-resistant colonies reflecting transduction and expression levels was slightly lower with MC. Molecular analysis by PCR evidenced 50% positive CFU with MC vs 62% with HCC. Again, the presence of the gene was detected in at least twice more colonies with respect to G418 resistance.

Exposure of CD34+ cells to haemopoietic growth factors prior to, or during infection may modify cells biology. We observed a small increase in recovery of CFU-GM with "Highly concentrated Cocktail" prestimulation, which could be explained by a wider stimulatory effect of this prestimulation mix. Conversely we did not find a significant difference in the percentage of neoR CFU-GM following either MC or HCC-prestimulation.

Cassel *et al* (15) have shown that 48 hours prestimulation followed by 72 hours infection gave the best results (17% G418R). Using a comparable protocol, Dunbar et al (10) found 21% G418R CFU-GM. A 6 hours exposure of CD34+ enriched PB cells to virus, without growth factors did not result in detectable gene transfer but other investigators have reported some success with this simplified regimen while transferring genes to unpurified BM cells (20). Our data do favour the use of MC prestimulation cocktail. The latter is more likely to spare pluripotentiality of (transduced) progenitors and avoid potentially contaminating tumour cells to proliferate.

6. Transduction length

PBSC from patients following chemotherapy might represent a population of already activated haematopoietic stem or progenitor cells. In order to avoid prolonged *in vitro* manipulation and subsequent cell loss we then, tested the performance of a single 12 hours exposure to virus supernatant (MOI 10) in comparison with a standard 48 hours virus exposure (four cycles MOI 10 each). Tranduction was performed in the presence of MC in both instances (Restrepo et al, submitted). Following 12 hours infection, CFU-GM numbers decreased to approximately 40% of input values. A longer transduction procedure (48 hours) further decreases CFU-GM number to 30% of input. One cycle of supernatant mediated transduction resulted in 35% G418-R colonies on average ; while four cycles of infection resulted 51% on average. In two samples a significant number of CFU-GM generating cells was lost in this second setting. PCR data proved rather similar in both cases 80% and 60% versus 90% and 50%, respectively. In two independent experiments performed on bone marrow derived CD34+ cells, similar transduction rates were evidenced whatever the length of tranduction.

A 12 hours exposure to virus resulted in a similar percentage of transduced cells as compared to 48 hours exposure to virus. Since stem cells might loose their potential during *in vitro* culture, reduction in duration of retroviral infection should be advantageous. The systematic two-fold increase in late CFU-output we observed following short transduction confirm the feasibility of such a strategy (Restrepo et al, submitted).

7. Insights into potential to initiate long term cultures and reconstitution

In order to evaluate transduction of LTC-IC and resulting expression, we first chose to initiate LTC on MS-5 cell line. The level of haemopoietic support provided by the MS-5-derivative we used proved limited in the case of cancer patient PBSC. Parallel experiments with the parental MS-5 cell line did not lead to detectable improvement (data not shown). These difficulties might relate to inherent characteristics of PBSC and/or to use of minimum concentrations of exogenous growth factors. In addition, mobilization protocols induce changes in the expression profile of adhesion molecules [23] enhancing the weak adhesion potential of cells already committed to circulate into blood flow.

Difficulties in performing LTC on neomycin resistant MS-5 xenogeneic stroma affected the readout of transduction of late CFU-GM. Still, in those experiments where late colonies could be analysed the potential of Friend virus to transduce hemopoietic progenitors and generate expression is of interest. Evaluation of transduction efficiency thus depends on the readout system applied. We further investigated suspension LTC, in the presence of haemopoietic growth factors as an alternative procedure. CFU-GM could be initated after at least 45 days in LTC and up to 55 in some instances. These data do compare favourably with Lu et al [14], who could maintain suspension LTC for 35 days at most.

Variability in both quality and *in vitro* survival potential of this cell source raise the issue of *in vivo* repopulating capacity.

In five additional samples investigated (data not shown) no CFU-GM colonies would grow neither before nor after transduction. Three samples generated a standard number of CFU-GM initally but post-transduction resistant colonies could not be evidenced. It seems unlikely that blood cell collection and CD34+ selection would play a critical role in sample heterogeneity. These procedures were highly standardized and gave reproducible results with other cell sources[13]. These data indicate that it is unlikely that CD34+ PBSC in cancer patients would always lead to haematopoietic reconstitution. Again these results are coherent with those reported by Ushida et al[22] whereby no cells harbouring 'a stem cell phenotype' (Thy-1+, Lin-, CD34+) can be detected in some cancer patients' samples.

8. Optimizing the virus backbone

Since presence of neoR-gene sequences in the retroviral context was detected by PCR in at least twice more colonies on average with respect to G418 resistance we chose to investigate whether this might relate to the use of too high doses of selective drug and/or poor gene expression resulting from poor translation of RNAs generated off the construct. The Fr-MuLV-FOCH29 vector backbone has thus been optimised towards improving translation/ expression features.

In order to improve expression and safety parameters of FOCH backbone two derivative constructs were designed (Cohen-Haguenauer et al, submitted). The first one, pFOCHmutGag carries a nonsense mutation close to the Gag-ATG in order to inhibit the putative synthesis of nucleocapsid proteins. This also results as expected in increased expression of the transgene while virus titers proved as high as with the first generation vector. The second derivative FOCHA assembles features of interest mapping to the FB29-LTR enhancer, the 5' leader and 3'UTR and potential to initiate expression off a spliced sub-genomic RNA. Highly titrating producer clones over 10^8 cfu/ml were obtained with the Green fluorescent Protein gene as reporter. Mo-MuLV based producer clones were generated in parallel with titers reaching 10^7 cfu/ml at most. We have thus delineated and assembled sequence features which make the FOCH-series of defective retrovirus vectors strong candidates for improved transduction of a variety of target cells based on high virus titers.

Acknowledgements

Brigitte TERNAUX, Marie-Noëlle LACASSAGNE, Isabelle ROBERT and Virginie MOUTON are warmly acknowledged for excellent technical assistance.
Work supported by grant from Fondation contre la leucémie, Ligue contre le cancer (Comité de Paris et Ligue Nationale), Association pour la Recherche contre le Cancer (ARC), and Associtation Française contre les Myopathies (AFM). Luz Marina Restrepo received fellowships from COLCIENCIAS (Instituto Colombiano para el avance de la Ciencia y la tecnologia) and Fondation de France ; Jan Bayer from A.N.R.S and Ligue National contre le Cancer.

References

1. Carter RF, Abrams-Ogg AC, Dick JE *et al.* Autologous transplantation of canine long-term marrow culture cells genetically marked by retroviral vectors. *Blood* 1992; **79**: 356-364.
2. Brenner MK, Rill DR, Holladay MS *et al.* Gene marking to determine whether autologous marrow infusion restores long-term haematopoiesis in cancer patients. *Lancet* 1993; **342**: 1134.
3. Stewart AK, Dubé ID, Kamel-Reid S, Keating A. A phase I study of autologous bone marrow transplantation with stem cell gene marking in multiple myeloma. *Human Gene Therapy* 1995; **6**: 107-119.
4. Udomsakdi C, Lansdorp PM, Hogge DE *et al.* Characterization of primitive hematopoietic cells in normal human peripheral blood. *Blood* 1992; **80**: 2513-21.
5. Henderson-Mannion J, Kemp A, Mohney T *et al.* Efficient retroviral mediated transfer of the glucocerebrosidase gene in CD34+ enriched umbilical cord blood human hematopoietic progenitors. *Experimental Hematology* 1995; **23**: 1628-1632.
6. Eipers PG, Krauss JC, Palsson B *et al.* Retroviral mediated gene transfer in human bone marrow cells grown in continuous culture vessels. *Blood* 1995; **86**: 3754-3762.
7. Bordignon C, Notarangelo LD, Nobilini N *et al.* Gene therapy in peripheral lymphocytes and bone marrow for ADA-immunodeficient patients. *Science* 1995; **270**: 470-475.
8. Blaese RM, Culver KW, Miller DA *et al.* T lymphocyte-directed gene therapy for ADA-SCID: initial trial results after 4 years. *Science* 1995; **270**: 475-480.
9. Van Beusechem VW, Kukler A, Heidt PJ, Valerio D. Long-term expression of human adenosine deaminase in rhesus monkeys transplanted with retrovirus infected bone-marrow cells. *Proc Natl Acad Sci USA* 1992; **89**: 7640.
10. Dunbar CE, Cottler-Fox M, O'shaughessy JA *et al.* Retroviral marked CD34-enriched peripheral blood and bone marrow cells contribute to long-term engraftment aftr autologous transplantation. *Blood* 1995; **85**: 3048-.
11. Kohn DB, Weinberg K, Nolta JA *et al.* Engraftment of gene-modified umbilical cord blood cells in neonates with adenosine deaminase deficiency. *Nature Medicine* 1995; **1**: 1017-1023.
12. Cohen Haguenauer O: Vecteur rétroviral pour le transfert et l'expression de gènes à vissée thérapeutique, dans les cellules eucaryotes. France, FR A, 2 707091, 1995
13. Cohen Haguenauer O, Restrepo LM, Masset M *et al.* Efficient transduction of haemopoietic CD34+ progenitors of human origin using an original retrovirus vector derived from Fr-MuLV: In vitro assessment. *Human Gene Ther, in press*
14. Lu M, Maruyama M, Zhang N *et al.* High efficiency retroviral-mediated gene transduction into CD34+ cells purified from peripheral blood of breast cancer patients primed with chemotherapy and granulocyte-macrophage colony-stimulating factor. *Hum Gene Ther* 1994; **5**: 203-208.
15. Cassel A, Cottler-Fox M, Doren S, Dunbar CE. Retroviral-mediated gene transfer into CD34-enriched human peripheral blood stem cells. *Exp Hematol* 1993; **21**: 585-591.

16. Danos O, Mulligan RC. Safe and efficient generation of recombinant retroviruses with amphotropic and ecotropic host ranges. *Proc Natl Acad Sci USA* 1988; **85**: 6460-6464.
17. Miller AD, Miller DG, Garcia JV, Lynch CM. Use of retroviral vectors for gene transfer and expression. *Methods in Enzymology* 1993; **217**: 581-99.
18. Miller D, Adam M, Miller A. Gene transfer by retroviral vectors occurs only in cells that are actively replicating at the time of infection. *Mol Cell Biol* 1990; **10**: 4239-4242.
19. Xu L, Stahl SK, Dave HPG *et al*. Correction of the enzyme deficiency in hematopoietic cells of Gaucher patients using a clinically acceptable retroviral supernatant transduction protocol. *J Exp Hem* 1994; **22**: 223-230.
20. Deisseroth AB, Zu Z, Claxton D *et al*. Genetic marking shows that Ph+ cells present in autologous transplants of chronic myelogenous leukemia (CML) contribute to relapse after autologous bone marrow transplantation n CML. *Blood* 1994; **83**: 3068.
21. Phillips, K, Gentry, T, McCowage, G, Gilboa, E, Smith, C. Cell-surface markers for assessing gene transfer into human hematopoietic cells. *Nat Med*1996; **2**:1154-6
22. Uchida N, He D, Friera AM *et al*. The unexpected G0/G1 cell cycle status of mobilized hematopoietic stem cells from peripheral blood. *Blood* 1997; **89**: 465-72.
23. Berthou C, Marolleau JP, Lafaurie C *et al*. Granzyme B and perforin lytic proteins are expressed in CD34+ peripheral blood progenitor cells mobilized by chemotherapy and granulocyte coloni-stimulating factor. *Blood* 1995; **86**: 3500-3506.
24. Bienzle, D, Abrams-Ogg, A. C, Kruth, S. A. *et al*. Gene transfer into hematopoietic stem cells: long-term maintenance of in vitro activated progenitors without marrow ablation. *Proc Natl Acad Sci U S A* 1994; **91**: 350-354.
25. Valera A, Perales, J. C.Hatzoglou, M.Bosch, F..Expression of the Neomycin-Resistance (*neo*) gene induces alterations in gene expression and metabolism. *Human Gene Ther* 1994; **5** : 449-456.

HIV Gene Therapy: Current Status and Its Role in Therapy

Ernst Bohnlein

Systemix, 3155 Porter Drive, Palo Alto, CA 94304

Introduction

In the early 1980s, surprisingly increasing numbers of patients with infections reminiscent of immune suppression were diagnosed predominantly in metropolitan US clinics. The common denominator appeared to be blood borne transmission of a virus as this patient group mostly comprised recipients of blood products (eg. factor VIII) and sexually active homosexual men. In 1983, a novel human retrovirus was isolated (2, 18, 24) and eventually named human immunodeficiency virus type I (HIV-1), the etiologic agent of the acquired immune deficiency syndrome (AIIDS). HIV- I disease is characterizedby an extended clinical latency period between the time of primary infection and the manifestation of the disease. IHV-1 infects cells of the hematopoietic system: CD4+ T lymphocytes, macrophages, dendritic cells, and microglial cells of the central nervous system (25). Recent studies suggest that HIV1 actively replicates throughout the "latency" period which eventually leads to depletion of immune effector cells (19, 48). This cell loss turns into a dysfunctional immune system characterized by a multitude of malignancies (ie. lymphomas, Kaposi sarcoma) and opportunistic infections (ie. candida, PCP, CMV, etc.), a hallmark of AIDS.

Shortly after the successful isolation of HIV, the scientific community was confident that effective therapies would become available shortly. This confidence was based on previously successful vaccination campaigns (ie. smallpox, polio). In addition, molecular biology techniques were available which led to the rapid determination of the complete HIV- 1 nucleotide sequence (37, 47). In a very short time period, HIV-1 became the best understood retrovirus to date (15) but this knowledge could not yet be translated into effectivetreatments for patients infected with HIV-1. In particular, clinical vaccination trials have failed until now. Monotherapies with antiviral drugs directed against critical retroviral enzymes [reverse transcriptase (RT), protease (PR)] have rapidly lead to a selective advantage of pre-existing minor HIV-1 variants resistant to the drug (14, 22). At present, triple-drug therapy (RT + PR inhibitors) looks promising (13), essentially reducing plasma viremia and in some patients even below detection levels. In most patients, triple drug therapy also results in increased peripheral CD4 T cell counts. It is too early to tell whether this treatment modality can prevent resistance development, reverse disease progression and restore immune function. Bio-availability, adverse effects and consequently patient compliance, the need to administer these anti-retroviral drugs for extended time periods and the actual cost of the drugs are issues which impact on the medical need for gene therapy. Nevertheless, the initial anti-retroviral drug studies support the notion that inhibition of HIV-1 replication is a critical step towards the restoration of immune function in HIV-infected individuals.

NATO ASI Series, Vol. H 105
Gene Therapy
Edited by Kleanthis G. Xanthopoulos

Gene Therapy for HIV Disease

The early experiences with vaccination and antiviral drugs led to the proposal to consider gene therapy ("intracellular immunization") as clinical alternative for HIV disease (1). The reverse transcribed HIV provirus integrates into the chromosome of infected cells and HIV-infected patients are mosaic (ie. contain both infected and non-infected cells). Hence, HIV infection can be considered an acquired genetic disease amenable to gene therapy. More importantly, understanding of HIV-1 molecular biology offered opportunities to interfere with the retroviral replication and gene therapy was becoming a clinical reality in late stage cancer patients (39) and children suffering from ADA immune deficiency (6).

HIV-1 is the most studied and, in molecular terms, the best understood retrovirus (15). Nevertheless, many open questions remain, in particular with respect to virus-host interactions and viral-induced pathology. For many years, HIV-1 was considered to be a "latent' virus. Recent data contradict this view and kinetic studies in peripheral blood demonstrate vigorous viral replication and massive T cell turnover indicative of an active and effective immune response early after infection (19, 48). HIV-1 infection has a clinically-latent phase but HIV is not a latent virus. The balance between newly generated CD4-positive T cells and cells lost to HIV pathology (T cell homeostasis) is critical and controlled by plasma viremia, a prognostic marker for disease progression (30).

EffectiveAIDS therapies must retain this peripheral T cell homeostasis for extended periods of time to arrest disease progression. This goal can be achieved in several ways: (i) Reduction in viral load will change the equilibrium in favor of the infected host and reverse disease progression. (ii) The number of peripheral T cell counts can be increased. In the absence of concomitant, effectiveantiviral therapy, this strategy could result in a "new homeostasis" at higher viral load and hence, more rapid disease progression. Genetic modification of "resistant" peripheral T cells could circumvent this problem. The peripheral T cell homeostasis comprises both T cell differentiation from hematopoietic stem/progenitor cells as well as expansion of peripheral T cells. Even early in the infection, generation of new T lymphocytes from progenitor cells might be required to completely restore immune function [eg. full T cell receptor (TCR) repertoire] in HIV-infected individuals. For these reasons, gene therapy approaches that can protect cells throughout the T cell differentiation pathway are ideal for "intracellular immunization" strategies.

In contrast to immuno-therapies and genetic vaccination strategies, the goal of intracellular immunization is the inhibition of HIV-1 replication at the cellular level. As a consequence, viral spread (cell-free and cell-associated) will be attenuated or ideally completely blocked, eventually resulting in reduced viral burden. Understanding of the viral replication cycle is critical towards developing effectiveas well as specific gene intervention strategies. The HIV replication cycle begins with the budding of immature virions. Subsequently, proteolytic processing of the gag-pol precursor proteins renders the viral particle infectious. The mature viral particles can then initiate another round of infection by binding to CD4, the primary HIV receptor. After penetration of the cell membrane, the reverse transcribed retroviral genome integrates into the cellular nucleic acid mediated by integrase. The number of proposed intracellular immunization strategies for HIV-1 disease is rather large and ranges from approaches specific for the HIV-1 virus (eg. RNA decoys for HIV-1 regulatory genes) to more broadly used methods (eg. ribozymes, antisense; see Table 1). For most strategies, feasibility has been demonstrated in transient systems, in many instances relying predominantly on surrogate experimental assays instead of directly measuring inhibition of HIV-1 replication. The variety of experimental systems used make it very difficult to compare the antiviral potency of any individual strategy. However, dominant-negative variants of the HIV-1 rev gene have been demonstrated to be effective by several research groups and, hence, can serve as referencepoints. In fact, the first approved HIV gene therapy clinical trial evaluates the safety and efficacy of RevM 10 in peripheral blood lymphocytes (34).

Approach	Genes	References
[illegible]	RRE, TAR	[illegible]
Antisense	Gag, Pol, Env, TAR	12, 11, 38, 41, 45
Ribozymes	Tat, Rev, 5'LTR	35, 40, 50
Trans-dominant proteins	Rev, Tat, Gag, Env, Protease, eIF5A	4, 5, 7, 20, 21, 28, 32, 36
Other Protein-based Strategies	Soluble CD4, IFN-α, IFN-γ,	3, 33, 46
Single-chain variable fragments (sFv)	Env, Rev, RT, Int, Tat	10, 17, 29, 31, 42, 44,
HIV-Inducible Toxins	HSV-Tk, Diphtheria A toxin	8, 16
[illegible]	Ribozyme, antisense, TAR decoy	[illegible]

Table 1: Proclinical HIV gene therapy approaches. This list is not comprehensive but tries to categorize the majority of reported preclinic proposals including key references

Results and Discussion

In this report, results on two antiviral gene therapy strategies as well as experiments addressing whether gene and drug therapies can work together will be described. First, our studies on intracellularly expressed antisense transcripts and their antiviral activity in human T cells will be described. In the second part, I will report on our experiments introducing RevM1O into hematopoietic progenitor cells and, subsequently, demonstrate antiviral efficacy in T cells derived from these cells. Phase I clinical gene therapy trials in HIV disease most likely will be performed in individuals concurrently receiving anti-retroviral therapy. In the last part, I will cover a series of experiments evaluating the antiviral efficacy of drug and gene therapy combinations.

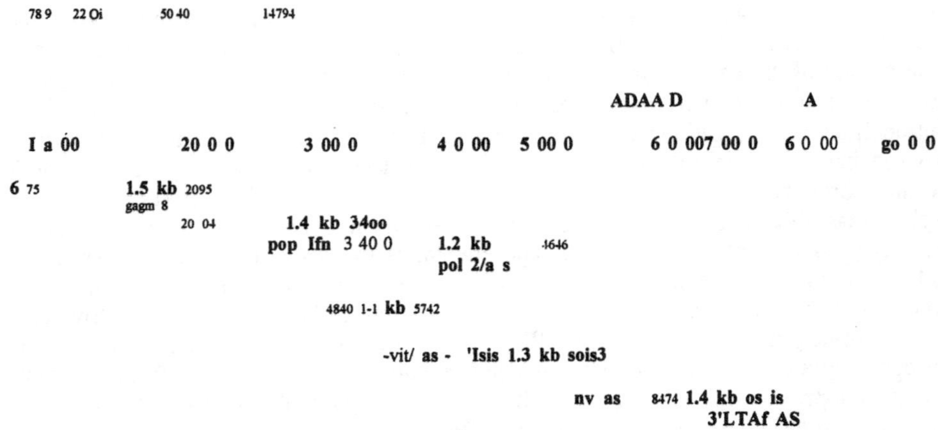

Fig. 1 On top is an outline of the HIV- I proviral genoniic organization. A indicates splice acceptor sites, D represents splice donor signals. The numbers indicate the Renome position in base pain. The individual antis -ense constructs are listed in the lower half indicating the name of the antisense fragment, the approximate length and the starting and ending nucleotides.

The small genome of HIV is an attractive target for antisense-based gene intervention strategies. Proof-of-principle was first demonstrated in 1990. Since then, many reports in the literature supported and expanded on this notion without demonstration of the molecular mode of action. It is generally assumed that complementary RNAs will form double-stranded RNA molecules intracellularly which could prevent translation of the target mRNA or could result in degradation of the annealed RNA transcripts. Expressing antisense transcripts directed against the HIV-1 Ψ-gag region, we could recently demonstrate the concomitant reduction of intracellular steady-state levels of both the target (HIV-1 genomic transcript) and the recombinant antisense RNA (45). We also studied parameters critical for intracellular heteroduplex formation. Using deletion mutants of the Ψ-gag RNA antisense transcript, we could delineate an inverse correlation between transcript length and antiviral efficacywith arttisense transcripts longer than 1,000 nucleotides yielding maximal effects. Two deletion mutants of comparable length without overlapping nucleotides yielded similar HIV inhibition suggesting that, at least in the case of the Ψ-gag antisense approach, no particular part of the transcript is required to be effective. This approach also demonstrated clearly selective degradation of the target full-length transcript. Next, we studied antisense transcripts directed against other parts of the retroviral genome including vif, pol, env and the 3' LTR (see Fig. 1). All vectors except the 3'LTR antisense construct were stable and inhibited replication of the HIV-1 HXB-3 strain, however with various efficacyin transduced human CD4-positive CEMss cells. The Ψ-gag and vif antisense transcripts were more effect:·et!·an RevM10 but tran·:ripts directed against the pol reading frame and the env gene were most effective. The molecular basis of these differencesis not understood and further studies to delineate the actual mode of action are required. Nevertheless, our studies clearly demonstrate that intracellular antisense transcripts longer than 1 kb are very effectivein inhibiting HIV replication and hence, are promising candidates to develop clinical gene therapy protocols for HIV disease.

HIV-1 critically depends on retrovirally-encoded gene products to complete the replication cycle, in particular the Tat and Rev trans-activating proteins (15). Both proteins exert their function interacting with a retroviral-encoded RNA *cis* element. This interesting molecular mode of action and their pivotal role in the HIV replication cycle resulted in intense studies to profile them as potential targets for intervention. Mutational analysis of the Rev protein revealed mutant proteins which were not only functionally inactive but also blocked wild-type protein function *in trans*. These trans-dominant (or dominant-negative) proteins could block Rev function and Rev-dependent gene expression in transient transfection studies. In 1992, we and Gary Nabel and colleagues independently demonstrated inhibition of HIV-1 viral replication in trans-duced CD4+ human T cells constitutively expressing RevM10 (4, 28). Subsequently, these studies were confirmed by others and it was shown that RevM10 can reduce HIV-1 replication in chronically-infected cells and in primary peripheral blood lymphocytes. These data formed the basis for the first RAC and FDA approved phase 1 clinical intracellular immunization study for HIV disease by Nabel et al (34). Preliminary results from this peripheral blood lymphocyte clinical study demonstrate longer persistence of RevM10 gene-modified T cells than control cells (49). The results from this pioneering study are exciting and provide clinical evidence that intracellular expression of therapeutic genes can provide a survival advantage in HIV-1 positive individuals. However, several shortcomings of the technology might prevent the clinical utility of this approach. First, transduction of T cells *in vitro* requires stimulation of the cells which also activates expression of HIV-1 proviral sequences. To overcome this problem, the current transduction protocols are performed in the presence of non-nucleoside reverse transcriptase inhibitors which selectively inhibit HIV-1 replication but do not prevent reverse transcription of the recombinant MLV vector genome required for successful transduction. Drug-resistance might become a practical hurdle which could preclude access to this technology for a sizable fraction of otherwise eligible patients. In addition, HIV infects many cell types other than CD4+ T cells including macrophages, dendritic cells and microglia cells of the CNS. For these reasons, we decided to deliver the antiviral RevM10 gene via hematopoietic stem cells which give rise to all HIV target cells. This approach also offersthe opportunity to restore immune function (ie. a complete T cell repertoire) which might already be depleted even in early stage patients.

Although the hematopoietic stem cell approach features several theoretical advancements over the T lymphocyte approach, it faces major technical hurdles. Many different research groups have made efforts to define human hematopoietic stem cells primarily by a combination of phenotypic marker genes (eg. CD34, CD38, Thy 1, HLA-DR, etc) and a variety of functional *in vitro* tests. The value of these *in vitro* studies is a continuous focus of scientific discussions but most experts agree that more stringent assays like competitive repopulation in the murine systems would be desirable. With respect to gene transfer into human stem/progenitor cell, the colony forming unit (CFU) assay in methyl-cellulose is the most widely used experimental setting. This assay relies on the proliferative expansion of individual "late" committed progenitor cells in semi-solid medium in the presence of cytokines. This assay is clonal in nature and hence, can be easily quantitated. The more sophisticated LTC-CFC assay requires a 5 week expansion of primary hematopoietic cells on a stromal cell layer followed by a methyl-cellulose assay. This assay reads out more primitive cells but is very time consuming and not quantitative. We have compared different transduction protocols using the CFU assay and our data suggests that triple cytokine combinations are required to achieve measurable retroviral gene transfer into these cells. We also compared retroviral supernatants from different packaging cell lines. These data clearly suggest that human cell-derived retroviral particles transduce human hematopoietic stem/progenitor cells more efficiently than supernatants derived from the murine PA317-packaging cell line. We also performed these experiments at large scale using up to $1x10^8$ cells and achieved essentially identical transduction rates. These transduction data were promising but we wondered whether we could demonstrate functional gene transfer into hematopoietic stem/progenitor cells. For this purpose, we developed retroviral vectors which encode a surface marker (eg. Lyt-2, NGF-R) in addition to the antiviral gene product. These vectors enable us to rapidly separate transduced from non-transduced cells which can then be differentiated into cell types relevant for HIV gene therapy. Based on this technology, we could demonstrate that CD4+ myeloid lineage cells can be derived which express the transgene. Subsequently, these cells were infected with macrophage-tropic HIV strains and we could demonstrate antiviral efficacy in these cells derived from transduced hematopoietic stem/progenitor cells. CD4+ T cells derived from transduced hematopoietic stem/progenitor cells represent the other relevant cell type and we adapted the SCID-hu Thy/Liv model and could generate CD3+CD4+ T cells carrying and expressing the recombinant retrovirus. We could also expand these cells *in vitro* and analyse them in more detail. Our results demonstrate that the transduction process and intracellular RevM10 expression have no negative impact on the proliferative potential of these cells. Further, we could show that these cells can respond to external stimuli like IL2, secrete cytokines comparable to control cells and are able to signal through their T cell receptor. By Southern Blot analyses, we could demonstrate that the gene-marked T cells are polyclonal in origin. These data strongly suggest that multiple hematopoietic stem cells with the potential to differentiate into CD4+ T lymphocytes were initially transduced using an amphotropic retroviral transduction protocol. Comparable to transduced primary T lymphocytes and CEMss cells, infection experiments of these cells demonstrated inhibition of HIV replication.

Taken together, these experimental results suggest that hematopoietic stem/progenitor-based gene therapy for HIV disease is feasible. Bone marrow is the "classical" clinically relevant tissue sources for hematopoietic stem cells. Umbilical cord blood has also been established as hematopoietic stem cell source but is almost exclusively used for pediatric applications. The discovery that cytokine-treatment results in drastically increased hematopoietic stem cell numbers in the peripheral blood made "stem cell transplantation" a much more feasible clinical therapy. The relative ease of this treatment [4 or 5 days of cytokine treatment (eg. G-CSF) followed by apherisis] made mobilised peripheral blood stem cells the clinical tissue source for the majority of clinical transplants. We collaborated with J. Zaia at City of Hope Medical Center to study the feasibility of G-CSF mobilisation of sufficient peripheral hematopoietic stem cells in HIV-infected individuals. The results of these preclinical experiments demonstrate that G-CSF administration can peripheralize hematopoietic stem/ progenitor cells in HIV-infected individuals in sufficient numbers. The biological potential of these cells as determined in the above described *in vitro* assays (CFU, LTC-CFC) is comparable to cells isolated from healthy volunteers. The HIV burden in these apheresis products is low but detectable by ELISA and DNA-PCR. The first enrichment step (CD34 selection) essentially eliminates all HIV-infected cells to the limits of the PCR assay sensitivity ($1/10^5$ cells). We have expanded this data set as part of our first clinical trial and the frequency of non-G-CSF responders is comparable to the situation in cancer patients.

These data together with our antiviral efficacyand toxicity data demonstrated the feasibility of hematopoietic stem cell based gene therapy for HIV disease. However, one clinical issue had not been adequately addressed: What patient population would be ideal for such a study and how will their current antiviral drug therapy influence the outcome of the gene therapy trial? We established *in vitro* HIV infection experiments using RevM10 and antisense-expressing cells as well as control cells and compared viral replication kinetics in the absence and presence of clinically relevant RT (AZT. ddC) and protease (Indinavir) inhibitors. These studies clearly demonstrated that drug and gene therapies can work together and are more effectivethan each therapy on its own. Similar to a previous study based on a dominant-negative protease mutant (20), we could also demonstrate that drug-resistant isolates were still inhibited by the gene therapyapproach.

Clinical Gene Therapy Protocols/Trials For HIV Disease

To date, several clinical intracellular immunization protocols have been initated (Table 2). The first trial performed at the University of Michigan Medical Center using RevM10-transduced peripheral blood lymphocytes. In this protocol, the RevM10 and a control gene are introduced by 2 differentways: a biolistics approach (gene gun) and amphotropic retroviral gene delivery. Preliminary results support the selective advantage of RevM10-transduced cells but also revealed very short persistence of the *ex vivo* manipulated T lymphocytes. Another study in T lymphocytes using 2 hairpin ribozymes expressed from Pol_{III} promoters in a retroviral vector has recently been initiated by F.Wong-Staal and colleagues at UCSD. This research group also plans a similar protocol based on CD34+ hematopoietic progenitor cells. In another T cell based trial at the NIH, a retroviral vector expressing a dominant-negative version of the Rev gene plus an antisense transcript has also been approved by the RAC. The most recent entrance is a T cell trial based on the intracellular expression of a single chain variable fragment (sFv) targeted at the HIV Rev protein. This trial will enter the clinical phase shortly at Thomas Jefferson University Medical Center. In addition to these T cell trials, several stem/progenitor protocols have entered the clinical arena. Rosenblatt and colleagues have a RAC-approved protocol based on a hammerhead ribozyme. Similarly, a collaborative effort between City of Hope and LA Cbildren's Hospital uses a retroviral vector to deliver a hammerhead ribozyme via CD34+ hematopoietic progenitor cells enriched from mobilized peripheral blood. At LA Children's, another approved protocol based on retrovirally-transducedbone marrow (BM) in pediatric patients will enter the clinical phase shortly. The protocol based on intracellular expression of the RevM10 gene delivered via retrovirally-transduced CD34+Thyl+ cells from MPB was the first clinical hematopoietic stem cell trial to enter the clinical phase.

PI	Phase	Target Cells	Antiviral Gene	DOA
G. Nabel	I	CD4+ PBLs	RevM10	9/93
F. Wong-Staal	I	CD4+ PBLs	Hairpin Rz	10/94
R. Morgan	I	Syngeneic lymphocytes	TAR-AS + TD Rev	4/95
W. Marasco	I	CD4+ PBLs	anti-Env sFv	7/95
J. Rosenblatt	I	CD34+ MPB	Hammerhead Rz	8/95 (FDA))
P. Greenberg	I	CD4+ T cells	RRE, RRE + Poly-TAR	11/95 (FDA)
R. Pomerantz	I	CD4+ PBLs	anti-Rev sFv	12/95 (FDA)
D. Kohn	I	CD34+ BM	RRE decoy	2/96 (FDA)
D. Kohn	I	CD34+ MPB	Hammerhead Rz	4/96 (FDA)
J. Belmont	I	CD34+ UCB	TD Trev	3/97 (FDA)
S. Deresinki, Carabasi	I	CD34+Thy+ MPB	TD Rev	FDA 1/96

Table 2: Approved gene therapy protcols. DOA: Date of approval. (FDA): RAC recommendation of sole FDA approval; indicated date reflects RAC recorrunendation; actual FDA approaval dates unknown. The CD34+Thy+ study has not been reviewed by RAC. Source: ORDA web site.

Outlook

Safety is the primary focus of these initial clinical trials. However, the T cell trials probably will provide experimental evidence to either support or negate the selective advantage of gene-modified cells in HIV-infected individuals. The stem/progenitor trials will provide insight into another critical scientific question: Is it actually possible to generate differentiated blood cells including T cells from *ex vivo* manipulated cells in an HIV-infected environment? Once these critical issues are positively resolved, gene therapy in combination with small molecule inhibitors can become a clinical reality for HIV disease.

Acknowledgments
I would like to thank all collegues at Systemix and Novartis who have contributed to the research, technical and clinical development of the HIV gene therapy project for their continuous support. I would also like to acknowledge our collaborators J. Mosca (Henry Jackson Foundation), J. Zaia (City of Hope) and C. Craik (UCSF).

References:

1. Baltimore D. Intracellular immunization. Nature 335:395-396; 1988
2. Barre-Sinoussi F, Cherinann JC, Rey F, Nugeyre MT, Chamaret J, Gruest J, Dauguet C, Axier-Blin C, Vezinet-Brun F, Rouzioux C, Rozenbaum W, and Montagnier L. Isolation of a T-lymphotropic retroviras from a patient at risk for acquired immunodeficiency syndrome (AIDS). Science 220:868-871; 1983
3. Bednarik DP, Mosca JD, Rai NBK, and Pitha PM. Inhibition of human immunodeficiency virus (HIV) replication by HIV-trans-activated a2-interferon. Proc Natl Acad Sci USA 86: 4958-4962; 1 989
4. Bevec D, Dobrovnik M, Hauber, J, and Böhnlein E. Inhibition of HIV-1 Replication in Human T-cells by Retroviral-Mediated Gene Transfer of a Dominant-Negative Rev Trans-Activator. Proc. Natl. Acad. Sci. USA 89: 9870-9874; 1992
5. Bevec D, Jaksche H, Oft M, Wöhl T, Himmelspach M, Pacher A, Schebesta M, Koettnitz K, Dobrovnik M, Csonga R, Lottspeich F, and Hauber J. Inhibition of HIV-1 replication in lymphocytes by mutants of the Rev cofactor eIF-5A. Science 271: 1858-1860; 1996
6. Blaese RM. Treatment of severe combined immune deficiency (SCID) due to adenosine deaminase (ADA) with autologous lymphocytes transduced with a human ADA gene. Hum Gene Ther. 1: 327-362, 1990
7. Buschacher GL, Freed EO, and Panganiban AT. Cells induced to express a human immunodeficiency virus type 1 envelope gene mutant inhibit the spread of wild-type virus. Hum Gene Thier. 3: 391-397; 1992
8. Caruso M, and Klatzmann D. Selective killing of CD4+ cells harboring a human immunodeficiency virus-inducible suicide gene prevents viral spread in an infected cell population. Proc. Natl. Acad. Sci. USA 89:182-186; 1992
9. Chang HK, Gendelman R, Lisziewicz J, Gallo RC , and Ensoli B. Block of HIV-1 infection by a combination of antisense tat RNA and TAR decoys: a strategy for control of HIV-1. Gene Ther, 1:208-216; 1994
10. Chen S-Y, Bagley J, and Marasco WA. Intracellular antibodies as a new class of therapeutic molecules for gene therapy. Hum Gene Ther 5:595-601; 1994
11. Chuah MK, Vandendriessche T, Chang HK, Ensoli B, and Morgan RA. Inhibition of human immunodeficiency virus type-1 by retroviral vectors expressing antisense-TAR. Hum Gene Ther. 5:1467-1475; 1994
12. Cohli H, Fan B, Joshi RL, Ramezani A, Li X, and Joshi S. Inhibition of HIV-1 multiplication in a human CD4+ lymphocytic cell line expressing antisense and sense RNA molecules containing HIV-1 packaging signal and rev response element(s). Antisense Res. and Dev. 4: 19-29; 1994
13. Collier AC, Coombs RW, Schoenfeld DA, Bassett RL, Timpone J, Baruch A, Jones M, Facey K, Whitacre C, McAuliffe VJ, Friedman HM, Merigan TC, Reichman RC, Hooper C, and Corey L. Treatment of human immunodeficiency virus infection with saquinavir, zidovudine, and zalcitabine. N Engi J Med 334: 1011-1017; 1996
14. Condra, JH, Schleif WA, Blahy OM, Gadryelski LJ, Graham DJ, Quintero JC, Rhodes A, Robbins HL, Roth E, Shivaprakash M, Titus D, Yang T, Teppler H, Squires KE, DeutschPJ, and Emini EA. *In vivo* emergence of HIV-1 variants resistant to multiple protease inhibitors. Nature 374:569-571; 1995
15. Cullen BR. Human Immunodeficiency Virus as a Prototypic Complex Retrovirus. J. Virol. 65:1053-1056; 1991
16. Curiel TJ, Cook DR, Wang Y., Hahn BH, Ghosh SK, and Harrison GS. Long-term inhibition of clinical and laboratory human immunodeficiency virus strains in human T-cell lines containing an HIV-regulated diphtheria toxin A chain gene. Hum Gene Ther. 4: 741-747; 1993
17. Duan L, Bagasra O, Laughlin MA, Oakes JW, and Pomerantz RJ. Comparison of Trans-Dominant Inhibitory Mutant Human Potent inhibition of human immunodeficiency virus type 1 replication by an intracellular anti-Rev single chain antibody. Proc Natl Acad Sci USA 91: 5075-5079; 1994

18. Gallo RC, Sarin PS, Gelmann EP, Robert-Guroff M, Richardson E, Kalyanaraman VS, Mann D, Sidhu GD, Stahl RE, Zolla-Pazner S, Leibowitch J, and Popovic M. Isolation of human T-cell leukemia virus in acquired immune deficiency syndrome (AIDS). Science 220:865-867; 1983
19. Ho DD, Neumann AU, Perelson AS, Chen W, Leonard JM, Markowitz M. Rapid turnover of plasma virions and CD4 lymphocytes in HIV-1 infection. Nature 373:123-126; 1995
20. Junker U, Escaich S, Plavec I, McPhee F, Rosé J, Craik CS, and Böhnlein E. Intracellullar Expression of HIV-1 Protease Variants Inhibits Replication of Wild-type And Protease Inhibitor Resistant HIV-1 Strains in Human T-Cell Lines. J. Virol.; in press; 1996
21. Junket U, Bevec D, Barske C, Kalfoglou C, Escaich S, Dobrovnik M, Hauber J, and Böhnlein E. Intracellular Expression of Cellular eIF5A Mutants Inhibits HIV- 1 Replication in Human T Cells: A New Approach for AIDS Gene Therapy. Hum Gene Ther., in press; 1996
22. Larder, BA, Kellam P, and Kemp SD. Convergent combination therapy can select viable multidrug resistant HIV-1 *in vitro.* Nature 365; 451-453; 1993
23. Lee T, Sullenger BA, Gallardo HF, Ungers GF, and Gilboa E. Overexpression of RRE-derived sequences inhibits HIV-1 replication in CEM cells. New Biol. 4: 66-74; 1992
24. Levy JA, Hollander H., Shimabukuro J, Mills, J, and Kaminsky L. Isolation of lymphdcytopathic retroviruses from San Francisco patients with AIDS. Science 225: 840-842; 1994
25. Levy JA. Pathogenesis of human immunideficiency virus infection. Microbiological Reviews 57(l): 183-289; 1993
26. Lisziewicz J, Sun D, Trapnell B, Thomson M, Chang HK, Ensoli B, and Peng B. An autoregulated dual-function antitat gene for human immnunodeficiency virus type 1 gene therapy. J. Virol. 69: 206-212; 1995
27. Lo KM, Biasolo MA, Dehni G, Palu G, and Haseltine WA. Inhibition of replication of HIV-1 by retrovirat vectors expressing tat-antisense and anti-tat ribozyme RNA. Virology 190: 176-183; 1992
28. Malim MH, Freimuth WW, Liu J, Boyle TJ, Lylerly HK, Cullen BR, And Nabel GJ. Stable expression of transdominant rev protein in human T cells inhibits human immunodeficiency virus replication. J. Exp. Med. 176: 1197-1201; 1992
29. Marasco WA, Hasletine Wa, and Chen SY. Design, intracellular expression, and activity of a human anti-human immunodeficiency virus type 1 gpl20 single-chain antibody. Proc. Natl. Acad. Sci. USA 90: 7889-7893; 1993
30. Mellors JW, Rinaldo CR, Gupta P, White RM, Todd JA, and Kingsley LA. Prognosis in HIV-1 infection predicted by the quantity of virus in plasma. Science 272: 1167-1170; 1996
31. Mhashilkar AM, Bagley J, Chen SY, Szilvay AM, Helland DG, and Marasco WA. Inhibition of HIV-1 tat-mediated LTR transactivation and HIV-1 infection by anti-Tat single chain intrabodies. EMBO J. 14:1542-1551; 1995
32. Modesti N, Garcia J, Debouck C, Peterlin M, and Gaynor R. Trans-dominant Tat mutants with alterations in the basic domain inhibit HIV-1 gene expression. New Biol. 3: 759-768; 1991
33. Morgan RA, Looney DJ, Muenchau DD, Wong-Staal F, Gallo RC, and Anderson WF. Retroviral vectors expressing soluble CD4: a potential gene therapy for AIDS. AIDS Res. Hum. Retroviruses 6:183-191; 1990
34. Nabel GJ, Fox BA, Post L, Thompson CB, and Woffendin C. A molecular genetic intervention for AIDS - effects of a transdominant negative form of Rev. Hum. Gene Ther. 5: 79-92; 1994
35. Ojwang JO, Hampel A, Looney DJ, Wong-Staal F, and Rappaport J. Inhibition of human immunodeficiency virus type 1 expression by a hairpin ribozyme. Proc. Nad. Acad. Sci. USA 89: 10802-10806; 1992
36. Pearson L, Garcia J, Wu F, Modesti N, Nelson J, and Gaynor R. A trans-dominant tat mutant that inhibits tat-induced gene expression from the HIV LTR. Proc. Natl. Acad. Sci. USA 87: 5079-5083; 1990
37. Ratner L, Fisher A, Jagodzinski LL Mitsuya H, Liou RS, Gallo RC, and Wong-Staal F. Complete nucleotide sequences of functional clones of the AIDS virus. AIDS Res. Hum. Retroviruses 3:57-69; 1987
38. Rhodes A, and James W. Inhibition of human immunodeficiency virus replication in cell culture by endogenously synthesized antisense RNA. J. Gen. Virol. 71: 1965-1974; 1990
39. Rosenberg SA. The treatment of patients with advanced cancer using cyclophosphamide, interleukin-2 and tumor infiltrating lymphocytes. Hum. Gene Ther. 1: 73-92; 1990
40. Sarver N, Cantin EM, Chang Pg,- Zaia JA, Ladne PA, Stephens DA, and Rossi JJ. Ribozymes as potential anti-HIV-1 therapeutic agents. Science 247: 1222-1225; 1990

41. Sczakiel G, and Pawlita M. Inhibition of human immunodeficiency virus type 1 replication in human T cells stably expressing antisense RNA. J. Virol. 65: 468-472; 1991
42. Shaheen F, Duan L, Zhu M, Bagasra O, and Pomerantz RJ. Targeting human immunodeficiency virus type 1 reverse transcriptase by intracellular expression of single-chain variable fragments to inhibit early stages of the vital life cycle. J. Virol. 70: 3392-3400; 1996
43. Sullenger BA, Gallardo HF, Ungers GE, and Gilboa E. Overexpression of TAR sequences renders cells resistant to human immunodeficiency virus replication. Cell 63: 601-608; 1990
44. Trono D, Feinberg MB, and Baltimore D. HIV-1 gag mutants can dominantly interfere with the replication of the wild-type virus. Cell 59:113-120; 1989
45. Veres G, Escaich S, Barske C, Baker J, and Böhnlein E. Constitutive intracellular Expression of Gag Antisense RNA Specifically and Effectively Inhibits HIV-1 Replication in Human CD4-Positive T Cells. Manuscript submitted.
46. Vieillard V, Lauret E, Rousseau V, and DeMayer E. Blocking of retroviral infection at a step prior to reverse transcription in cells transformed to constitutively express interferon beta. Proc. Natl. Acad. Sci. USA 91: 2689-2693; 1994
47. Wain-Hobson S. Sonigo P., Donas O, Cole S. and Alizon M. Nucleotide sequence of the AIDS virus LAV. Cell 40:9-17; 1985
48. Wei X, Ghosh SK, Taylor ME, Johnson, VA, Emini EA, Deutsch P, Lifson JD, Bonhoeffer S, Nowak MA, Hahn BH, Saag MS, Shaw GM. Viral dynamics in human immunodeficiency virus type 1 infection. Nature 373:117-122; 1995
49. Woffendin C, Ranga U, Yang Z-Y, Xu L, and Nabel GJ. Expression of a protective gene prolongs survival of T cells in human immunodeficiency virus infected patients. Proc. Natl. Acad. Sci. USA 93: 2889-2894; 1996
50. Zhou C, Bahner I, Larson GP, Zaia JA, Rossi JJ, and Kohn DB. Inhibition of HIV-1 in human T lymphocytes by retrovirally transduced anti-tat and rev hammerhead ribozymes. Gene 149: 33-39; 1994
51. Zimmermann K, Weber S, Dobrovnik M, Hauber J, and E. Böhnlein E. Expression of Chimeric Rev Response Element Sequences Interferes with HIV-1 Rev Function, Hum Gene Ther. 3: 155-161; 1992

Adeno-Associated Viral Vectors: Principles and in vivo Use

Olivier Danos, Gene Therapy Program. Généthon, CNRS URA 1922, 1bis rue de l'Internationale, 91002 Evry, France.

I. Biology of Adeno-Associated Viruses

Small non-enveloped viruses with an icosahedral capsid and a single-stranded DNA genome are grouped in the family of Parvoviridae. The virions are usually around 20 nm in diameter and are extremely resistant to a variety of harsh physical conditions (low pH, heat, presence of detergent). Within this family, Dependoviruses where first identified as contaminants of human adenovirus preparations and shown to be strictly dependent on the presence of the helper adenovirus for their replication. These Adeno-associated viruses (AAV) are found in a variety of birds and mammals including humans. Five human serotypes have been identified, and around 70 % individuals are seropositive in the general population. However, no pathology has been associated with AAV infection and although the AAV genome is known to integrate the host cell genome under certain conditions, no vertical transmission has been documented. For a complete review on Parvoviruses and AAV, see (1).

AAV particles are able to infect a variety of cell type in culture. Viral capsids entering the cytosol are rapidly transported to the nucleus were the single stranded genome is delivered. In the presence of a helper virus, such as Adeno, Herpes or Vaccinia virus, the genome is actively transcribed and replicated. This step is performed mostly by the cellular DNA and RNA synthesis machineries and requires the presence of AAV proteins encoded by the rep gene. It is most efficiently activated by a co-infecting helper virus, but more generally, it can be seen in response to a variety of cellular stresses (2). Viral capsids assemble in the nucleus, package the replicated genome and virions are liberated upon cell lysis (3).

In the absence of helper functions, the AAV genome is not productively transcribed or replicated. The single stranded (ss) DNA genome is converted into a double stranded (ds) form and eventually integrates into the cellular DNA. When human cells are infected with helper free AAV, tandem arrays of the genome are found at a preferred location on chromosome 19. This site specific integration requires the presence of Rep proteins (4). The latent genomes can be reactivated and rescued by an infection with a helper virus (5).

The 4679 nucleotide genome packaged into AAV virions can be of plus or minus polarity. It contains a 145 nt inverted terminal repeat (ITR) of which the first 125 nt can form a T-shaped hairpin structure. The ITR contains all the information needed in

NATO ASI Series, Vol. H 105
Gene Therapy
Edited by Kleanthis G. Xanthopoulos

cis for replication, packaging and integration of the genome (6). Viral proteins are encoded by two genes, rep and cap. Multiple proteins are translated from differentially initiated or spliced transcripts. There are 4 Rep (MW 78, 68, 52 and 40) and 3 Cap proteins (VP1, VP2, VP3). The protein composition of AAV virions is 80% VP3.

The smaller Rep proteins are involved in viral RNA maturation and packaging. The two larger Rep proteins display site specific endonuclease, ATPase and helicase activities (7) and bind repeated copies of the motif GAGC on double stranded DNA. Binding sites are found: i) on the A stem of the ITR; ii) within the three AAV promoter regions (8) and iii) at several loci the human genome, including one close to the AAV preferential integration site on chromosome 19 (4, 9). The current view of the role of the large Rep proteins is that they: i) resolve the closed T-shaped end of the double stranded replication intermediates and allow for the formation of monomeric single stranded genomes ; ii) modulate AAV gene expression through transcriptional regulation (10); iii) mediate site specific integration, possibly by bridging the viral and cellular DNA binding sites (11).

II. AAV-derived Vectors

A. *Principle and Methods*

AAV vectors only retain the duplicated 145 nt ITR from the viral genome, the rep and cap genes being removed and replaced by the sequences of interest. Vector particles can theoretically be produced in the absence of wild type AAV, if the rep and cap gene products are provided in trans (6). The initial method to achieve this was to transfect cells with the vector and a rep-cap expression cassette on two separate plasmids (12, 13). Human 293 cells, which are highly permissive to both AAV and adenovirus replication and can be efficiently transfected, are of general use for vector production. Following transfection, helper functions are provided by infecting the cells with Adenovirus. Since the rep and cap sequences are not linked to the AAV ITR, only the recombinant construct is replicated and packaged. Recombinant virions, free of wt AAV can be purified from cell lysates and separated from the Adenovirus particle by density centrifugation. This procedure can be optimized to yield 1-5 x 103 recombinant AAV (rAAV) particles per transfected cell, which is within the range of wild type AAV production (13, 14).
Still, the current method presents several limitations : i) a typical prepartion is very work intensive, since it involves the transfection of 1-5 x 108 cells, followed by week-long purification and assay procedure (13); ii) each viral preparation is made from a new transfection, and reproducibility problems arise ii) a low level contamination of the final stock by the helper adenovirus is unavoidable, following physical separation (14); iii) illegitimate recombination, which may occur during the transfection, or under the influence of Rep results into structures where ITR sequences are linked to rep and cap. These chimeric forms can subsequently evolve into replication competent AAV genomes, through additional recombination (15).

Several significant improvements have been made to the transient transfection method. Adenovirus infection can now be replaced by the cotransfection of a cloned Adenovirus genome deleted for the packaging sequence (() and part of the E1 region (unpublished observation). This allows for the production of rAAV particles in the absence of functional Adenovirus. Further deletions in the Adenovirus genome, eliminating most of the late (L) genes, have also been shown to provide efficient help for rAAV production (X Xiao and RJ Samulski, personal communication). Following this method, adenoviral capsids are eliminated and cleaner prepartions should be obtained. It was also observed that the level of Rep proteins produced after transfection was critical for obtaining high titer rAAV. Surprisingly, low amounts of rep expression result in higher rAAV production, probably because of the negative influence of Rep on cap expression. Helper (rep-cap) plasmids designed to express low levels of Rep allow for the production of 5-10 fold more rAAV (16). In achieving high titer preparations, the ability to synthetize large amounts of Cap proteins is critical (14).

One obvious goal is to obtain stable packaging systems for the production of rAAV. This would not only simplify the preparation method, but also facilitate the definition of specifications for the production of vectors to be used in pre-clinical and clinical experiments. The development of a universal packaging line is underway. Clones stably expressing rep and cap as well as a vector construct have been used to produce helper free rAAV, following Adenovirus infection (17). The isolation of « proto-producer » clones in which rAAV production is triggered by Adenovirus infection, requires more work initially, but they result in permanent sources of rAAV that can be mobilized by Ad infection. This approach is likely to become of general use, once its reproducibility will be established.

B. Factors influencing AAV-mediated gene transfer efficiency:

The understanding of the replication mechanism of AAV is only partly relevant to the situation encountered with vectors. Like the parental wild-type (wt) AAV, recombinant genomes entering the cell nucleus must be converted to a dsDNA in order to be transcribed. In wt AAV the Rep proteins either directly facilitate second strand synthesis, or can trigger the (still undefined) cellular functions involved. On the other hand, vectors do not retain the rep gene and the Rep proteins are not detectably associated with the released viral particles. Thus the recombinant genomes uniquely depend on the cell DNA synthesis machinery for second strand synthesis. This implies that the expression of sequences transferred by an AAV vector will be highly dependent on the presence of an appropriate cellular environment. Different situations have been shown to enhance rAAV-mediated gene transfer by allowing for a more efficient second strand synthesis. One is the expression in the target cell of the open reading frame 6 from the E4 region of Adenovirus (18, 19), a protein that can bind p53 (20), and may therefore interact with cell cycle signals. More generally, drug or radiation induced genotoxic stresses, have an enhancing effect on rAAV-mediated gene transfer in cultured cells, or in vivo (20-24). Although a unifying explanation for these observations is still lacking, the initiation of unscheduled DNA repair synthesis is probably central for rAAV second strand synthesis.

III. AAV vectors for direct gene transfer in vivo:

A number a features of rAAV make them well suited for direct gene transfer in vivo. Compared to retroviruses or adenoviruses, the rAAV particles are small and very resistant, and may therefore remain in the circulation or in an injected tissue for longer times, and diffuse more efficiently across some the natural barriers within the organism. In addition, rAAV should, in principle, be able to transduce non-dividing cells, which constitute the overwhelming majority of targets encountered in vivo. Limitations include, the size of the transferred sequences and, in the prospect of human clinical trials, the widespread seropositivity for AAV in human populations. Over the past couple of years, the methods for rAAV production, have become efficient enough to yield preparations with titer and quality compatible with in vivo experimentation. A number of in vivo targets have been explored, including central and peripheral nervous system (23, 25-30), repiratory epithelium (31), vascular, cardiac and skeletal muscle (32-36) and liver (24). Here we discuss our experiments with skeletal muscle and liver (37, 38).

A. Gene Transfer to the Skeletal Muscle

Our interest was to further document the remarkable efficiency of rAAV for gene transfer into the skeletal muscle and to analyze the structure of the recombinant genome in the transduced muscle fibers. A secondary goal was to start to analyse the contribution of the post-mitotic environment found in the terminally differentiated muscle fiber, to the high-efficiency of AAV-mediated gene transfer.

The efficiency of AAV-mediated gene transfer into the skeletal muscle of mice was evaluated by performing a single injection of purified recombinant AAV (rAAV) particles encoding a modified E. coli ß-galactosidase with a nuclear localization signal, under the control of the CMV promoter. The rAAV preparations were titered by dot blot hybridization (1012 physical particles / ml) and by limiting dilution infections and X-gal staining of 293 cells (5.6 x 108 pfu / ml and 9.8 x 109 pfu / ml in the absence or presence of adenovirus co-infection, respectively). The stocks were determined to be free of detectable contamination by adenovirus and by replication competent AAV (less than 1 wt AAV / 109 rAAV). Young adult Balb/c mice were injected into the quadriceps muscle with 30 μl of rAAV preparation containing 1.7 x 107 ß-gal infectious units and animals were sacrificed between 2 and 28 weeks later. Histological sections of the injected muscles, stained for ß-galactosidase activity revealed between 10 and 70% positive fibers on more than 75% of the muscle length.

Two weeks after gene transfer, the muscle had a normal histological aspect, whereas on sections taken at four and eight weeks, infiltrates became conspicuous in the areas containing positive fibers. Regenerated fibers with centrally located nuclei were also noted in the infiltrated areas. Remarkably, these fibers still expressed the transgene and may therefore have arisen through the proliferation and fusion of transduced satellite myoblasts containing the recombinant AAV genome. The cellular response was not observed at later time points. In conclusion, this gene transfer procedure is associated with a transient and limited immune response, without any major impact on the long term genetic modification of muscle cells.

We then asked whether an enhancement of gene transfer is observed when muscle satellite cells are induced to proliferate as a consequence of muscle injury. The same gene transfer procedure was applied to animals pretreated 48 hours before by an intra-muscular injection of barium chloride ($BaCl_2$) which provoks a rapid necrosis of the muscle fibers, followed by tissue regeneration. In animals analysed 2 weeks after gene transfer, over 85% of the muscle fibers had been regenerated and displayed centrally located nuclei.

Surprisingly, most of the ß-galactosidase positive fibers were confined to the remaining, non regenerated, areas and accounted for less than 20% of all fibers. At this time point, cellular infiltrates were observed in the regenerated areas, but not in the intact tissue. At late time points, the infiltrates became much less pronounced, and only limited areas of positive and mostly regenerated fibers remained. We concluded that there is no obvious enhancement of gene transfer when the AAV vector is applied to a regenerating muscle containing proliferating cells. On the contrary, the terminally differentiated, post-mitotic, muscle fibers appear to be preferentially permissive to AAV vectors.

Another group of mice received a single intra-muscular injection of a rAAV encoding murine erythropoietin (Epo, the hormone regulating erythopoiesis), under the control of the CMV promoter. Each animal was injected with 2.7x1010 total vector particles, blood samples were collected over time and Epo production was monitored either directly by a radio-immunoassay on plasma, or indirectly by measuring the hematocrit which reflects the number of circulating erythrocytes. The hematocit of every animal increased during the first four weeks following gene transfer, and reached a plateau value of 80 to 90%. These high hematocrits were observed until the animals were sacrificed after 2 to 4 months. They corresponded to 10 to 20 fold increase in serum Epo levels.

The structure of the rAAV genomes was analyzed by Southern blot on high molecular weight DNA prepared at different time points from the injected muscles. Eight weeks after injection, 1 to 3 copies of rAAV per haploid genome were measured. This material was associated with high molecular weight DNA, under the form of concatemers. No proof of integration, like the characterisation of junction between cellular and vector genome, has been obtained. It is possible that the rAAV genome remains extra-chromosomal, under the form of either head-to-tail tandem repeats or interlocked circles.

The status of the vector DNA was also analyzed at early time points following injection. After one and two days, a strong signal is found corresponding to the input ssDNA genome. This signal progressively decreases until 2 weeks. Double stranded monomers of the genome are also observed during the first days following gene transfer. Although this could represent an artefactual reassociation of the complementary strands of DNA delivered by the rAAV particles, it may also reflect the high permissivity of the muscle fiber. This ds monomers progressively disappear during the first two weeks, and are chased into the high molecular weight forms. In conclusion, we observe two successive transformation of the recombinant genome upon entry into the muscle fiber nuclei: a conversion into dsDNA monomer, followed by a con-catemerisation.

B. Gene Transfer into the Liver

The steady state liver contains mostly non dividing cells, 90% of which are hepatocytes. These hepatocytes are normally quiescent but they are induced to divide in response to liver injury.

The quiescent or regenerating liver can be subjected to gene transfer using a variety of vectors by portal infusion. We have examined the potential of rAAV in this system by administrating a preparation of rAAV carrying the human Factor IX cDNA to adult C57Bl/6 mice (37). Factor IX is a component of the blood coagulation cascade, normally produced by the liver. Its deficiency results in Hemophilia B. Beetween 2 and 8 x 1010 vector particles were infused in the portal vein and the presence of human Factor IX was periodically measured in plasma samples. The human protein was detectable at low levels during the first week and increased to steady state concentrations of 250 to 2000 ng/ml. The secreted Factor IX was active in coagulation assays and the serum levels were dose dependent and persisted for at least 36 weeks (the duration of the experiment). When extrapolated to a clinical situation, such concentrations would be relevant for the treatment of Hemophilia B.

The amount of rAAV genomes present in the infused livers was measured by Southern blot to be between 1 and 4 copies per haploid cellular genome. In situ hybridization showed that human Factor IX mRNA could be found in only 1 to 5 % of the hepatocytes, implying that either a few cells contain and express multiple (30 to 50) copies of the rAAV genome, or that most cells contain vector genomes and only a minority expresses them. Considering the number of target cells in the mouse liver (108) and the estimated amount of infused infectious particles (1-5 x 108), it is unlikely that most cells can be transduced. As a comparison, over 5 x 109 adenoviral particles are needed to transduce most cells in the liver . We therefore favor the hypothesis where the gene transfer procedure would deliver the ssDNA genome to 1 to 5 % of the liver cells, and after conversion into the ds form, an amplification step would take place.

IV. Conclusions

AAV vectors have now proven to be very efficient for stable gene delivery into a number of in vivo targets. At this time, preparation of high quality and high titer rAAV, although dramatically improved over the past couple of years, remain a significant bottleneck. Yet, specifications for clinical grade material are starting to be defined and phase I clinical trials involving rAAV as vector are now underway.

In animal experiments, high gene transfer efficiency is seen in myotubes and neurones, which are post-mitotic cells, and possibly in a sub-population of hepatocytes for which the cell cycle status is unknown.

Not all arrested primary cells, however are permissive to rAAV gene transfer. Bone marrow cells enriched in non cycling CD34+ hematopoietic progenitors will accumulate rAAV DNA, but mostly fail to express the transfered gene (C. Jordan, personal communication). Understanding the determinants of cellular permissivity to AAV will have important implications for the definitions of optimal gene transfer targets in vivo.

V. References

1. Berns, K. I. (1996) in Fields Virology, eds. Fields, B., Knipe, D. & Howley, P. (Lippincott-Raven, Philadelphia), Vol. 2, pp. 2173-2220.

2. Schelehofer, J. R., Matthias, E. & Zur Hausen, H. (1986) Virology 152, 110-117.

3. Wistuba, A., Kern, A., Weger, S., Grimm, D. & Kleinschmidt, J. (1997) J. Virol 71, 1341-1352.

4. Linden, R. M., Winocour, E. & Berns, K. I. (1996) Proc. Natl. Acad. Sci. U.S.A. 93, 7966-7972.

5. McLaughlin, S. K., Collis, P., Hermonat, P. L. & Muzyczka, N. (1988) J. Virol. 62, 1963-1973.

6. Muzyczka, N. (1992) Curr. Top. Microbiol. Immunol. 158, 97-129.

7. Im, D. & Muzycska, N. (1990) Cell 61, 447-457.

8. Mc Carty, D. M., Pereira, D., Zolotukhin, I., Zhou, X., Ryan, J. H. & Muzyczka, N. (1994) J. Virol. 68, 4988-4997.

9. Wonderling, R. & Owens, R. (1997) J. Virol. 71, 2528-2534.

10. Pereira, D., McCarty, D. & Muzyczka, N. (1997) J. Virol 71, 1079-1088.

11. Weitzman, M. D., Kyostio, S. R. M., Kotin, R. M. & Owens, R. A. (1994) Proc. Natl. Acad. Sci. U.S.A. 91, 5808-5812.

12. Samulski, R. J., Chang, L. S. & Shenk, T. (1989) J. Virol. 63, 3822-3828.

13. Snyder, R., Xiao, X. & Samulski, R. J. (1996) in Current Protocols in Human Genetics, ed. Haines, J., pp. 12.1.1-121.1.23.

14. Vincent, K., Piraino, S. & Wadsworth, S. (1997) J. Virol. 71, 1897-1905.

15. Allen, J., Debelak, D., Reynolds, T. & Miller, A. (1997) J. Virol 71, 6816-6822.

16. Li, J., Samulski, R. J. & Xiao, X. (1997) J Virol 71, 5236-5243.

17. Clark, K., Voulgaropoulou, F., Fraley, D. & Johnson, P. (1995) Hum. Gene Ther. 6, 1329-1341.

18. Ferrari, F., Samulski, T., Shenk, T. & Samulski, R. (1996) J. Virol. 70, 3227-3234.

19. Fisher, K. J., Gao, G. P., Weitzman, M. D., De Matteo, R., Burda, J. F. & Wilson, J. M. (1996) J. Virol. 70, 520-532.

20. Dobner, T., Horikoshi, N., Rubenwolf, S. & Shenk, T. (1996) Science 272, 1470-1473.

21. Alexander, I., Russell, D. & Miller, A. (1994) J. Virol. 68 N°12, 8282-8287.

22. Russell, D. W., Alexander, I. E. & Miller, A. D. (1995) Proc. Natl. Acad. Sci. U.S.A. 92, 5719-23.

23. Alexander, I., Russell, D., Spence, A. & Miller, A. (1996) Hum. Gene Ther. 7, 841-850.

24. Koeberl, D. D., Alexander, I. E., Halbert, C. L., Russels, D. W. & Dusty Miller, A. (1997) Proc. Natl. Acad. Sci. U.S.A. 94, 1426-1431.

25. Kaplitt, M., Leone, P., Samulski, R., Xiao, X., Pfaff, D., O'Malley, K. & During, M. (1994) Nat. Genet. 8.

26. Ali, R. R., Reichel, M. B., Thrasher, A. J., Levinsky, R. J., Kinnon, C., Kanuga, N., Hunt, D. M. & Battacharya, S. S. (1996) Hum. Mol. Genet. 5, 591-594.

27. Lalwani, A. K., Walsh, B. J., Reilly, P. G., Muzyczka, N. & Mhatre, A. N. (1996) Gene Therapy 3.

28. McCown, T. J., Xiao, X., Li, J., Breese, G. R. & Samulski, R. J. (1996) Brain Res. 713, 99-107.

29. Peel, A., Zolotukhin, S., Schrimsher, G. W., Muzyczka, N. & Reier, P. J. (1997) Gene Therapy 4, 16-24.

30. Flannery, J., Sergei Zolotukhin, S., M. Isabel Vaquero, M., Matthew M. LaVail, M., Muzyczka, N. & Hauswirth, W. (1997) Proc. Natl. Acad. Sci. USA 94, 6916-6921.

31. Afione, S. A., Conrad, C. K., Kearns, W. G., Chunduru, S., Adams, R., Reynolds, T. C., Guggino, W. B., Cutting, G. R., Carter, B. J. & Flotte, T. R. (1996) J.Virol. 70, 3235-3241.

32. Kaplitt, M. G., Xiao, X., Samulski, R. J., Li, J., Ojamaa, K., Klein, I. L., Makimura, H., Kaplitt, M. J., Strumpf, R. K. & Diethrich, E. B. (1996) Ann Thorac Surg 62, 1669-1676.

33. Xiao, X., Li, J. & Samulski, R. J. (1996) J. Virol. 70, N°11, 8098-8108.

34. Kessler, P. D., Podsakoff, G. M., Chen, X., McQuiston, S. A., Colosi, P. C., Matelis, L. A., Kurtzman, G. J. & Byrne, B. J. (1996) Proc. Natl. Acad. Sci. U.S.A. 93, 14082-14087.

35. Fisher, K. J., Jooss, K., Alston, J., Yiping , Y., Haecker, S. E., High, K., Pathak, R., Raper, S. E. & Wilson, J. M. (1997) Nature Medicine 3, 306-312.

36. Herzog, R., Hagstrom, J., Kung, S., Tai, S., Wilson, J., Fisher, K. & High, K. (1997) Proc. Natl. Acad. Sci. USA 94, 5804-5809.

37. Snyder, R., MIao, C., Patijn, G., Spratt, K., Danos, O., Nagy, D., Gown, A., Winther, B., Meuse, L., Cohen, L., Thompson, A. & Kay, M. (1997) Nat. Genet. 16, 270-276.

38. Snyder, R., Spratt, S., Lagarde, C., Bohl, D., Kaspar, B., Sloan, B., Cohen, L. & Danos, O. (1997) Hum. Gene Ther. , in press.

Genetic Correction for Gene Therapy

Eric B. Kmiec[1], Allyson Cole-Strauss, Michael C. Rice and Pamela Havre

Department of Microbiology
Kimmel Cancer Center
Thomas Jefferson University
233 South 10th Street
Philadelphia, PA, 19107
Phone: 1-215-503-4618
Fax: 1-215-923-1098

LECTURE I

The development of gene targeting systems has been enabled by the great advances in molecular genetics and cell culture technology. The success of producing genetic knock-outs in mice through the use of embryonic stem cells allowed the conception of efficient targeting in mammalian cells to become a distinct possibility. The availability of cloned genes and DNA sequences, combined with the ability to transfer and express genes in mammalian cells, forms the basis of gene targeting strategies. The challenges of gene targeting in mammalian cells are enormous, however, and they fall into three general categories.

The first centers around the vector itself and choices as to the possibility of using a synthetic molecule, a DNA fragment, or, in some cases, plasmid DNA. The next category of concern centers around the choice of the cell type to transfect. This is governed by the endpoint an investigator wishes to achieve. Somatic cells are available primarily through the use of tissue culture cells or transformed cell lines, making experimentation easier. However, in many diseases the choice has been made to focus on primary cells such as hemopoietic stem cells, wherein a single targeting event in the primitive cell will lead to the genesis of continually modified daughter cells. The step-up to primary cells is often the Achilles Heel of promising therapies

The third category is based on the design of the molecule and goes to the center of the argument about gene targeting. If one considers what gene targeting is all about at the mechanistic level, one must revisit models of homologous recombination. Homologous recombination is a process that leads to the permanent genetic exchange of DNA or the alteration of DNA in a particular gene. Frequencies of recombination, even in mammalian cells, vary widely and so the charge to people trying to improve

[1] Corresponding author

NATO ASI Series, Vol. H 105
Gene Therapy
Edited by Kleanthis G. Xanthopoulos

gene targeting events is to clearly influence the rate-limiting step of the reaction. We believe that the rate-limiting step is the pairing of the target strand, or DNA molecule, with the vector. Successful targeting should depend, therefore, on the ability of the system to influence the rate-limiting step of the reaction. In eukaryotic cells, factors affecting the rate-limiting step, are the position of the target within the chromosome, and most importantly, the availability of enzymes present to catalyze the pairing reaction.

There are two ways in which the enzymatic activity of DNA pairing/gene targeting can be influenced. The first is by making a vector or utilizing a vector that is acted on efficiently by the low-level of enzymatic activity present in the cell. The second way is to increase the enzymatic activity of the cell directly by complementing the cell's allotment of this activity using the gene-addition approach. In this strategy, genes are added to the cell to increase the presence of important enzymes. Our laboratory, over the last several years, has taken both approaches. First, we have designed a hyperactive oligonucleotide that facilitates increased gene targeting frequency in mammalian cells. Second, we isolated, cloned and characterized candidate genes whose expression in mammalian cells augments activity and improves the frequency of gene targeting events.

Our studies began using the lower eukaryote *Ustilago maydis* as a fungal, genetic prototype system. We had previously identified a gene known as REC2 which was responsible for 90% of the plasmid-chromosomal targeting events occurring in this fungus. Based on these genetic studies and the biochemistry of the purified Rec2 protein, we concluded that this gene may be important in the pathway to improving gene targeting in all eukaryotes. The Rec2 protein catalyzed many of the biochemical reactions established for the RecA protein of prokaryotes, a protein responsible for the vast majority of homologous recombination events in prokaryotes. We utilized a series of recombinogenic assays to evaluate the Ustilago Rec2 protein. Results of these assays indicated that Rec2 could catalyze strand transfer events and these results were used as a barometer to measure recombination efficiency.

Although DNA/recombinase assays are important to identify activities, in eukaryotes the cellular target for recombination is more complex. Therefore, we conducted a series of biochemical studies in which we measured the recombination activity of the Ustilago Rec2 protein on DNA packaged into chromatin. Such nucleosomal DNA reflects more precisely the status of the DNA in the chromosome. We reasoned that the environment of the chromosome itself was somehow important to targeting and the environment in which our vector would be operational was important to understand.

A long series of studies revealed the following facts: *A.)* Strand transfer events promoted by Rec2 or RecA protein were completely blocked when the target DNA was packaged into chromatin or was in the form of nucleosomal DNA; *B.)* The activation of nucleosomal DNA to be receptive to gene targeting or homologous recombination events, relied on the active transcription of the template by RNA polymerase; and *C.)* The RNA itself was caught up, or became part, of the product recombinant molecule.

These three observations led us to consider what the role of RNA might be in recombination. We designed experiments to test the hypothesis that RNA reduced the length of homology that was required for successful joint molecule formation. We were measuring joint molecules as a representative of homologous recombination events promoted by Rec2 or RecA. The results of these experiments clearly demonstrated that as the target size was reduced from 170 to 70 and on to 25 bases of homology between the pairing partners, one of the two partners must be transcriptionally active. We concluded from these studies that part of the vector for gene targeting events in mammalian cells would be more efficient if it contained stretches of RNA.

Our first attempt to create such a vector was to transcriptionally activate plasmids of DNA as one of the pairing partners. This did not reproducibly improve targeting events in mammalian cells. To create a stable RNA molecule, we designed an oligonucleotide that consisted of both DNA and RNA residues and named it the *chimeric oligonucleotide* to reflect the heterogeneity of bases in the structure. Direct tests of the recombination capacity of RNA were conducted using these molecules. The results of these studies indicated directly that oligonucleotides containing RNA residues were able to pair with the DNA target containing short stretches of homology. Oligonucleotides lacking RNA were unable to pair at the same frequency when the target size was reduced. To understand the parameters of these results, the physical properties of the chimera were examined first. The importance of purity and melting temperatures were evaluated and association kinetics between the chimeric oligonucleotide and the DNA target were measured. We also established that the chimeric oligonucleotide could pair with nucleosomal DNA and produce significant levels of joint molecules *in vitro*.

Based on information regarding the use of oligonucleotides in *Saccharomyces cerevisiae*, we attempted to mutate or correct specific point mutations in episomal or genomic targets using the oligonucleotide. This goes to the foundation of the approach known as site-specific mutagenesis *in vivo*.

Three levels of targeting using chimeric oligonucleotides were tested. The first system, based on evolutionary progression, is bacterial. Experiments have been designed to test the chimeric oligonucleotide in prokaryotic gene targeting by using a tetracycline gene that is missing a base in the coding region. Chimeric oligonucleotides were designed to engineer the insertion of the nucleotide at the specific site and reconstruct an intact coding region. The coding region would then permit the expression of tetracycline resistance in *E. coli* and colonies would be evident on plates containing tetracycline.

Such a genetic readout in bacterial cells has enormous implications both from possible therapeutic uses to the development of a system to evaluate the mechanism of correction or mutagenesis. The results indicated that indeed chimeric oligonucleotides and not DNA oligonucleotides enabled the specific insertion of a base into the coding region of the tetracycline gene and the onset of gene expression enabled resistance to tetracycline. The frequency of such events varies between one and five percent.. In addition, the RecA protein is required for this type of gene targeting in bacterial cells and, importantly, the Ustilago and human REC2 genes can complement a *recA*-deficient cell line. We have now extended the work into genomic targets in bacterial cells using a LacI gene or a Tet^S gene by measuring for color changes in the colonies or conferring tetracycline resistance to a tetracycline-sensitive strain.

The next system in which chimeric oligonucleotides have been used is *Ustilago maydis* itself. In this system, the genomic copy of the PYR6 gene was corrected (in which a codon which normally is read as lysine, is converted to a stop codon). The result of this mutation is to essentially destroy the coding region of the PYR6 gene. Once destroyed, the PYR6 gene is unable to be expressed and becomes resistant to the toxic chemical, 5-FOA. Successful completion of these experiments demonstrated that the chimeric oligonucleotide is effective in converting or knocking-out genes in lower eukaryotic cells.

The next step was to test the chimeric oligonucleotide on an episomal target in mammalian cells. The episomal target chosen was an alkaline phosphatase gene containing a point mutation in the coding region. The genetic readout occurs even in the mutated form of the gene but the translated protein folds improperly and is unable to be detected by a zinc-based stain. Chimeric oligonucleotides were found to be effective in correcting this mutation in Chinese Hamster ovary cells evidenced by the appearance of red-staining cells. These results were confirmed at both the genetic level and at the protein level since wild-type alkaline phosphatase was produced.

An independent group working at the University of Minnesota repeated and confirmed the alkaline phosphatase targeting and extended it to the genomic level using the liver cell line, HUH7. These cells were found to be amenable to chimeric-based oligonucleotide targeting by mutating a wild-type alkaline phosphatase gene at frequencies approaching 20%. Efficient transfection conditions of mammalian cells are of critical importance for the success of this technique and liver cells tend to be the most amenable to liposome-mediated uptake. Routinely, we observe between 50 to 70% of the cells being positive for the chimeric oligonucleotide which localizes to the nucleus. Importantly, the ratio of liposome and chimeric oligonucleotide often determines the successful transfection of the cell and its cellular compartmentalization, i.e. nuclear or cytoplasmic. The design of the specific chimeric oligonucleotides is, in the HUH7 setting, (genomic targeting) identical to those used in the episomal targeting. The molecular analysis confirms that the specific mutation was made at the designated site. This work was recently featured on the cover of the June issue of *Hepatology*.

LECTURE II

We believe that the chimeric oligonucleotide pairs with its target site and induces a mutation or correction by the nuclear excision repair pathway or by mismatch repair. After enabling repair, the chimeric oligonucleotide dissociates and is eliminated by the cell. There are several issues regarding the chemistry of the chimeric oligonucleotide. First, the purity of the molecule is essential in order to obtain successful targeting results. Failures of synthesis in the preparation of the oligonucleotide can often lead to failure to observe significant changes. To ensure that the oligonucleotide contains high purity, we have employed melting temperature curves to distinguish between failed synthesis and full-length molecules. S1-nuclease has also been used as a probe to assess the structure of the oligonucleotide. These structural studies have revealed that the chimeric oligonucleotide is in a double hairpin conformation. Our general strategy is, again, to use this molecule as a mutator or corrector of genetic defects in chromosomal targets.

To evaluate other mammalian genes that may be amenable to this targeting event, we chose the ß-globin gene in Sickle Cell Anemia. Using a chimeric oligonucleotide we hoped to change the T-A base pair in the β^S globin to an A-T base pair, regenerating the wild-type β^A genotype. This particular mutation was chosen because it is uniform among the affected population. It also provides a convenient restriction polymorphism cleavage site that enables facile detection of successful conversion. Although it is clear the hematopoeitic stem cells would be the chosen cell type for treatment of Sickle Cell Anemia, we used EBV-transformed B cells as the prototype system. These cells grow readily in culture but have a number of disadvantages. First, they are basically primary cells and we have found that they must be transformed routinely in order to generate cells receptive to chimeric oligonucleotide-mediated conversion. Second, it is also possible that the EBV may contribute to some parts of the reaction. The protocol for transfer of the oligonucleotide into the B cells was designed to employ liposomes. After successful transfection we were able to detect significant levels of correction from the β^S to the β^A genotype. This exchange was detected by RFLP, by Southern blots and by direct DNA sequencing. There are, however, indications that PCR may be artefactual in some systems. So we conducted a series of detailed experiments to ensure that we were measuring what we thought we were. In no case did the reaction buffer, the water, or the oligonucleotide serve as a source for PCR priming. In addition, the chimeric oligonucleotides do not integrate into cell lines. In spite of this, criticism has arisen as to the methods of detection.

To address these issues, we now employ Allele-Specific Polymerase Chain Reactions, or ASPCR. This technology allows fast detection of specific base changes down to 0.5 - 1%. It also avoids questions of manipulation of PCR-generated fragments. Other concerns have been raised regarding the cloning of the B cells after exposure to chimeric oligonucleotides. At this point, we have been unsuccessful in cloning liposome-treated EBV-transformed B cells. To ensure that the genetic change mediated by the oligonucleotide was permanent, we reproduced the data using HUH7 cells and HeLa cells. The same strategy was employed, both in the transfection and in the conversion analysis. Our results indicate that conversion occurs at a 2 - 8% level routinely. In addition, both types of cells have been cloned and individual cell lines have now given rise to heterologous ($\beta^S\beta^A$) hemoglobin genes.

The next gene that has been targeted is Factor IX, a mutation in which is responsible for Hemophilia B. This X-linked gene is attractive because there is only one copy of it in the human genome. Primary human hepatocytes were isolated and chimeric oligonucleotides were transfected and successful conversion of the A $\rightarrow$ C nucleotide at position 935 in the Factor IX gene was observed at a 10 - 15% frequency. It is noteworthy to remember that 2 - 5% conversion will cure the disease. Genetic conversion was measured by both hybridization of colony blots and by direct DNA sequencing. It is the first evidence of primary cell targeting using a gene conversion method.

These results allowed for the development of an animal model to test Factor IX targeting. In this case, a normal Sprague-Dawley rat was used and we attempted to mutate the Factor IX gene. Direct injection of the chimeric oligonucleotide mixed with liposomes was delivered into the caudal lobe of the liver. The lobe was then clamped off to allow ballooning to occur, thereby increasing the blood flow into the local area. After four hours, the caudal lobe block was released and conversion frequencies were measured 48 hours later. Colony hybridization blots and Southern blots have indicated that conversion rate appears to be 2 - 7%. This is the first demonstration of gene conversion in an animal model.

The original purpose of our work has been to study the effect of DNA recombination genes and proteins on gene targeting in humans. We set out to clone, therefore, the human analog of the Ustilago REC2 gene. This successful effort was published in the Proceedings of the National Academy of Science, July 8th issue (1997). The amino acid sequence of the protein, predicted from the nucleotide sequence, was found to be to be highly similar to the RecA protein of *E. coli.* It is also noteworthy that the size of the human REC2 gene, hREC2, is identical in kilodaltons, to the RecA protein, yet analysis of the human REC2 gene found a close homology to several human repair genes. Therefore, based on its analogous activity of gene targeting in Ustilago, we wondered whether the human repair gene, hREC2, was involved in mediating gene targeting of oligonucleotides. It also led to two other important questions: *1.)* Is repair more important than recombination in gene targeting events using chimeric oligonucleotides?; and *2.)* Are we truly measuring a recombination event or is repair the dominant process?

The human REC2 gene is expressed in all mammals tested thus far and is widely evident in most tissues. However, in cell culture, very few established cell lines produce detectable levels of the human REC2 gene transcript. We discovered that the human REC2 transcript can be induced by both ultraviolet light and by X-ray irradiation. This observation places the human REC2 gene in a unique category because very few radiation-inducible genes have been identified in human cells.

To study the effect of human REC2 on DNA repair, we created integrated cell lines of Chinese Hamster ovary cells containing full length copies of the REC2 gene, a truncated version of the REC2 gene, a phosphorylation mutation of the REC2 gene, and the inverse orientation of the REC2 gene. All of these constructs were expressed at high levels in Chinese Hamster Ovary cells, driven by a CMV promoter. Chinese Hamster Ovary cells were then exposed to ultraviolet light, using a traditional repair assay. After irradiation with UV light, the cell lines were placed in a selection media containing 6-thioguanine. This assay measures the repair efficiency at the HPRT locus. The only evidence of DNA repair was provided in cells containing the full length copy of the wild type REC2 gene. All other variations of the REC2 construct were found to induce repair at controlled levels. Furthermore, the REC2 gene was found to be associated with p53, cdc2, and pCNA; all proteins involved in the control of cell division or DNA replication. This observation was determined by both amino precipitation and by direct protein association. Interestingly, the protein p53 has been implicated in the process of DNA repair, as well as having enormous influence on cell cycle/division rates. REC2-integrated CHO cells exhibited a delay in S-phase as the cells progressed through the cell cycle. We have observed a retardation of the progression through S-phase from G1 to G2. This effectis completely lost when a tyrosine residue inside the coding region of the REC2 gene is changed to an alanine. We therefore believe, that this tyrosine residue must be phosphorylated in order to have REC2 exhibit its retardation of the cell cycle.

A knock-out mouse and a transgenic animal have been generated using traditional methods of molecular biology. The radiation-inducible promoter of REC2 has also been isolated. If the REC2 gene is over-expressed in cell lines and a low dose of X-ray irradiation or UV irradiation is given to cells, the vast majority of cells progress to apoptosis. It is our belief that this can be used as a cancer gene therapy in a strategy in which the REC2 gene is introduced into cells and then a low dose of radiation enables the activation of the cells toward apoptosis. We have also observed an increase in the targeting frequency of the chimeric oligonucleotide in cells containing the human REC2 gene. So far, the ß-globin gene and the HPRT have been successfully targeted under conditions where the targeting event is completely dependent on expression of human REC2. Success has also been achieved in plants. In summary, we believe that the combination of oligonucleotides and human repair genes enables gene therapy of certain types of radiation-resistant tumors.

Retroviral Vectors for Human Gene Therapy

Giuliana Ferrari and Fulvio Mavilio

The H.S.R.-Telethon Institute for Gene Therapy of Genetic Diseases, Istituto Scientifico H.S.Raffaele, Via Olgettina 58, 20132 Milano, Italy.

1. Retroviral vectors

Many different technologies are currently utilized to transfer new genetic information into animal cells. Although the gene transfer techniques used in experimental settings imply variable chemical, physical, mechanical and biological tools, only a few of them are practically useful for human gene therapy applications. Each technique offers unique advantages and suffers from some problems, and the development of a vector combining all the strong points and the benefits of the different systems is presently a hard task. To be relevant for clinical application, a gene transfer method should provide high efficiency of gene transfer, stability of integration in the human genome ensuring long-term expression of the transduced gene, safety and feasibility. Some of these important features denoting the most commonly used viral and non-viral techniques are listed in table 1. Among all these systems, viral vectors provide most of the features required for the treatment of human diseases and they are utilized in over 85% of the ongoing clinical protocols[1]. Many features make retrovirus vectors a good choice for gene transfer into human cells. Most importantly, these vectors integrate efficiently into the target cell genome resulting in unrearranged transfer of the desired genes. In general, long-term transgene expression, is obtained both *in vitro* and *in vivo*.

An important point to be considered in the choice of a vector is also the potential immunogenicity of viral-encoded proteins. Typical examples of induction of immune response to viral products come from the use of adenoviral-derived vectors, both in the pre-clinical and clinical studies [2,3] such as the cystic fibrosis gene therapy trials. Since no viral proteins are coded by replication-defective retroviral vectors, host immune response against these products has never been observed.

The major disadvantages related to retroviral vectors are the constraints in the size of the DNA that they can accommodate, and the dependence on active DNA replication for efficient integration[4]. New technologies based on the development of lentivirus-based vectors, transducing both dividing and resting cells, may circumvent some of these limitations [5]. Two other potential problems with retroviral vectors are those of insertional mutagenesis and risk of

NATO ASI Series, Vol. H 105
Gene Therapy
Edited by Kleanthis G. Xanthopoulos

helper virus production. Problems with insertional mutagenesis, such as activation of cellular oncogenes, are shared with any transfer technique resulting in random integration of new sequences into the cellular genome. Retroviral activation of oncogenes in mice was reported in the context of a spreading infection by wild-type viruses and whether such event can occur at appreciable rates after infection by replication-defective retroviral vectors remains to be seen. The issue of helper viruses generation will be discussed in the section related to the packaging cell.

Overall, the most efficient way to stably introduced genes into target cells employs retroviral vectors. This is reflected in the ongoing gene therapy clinical trials, in which retroviruses remain the most frequent vector (63% of all patients)[6,7]. For gene therapy of inherited diseases, which require the permanent correction of a defective gene product, stability of integration into the genome combined with the long-term transgene expression accomplished by retroviral vectors make this system the most eligible one.

Retroviral vectors are derived from retroviruses by deletion of all viral genes which are replaced by the cDNA to be transferred. Modifications of wild type retroviruses result in replication-defective recombinant vectors retaining the viral information required to package virions (psi sequence) and integrate viral genome into host DNA (long terminal repeats). The most commonly used vectors derive from murine leukemia viruses (MLV), a well characterized virus which can be easily manipulated in order to construct vectors. Retrovirus particles consist of a 9-10 kb RNA genome enclosed within a protein capsid and an outer envelope derived from the plasma membrane of the infected cell. Usually, two copies of the coding strand of the RNA genome are packaged into each viral capsid, along with reverse transcriptase (RT) and a cellular tRNA primer that hybridizes to the primer binding site present near the 5'-end of the viral RNA. A transmembrane envelope glycoprotein (env) encoded by the virus serves as a ligand for a specific receptor on the recipient cell, which allows attachment and entry of the viral particle. By fusion with the plasma membrane, the viral RNA is released inside the cytoplasm and converted to DNA by the RT enzyme. Integration of viral DNA as a provirus into the host cell genome is mediated by the integrase activity of RT. Transcription of viral sequences by the cell transcriptional machinery leads to production of both viral protein and genomic RNA, resulting into the assembly and budding of new viral particles. The genetic sequences necessary for the life cycle, present into the retrovirus genome are the following:

a. long terminal repeats (LTR), containing sequences controlling proviral integration and transcription, i.e. promoter, enhancer and poly(A) signal;

b. psi (Ψ) signal, required for the packaging of RNA viral genome into complete infectious particles;

c. *gag*, *pol* and *env* genes, coding for structural proteins (gag), reverse transcriptase (pol), and envelope (env) proteins;

d. splicing signals, donor (SD) and acceptor (SA), controlling the generation of genomic and subgenomic RNA, translated into *pol* and *env* products.

Construction of a retroviral vector involves creation of a plasmid containing the LTRs, the Ψ sequence and heterologous cDNAs which replace the *gag, pol* and *env* genes. All the deleted viral functions are provided in *trans*, utilizing approaches that involve complementation of defective vectors by genes inserted into "helper" cell lines to generate the infectious particles. A detailed description of "helper" or packaging cell lines will be provided in the next section. In this way, the production of infectious viral particles occurs in a single round, after transfection of retrovirus vector DNA into the packaging cell, and no further virus spreading will take place once the target cells are infected. The most widely used vectors are those made from MoMLV backbones. The *pol, env* and most of the *gag* genes are removed. The residual part of the *gag* gene is necessary since it overlaps with the Ψ sequence required for the virus packaging. A cryptic splice acceptor site within *gag* allows generation of subgenomic RNA by splicing the primary transcript from LTR promoter.

The simplest and most classical vector contains a gene conferring resistance to a drug, e.g. the neomycin phosphotransferase gene (NeoR). Since the transduction efficiency by retroviral vectors, although usually very high, does not allow gene transfer into 100% of the target cell, the presence of a selectable gene is a useful advantage. In addition to the NeoR gene, conferring resistance to the neomycin analog G418, other selectable genes were recently cloned into retroviral vectors allowing positive selection of the transduced cells. Strategies implying vectors coding for cell surface proteins, easily detectable and allowing rapid and efficient immunoselection, offer an extremely powerful tool to obtain homogeneously transduced cell populations for clinical applications [8,9].

The requirement for expression of both marker/reporter genes and therapeutic genes makes the use of two transcriptional promoters necessary. The first promoter is usually the vector LTR, while the second one may be of cellular gene or viral origin, e.g. CMV (cytomegalovirus), SV40 (simian virus 40), RSV (Rous sarcoma virus), HSV TK (Herpes simplex virus thymidine kinase), MMTV (murine mammary tumour virus) promoters. Depending on the strategy, the transcription of the cDNA of interest may start by the LTR promoter or by an internal promoter. Constructs of this kind suffer from a major drawback due to the proximity of two transcriptional units. In particular the efficiency of internal promoters is often greatly reduced by transcriptional interference exerted by the potent viral promoter/enhancer contained in the LTR. Nevertheless, transcription of a cDNA from the LTR promoter may be highly compromised because of the inactivation of the viral promoter by *in vivo* methylation.

In order to avoid these limitations, different vectors constructions were developed. In the so-called "splicing" vectors it is possible to control two genes with a single LTR by means of alternative splicing. Usually the first gene is product encoded by the LTR primary transcript, whereas the second one is encoded by the spliced subgenomic one. This strategy is similar to that utilized by the wild-type retrovirus to produce *gag*, *pol* and *env* gene products from the same promoter. Splicing vectors are not very commonly used, since it is difficult to envisage splicing efficiency and consequently the relative rate of production of the two proteins. On the other hand, splicing increases the efficiency by which a viral genomic RNA is translated into a protein product. In the MFG vector [10,11] the introduction of an acceptor splicing site derived from the MoMLV *env* gene 5' to a cloned cDNA provide efficient transgene expression from a spliced message in combination with high viral titer.

An alternative strategy to avoid the problem of transcriptional interference is the one used in the so-called self-inactivating vector, obtained by deletion of the viral enhancer sequence from the U3 region of the 3' LTR. However this modification causes a severe drop in viral titer[12].

An interesting approach was utilized to construct the "double-copy" vector, aimed to insert a minigene outside the transcriptional unit of the vector in the 3' LTR[13]. During proviral integration into the host genome, duplication and transfer of the U3 region in the 3' LTR to 5' position, cause also transfer of the minigene. In this configuration the 5' copy of the minigene lies upstream from, and is not influenced by the viral promoter.

In the most recent retroviral vectors designs, transcription of two or more genes is allowed by the use of bi- or poly-cystronic mRNA. These vectors contain poliovirus or encephalomyocarditis virus sequences responsible for a conserved secondary structure called internal ribosome entry site (IRES). The presence of an IRES within a RNA bypasses the constraint of ribosome scanning mechanism and permits direct entry of ribosomes downstream from the ATG start codons. An IRES provides cap-independent translation initiation at internal initiation codons (reviewed in ref.[14]). When used in the context of a retroviral vectors, this may alleviate some of the problems associated with promoter interference[15,16]. An interesting application of this type of vectors is in the production of multiple proteins or different subunits of multimeric proteins under the control of the same promoter[17].

2. Packaging cells

A packaging or "helper" cell is a cell that produces retroviral proteins in the absence of transmissible RNA genome. Thus, it allows to package a replication- defective retrovirus genome by complementing in *trans* all the missing viral functions. Packaging cells provide a way to introduce genes permanently into the DNA of somatic cells without introducing infectious virus or viral genes.
There are different packaging cells derived by a variety of cell lines stably transfected with the sequence of *gag, pol* and *env* genes (for a detailed list, see ref.[18]). In contrast to retroviral vectors, the viral RNA encoding the structural proteins lacks the Ψ signal. Thus, the packaging cells do not package RNA that encodes the *gag, pol* and *env* genes, while, after transfection of retroviral vector DNA, it packages vector RNA. As a result, infectious viral particles carrying only the vector genome are released by the packaging cell.

As with wild-type retroviruses, recombinant retroviral vectors enter the host cell via interaction of a viral envelope glycoprotein, the product of the *env* gene, with a cellular receptor [19]. The host range of a retrovirus or a packaging cell line is defined primarily by the *env* gene specificity. The commonly used murine virus subtypes have several classes of envelope glycoproteins which interact with different host-cell receptors [20]. The most widely used types of envelope glycoproteins are ecotropic and amphotropic envelopes, binding to specific cell receptors. The ecotropic env receptor is present on mouse cells and is a cationic amino acid transporter [21]. The amphotropic receptor (Ram 1), present on cells derived by many species, including humans, mice, rats, chickens, monkeys and dogs, is a sodium-dependent phosphate transporter [22]. Similarly, the receptor for the gibbon ape leukemia virus (GALV) envelope, that is expressed at higher level than Ram1 in human hematopoietic stem cells [23,24], is a sodium-dependent phosphate symporter [25,26].

One main disadvantage associated with the usage of retroviral vectors relies in the risk of generation of replication-competent (helper) viruses. The potential for production of helper virus during the preparation of viral stocks remains a concern, although for practical application this problem has been solved by introducing extensive, accurate and very sensitive monitoring. So far, no occurrence of helper viruses in patients involved in gene therapy trials has ever been reported. The potential for helper virus production depends on viral sequences present both in the retroviral vector and in the packaging cells. Recombination between these sequences could produce replication-competent retroviruses (RCR). Another source for recombination events is represented by the presence of endogenous MLVs sequences in murine cells, and therefore in packaging cells of murine origin. In order to minimize the risk of generating RCRs into the packaging cell, it is particularly important to avoid homologous overlap between the separate viral sequences used for construction of packaging cell lines and retrovirus vectors. Several generation of helper cell lines have been successively devised to provide greater margins of safety against RCR formations. One of the most widely used packaging cell line, PA317 [27], contains DNA encoding viral structural proteins as a contiguous sequence, but presents also deletions in the 5' LTR, packaging signals and 3' LTR. Use of vectors having little or no overlap with viral sequences in the PA317 [28], drastically reduce the risk of generation of RCR, since a minimum of two recombination events are required.

Another generation of packaging cell lines (GP+E86 and GP+*env*Am12) further reduces the chances of RCR formation by physically separating the *gag-pol* and *env* genes on two plasmids prior to their introduction into cells. As a consequence, at least three recombination events would be required to generate RCR by recombination between vector and these helper cell lines [29,30].

While generation of helper virus appears to be preventable and can be monitored appropriately, even defective retroviruses could in theory lead to malignant transformation by insertional mutagenesis. Such an event has not yet been observed in clinical trials, nor in preclinical animal models in the absence of replication competent virus. However, the report of thymic lymphoma developing in three out of eight rhesus monkeys transplanted with bone marrow that was contaminated with large amounts of replication-competent virus [31] attest to the importance of stringent monitoring for helper virus in human gene transfer experiments.

3. Retroviral cell targeting

Retroviral-mediated gene transfer is the widest used technologies in all the *ex vivo* gene therapy approaches. Almost all kinds of replicating cells are susceptible to retroviral vector transduction, including clinical relevant types like hematopoietic cells, smooth and skeletal muscle cells, fibroblasts, keratinocytes, endothelial cells, hepatocytes, sinoviocytes, neoplastic cells. The *ex vivo* strategy implies accessibility and removal from the patient of the target tissue, *in vitro* expansion and retroviral-mediated transduction, and re-implantation of the genetically-corrected cells back into the patient. This procedure restricts the application of gene therapy to particular indications. For many diseases, the *in vivo* direct transduction of the target cells would be preferable and in some cases the only choice strategy. Targeting retroviral vectors to specific cell types would provide many advantages both to *ex vivo* and *in vivo* gene transfer technologies. Depending on the strategy, molecular engineering may be used at two stages to achieve targeted gene therapy (reviewed in [32,33,34]). The expression of the therapeutic gene can be regulated on the transcriptional level through tissue-specific or inducible promoters. The second possibility for targeting is provided by the infection event. Targeting at the level of infection limits the transduction of the transgene to a desired cell type. Transcriptional targeting implies no selective target transduction but activation of transgene expression to a desired cell type. A targeting approach based on the susceptibility of dividing cells to viral infection was used to transduced brain tumours *in vivo.* The herpes thymidine kinase gene was introduced into tumour cells to provide sensitivity towards ganciclovir and as a consequence selective killing of transduced tumour cells35.

Since the host range determinants on the virus are located in the envelope proteins, the most promising strategies that have been used for targeting retrovirus entry are based on the modification of the retroviral envelope gene. The sequence of the *env* gene responsible for recognition of the cellular receptors can be replaced by sequences encoding new non-viral ligands. A major drawback in this strategy is due to the fact that the envelope protein participates also in virus fusion and entry. Thus, a molecular modification producing the desired effect of altering the tropism of the viral binding, might be detrimental to subsequent step of viral-mediated gene transfer, such as internalization [36]. Despite these problems , there have been several reports of successful reshaping of *env* protein. Replacement of a portion of the envelope protein with part pf the erythropoietin (EPO) protein led to generation of chimeric viral particles able to specifically transduced EPO-receptor-bearing cells [37].

Other approaches employing creation of fusion products between single chain antibody specific for a cell surface antigen and an ecotropic MoMLV envelope protein [38,39] revealed an intrinsic limit of this approach. Chimeric envelopes were incorporated into viral particles and were shown to bind to ligand-expressing cells, but the subsequent infection step was occurring at very low efficiency. It seems that although modifications in the ligand domains allow high affinity and binding to selective receptors, the subsequent events required for efficient viral entry resulted impaired. Insertion of a cleavable linker to fused epidermal growth factor to an amphotropic retroviral envelope protein was adopted to release the virus, after the binding to the specific receptor, allowing cell transduction through its own natural receptors [40].

A different approach utilizes modification of pre-formed vector particles by coupling agents to the envelope protein. This "molecular bridge " strategy lies in the conjugation of an antibody to the retroviral envelope protein, and linkage of this antibody to a second cell-surface-specific antibody using streptavidin. In all studies, using antibodies against major histocompatibility complex antigens [41], EGF [42] and transferrin receptors [43], internalization of the virus by these routes occurred but was not followed by establishment of the proviral state. This is probably because the receptor (and the agent bound to it) enters a cellular compartment that does not allow access to the nucleus, because it is normally degraded or recirculated to the cell membrane.

A more successful result was obtained by chemical coupling of lactose to the envelope glycoproteins to induce retroviral particles to bind specifically to asialoglycoprotein receptors like those expressed by hepatocytes [44].

4. Targeting of gene expression

A different way to achieve targeted gene therapy is based on the transcriptional control of transgene expression. For certain strategies, it is necessary to deliver genes whose expression is under the control of appropriate tissue-specific or cell-specific promoters. Gene therapy for thalassemia is the most typical case of requirement for stringent regulation of gene expression since the ultimate goal of this kind of approach is the regulated correction of the globin deficiency. In experiments with the human ß-globin gene, retroviral-mediated transduction of hematopoietic stem cells lead to erythroid-specific gene expression, but only at very low level [45,46]. The finding that additional sequences located 20 kb upstream of the ß-globin gene (termed LCR, i.e. locus control region, sequences) are necessary for high levels of ß-globin transcription [47] suggested a mean for improving ß-globin gene expression in retroviral vectors. However, development of retroviral vectors carrying ß- or g-globin genes including elements derived from the LCR, and able of providing efficient gene transfer and expression, has proven difficult. Position effects cause considerable clonal variation in expression levels among constructs, suggesting that small ß-globin LCR sequences may not provide for a such a strict position-independent expression of ß-globin, at least in the context of retroviral-mediated gene transfer[48].

Tissue-specific and differentiation-specific transgene expression in the context of retroviral-mediated gene transfer was successfully achieved by insertion of a muscle creatine kinase (MCK) enhancer element in the U3 region of the viral LTR [49]. The MCK enhancer was shown to exert an active negative effect on transcription from the viral promoter, so that transgene expression was induced only in cells undergoing muscle differentiation. In a similar study, the murine tyrosinase promoter was used to replace the MoMLV enhancer [50]. Although virus titer were very low, a strong transgene expression was obtained only in melanoma cells, while expression in fibroblasts was barely above background level.

To improve tissue and lineage specificity of retroviral vectors, promoters sequences derived from the integrin CD11a, CD11B and CD18 subunits genes have been incorporated into a retroviral construct [51]. Although tissue-specific expression was obtained in transient assay, only low levels of expression were observed in stable cell lines.

5. References

1 Ross G., Erickson R., Knorr D., Motulsky A. G., Parkman R., Samulski J., et al. Gene therapy in the United States: a five-year status report. Hum. Gene Ther. 1996; 7: 1781-1790.

2 Yang Y., Nunes F. A., Berencsi K., Furth E. E., Gonczol E., Wilson J. M. Cellular immunity to viral antigens limits E1-deleted adenoviruses for gene therapy. Proc. Natl. Acad. Sci. USA 1994; 91: 4407-4411.

3 Crystal R. G., McElvaney N. G., Rosenfeld M. A., Chu C. S., Mastrangeli A., Hay J. G., et al. Administration of an adenovirus containing the human CFTR cDNA to the respiratory tratc of individuals with cystic fibrosis. Nature. Gen. 1994; 8: 42-51.

4 Miller D. G., Adam M. A., Miller D. Gene transfer by retrovirus vectors occurs only in cells that are actively replicating at the time of infection. Mol. Cell. Biol. 1990; 10: 4239-4242.

5 Naldini L., Blomer U., Gallay P., Ory D., Mulligan R., Gage F. H., et al. In vivo gene delivery and stable transduction of nondividing cells by a lentiviral vector. Science 1996; 272: 263-267.

6 Marcel T., Grausz D. J. The TMC worldwide gene therapy enrollment report. Hum. Gene Ther. 1996; 7: 2025-2046.

7 Anderson F. A. End -of-the-year potpourri-1996. Hum. Gene Ther. 1996; 7

8 Mavilio F., Ferrari G., Rossini S., Nobili N., Bonini C., Casorati G., et al. Peripheral blood lymphocytes as target cells of retroviral vector-mediated gene transfer. Blood 1994; 83: 1988-1997.

9 Pawliuk R., Kay R., Lansdorp P., Humphries R. K. Selection of retrovirally transduced hematopoietic progenitor cells using CD24 as a marker of gene transfer. Blood 1994; 84: 2868-2877.

10 Riviere I., Brose K., Mulligan R. C. Effects of retroviral vector design on expression of human adenosine deaminase in murine bone marrow transplant recipients engrafted with genetically modified cells. Proc. Natl. Acad. Sci. U.S.A. 1995; 92: 6733-6737.

11 Krall W. J., Skelton D. C., Yu X.-J., Riviere I., Lehn P., Mulligan R. C., et al. Increased levels of spliced RNA account for augmented expression from the MFG retroviral vector in hematopoietic cells. Gene Therapy 1996; 3: 37-48.

12 Yu S., von Ruden T., Kantoff P. W., Garber C., Sciberg M., Rather U., et al. Self-inactivating retroviral vectors designed for transfer of whole genes into mammalian cells. Proc. Natl. Acad. Sci. USA 1986; 83: 3194-3198.

13 Hantzopoulos P. A., Sullenger B. A., Ungers G., Gilboa E. Improved gene expression upon transfer of the adenosine deaminase minigene outside the transcriptional unit of a retroviral vector. Proc. Natl. Acad. Sci. USA 1989; 86: 3519-3523.

14 Boris-Lawrie K. A., Temin H. M. Recent advances in retrovirus vector technology. Current Opinion Genet. and Dev. 1993; 3: 102-109.

15 Ghattas I. R., Sanes J. R., Majors J. E. The encephalomyocarditis virus internal ribosome entry site allows efficient coexpression of two genes from a recombinant provirus in cultured cells and in embryos. Mol. Cell. Biol. 1991; 11: 5848-5859.

16 Morgan R. A., Couture L., Elroy-Stein O., Ragheb J., Moss B., Anderson W. F. Retroviral vectors containing putative internal ribosome entry sites: development of a polycistronic gene transfer system and applications to human gene therapy. Nucl. Acid Res. 1992; 20: 1293-1299.

17 Zitvogel L., Tahara H., Cai Q., Storkus W. J., Muller G., Wolf S. F., et al. Construction and characterization of retroviral vectors expressing biologically active human interleukin-12. Hum. Gene Ther. 1994; 5: 1493-1506.

18 Cepko C., Pear W. Overview of the retrovirus transduction system. In: Current protocols in molecular biology, John Wiley & Sons, 1996; Suppl.36: 9.9.1-9.9.16.

19 Albritton L. M., Tseng L., Scadden D., Cunningham J. M. A putative murine ecotropic retrovirus receptor gene encodes a multiple membrane-spanning protein and confers susceptibility to virus infection. Cell 1989; 57: 659-666.

20 Miller A. D. Cell-surface receptors for retroviruses and implications for gene transfer. Proc. Natl. Acad. Sci. USA 1996; 93: 11407-11413.

21 Wang H., Kavanaugh M. P., North R. A., Kabat D. Cell-surface receptor for ecotropic murine retroviruses is a basic aminoacid transporter. Nature 1991; 352: 729-731.

22 Miller D. G., Miller A. D. A family of retroviruses that utilize related phosphate transporters for cell entry. J. Virol. 1994; 68: 8270-8276.

23 Kavanaugh M. P., Miller D. G., Zhang W., Law W., Kozak S. L., Kabat D., et al. Cell-surface receptors for gibbon ape leukemia virus and amphotropic murine retrovirus are inducible sodium-dependent phosphate symporters. Proc. Natl. Acad. Sci. U.S.A. 1994; 91: 7071-7075.

24 Bauer T. R. J., Miller A. D., Hickstein D. D. Improved transfer of the leukocyte integrin CD18 subunit into hematopoietic cell lines by using retroviral vectors having a gibbon ape leukemia virus envelope. Blood 1995; 86: 2379-2387.

25 Miller D. G., Edwards R. H., Miller A. D. Cloning of the cellular receptor for amphotropic murine retroviruses reveals homology to that for gibbon ape leukemia virus. Proc. Natl. Acad. Sci. USA 1994; 91: 78-82.

26 van Zeijl M., Johann S. V., Closs E., Cunningham J., Eddy R., Shows T. B., et al. A human amphotropic retrovirus receptor is a second memeber of the gibbon ape leukemia virus receptor family. Proc. Natl. Acad. Sci. USA 1994; 91: 1168-1172.

27 Miller A. D., Buttimore C. Redesign of retrovirus packaging cell lines to avoid recombination leading to helper virus production. Mol. Cell. Biol. 1986; 6: 2895-2902.

28 Miller A. D., Rosman G. J. Improved retroviral vectors for gene transfer and expression. BioTechniques 1989; 7: 980-990.

29 Markowitz D., Goff S., Bank A. Construction and use of a safe and efficient amphotropic packaging cell line. Virology 1988; 167: 400-406.

30 Markowitz D., Goff S., Bank A. A safe packaging line for gene transfer: separating viral genes on two different plasmids. J. Virol. 1988; 62: 1120-1124.

31 Donahue R. E., Kessler S. W., Bodine D., Mc Donagh K., Dunbar C., Goodman S., et al. Helper virus induced T cell lymphoma in nonhuman primates after retroviral mediated gene transfer. J. Exp. Med. 1992; 176: 1125-1135.

32 Cosset F.-L., Russell S. J. Targeting retrovirus entry. Gene Therapy 1996; 3: 946-956.

33 Harris J. D., Lemoine N. R. Strategies for targeted gene therapy. Trends in Genet. 1996; 12: 400-405.

34 Schnierle B. S., Groner B. Retroviral targeted delivery. Gene Therapy 1996; 3: 1069-1073.

35 Oldfield E. H. Gene therapy for the treatment of brain tumors using intra-tumoral transduction with the thymidine kinase gene and intravenous ganciclovir. Hum. Gene Ther. 1993; 4: 531-533.

36 Salmon B., Gunzburg W. H. Targeting of retroviral vectors for gene therapy. Hum. Gene Ther. 1993; 4: 129-141.

37 Kasahara N., Dozy A. M., Kan Y. W. Tissue-specific targeting of retroviral vectors through ligand-receptor interaction. Science 1994; 266: 1373-1376.

38 Russell S. J., Hawkins R. E., Winter G. Retroviral vectors displaying functional antibody fragments. Nucleic-Acids-Res. 1993; 21: 1081-1085.

39 Chu T.-H., Martinez I., Sheary W. C., Dornburg R. Cell targeting with retroviral vector particles containing antibody-envelope fusion protein. Gene Therapy 1994; 1: 292-299.

40 Nilson B. H., Morling F. J., Cosset F. L., Russell S. J. Targeting of retroviral vectors through protease-substrate interactions. Gene Therapy 1996; 3: 280-286.

41 Roux P., Jeanteur P., Piechaczyk M. A versatile and potentially general approach to the targeting of specific cell types by retroviruses: application to the infection of human cells by means of major histocompatibility complex class I and class II antigens by mouse ecotropic murine leukemia virus-derived viruses. Proc.Natl.Acad.Sci.U.S.A. 1989; 86: 9079-9083.

42 Etienne- Julan M., Roux P., Carillo S., Jeanteur P., Piechaczyk M. The efficiency of cell targeting by recombinant retroviruses depends on the nature of the receptor and the composition of the artificial cell-virus linker. J. Gen. Virol. 1992; 73: 3251-3255.

43 Goud B., Legrain P., Buttin G. Antibody-mediated binding of a murine ecotropic Moloney retroviral vector to human cells allows internalization but not the establishment of the proviral state. Virology 1988; 163: 251-254.

44 Neda H., Wu C. H., Wu G. Y. Chemical modification of an ecotropic murine leukemia virus results in redirection of its target cell specificity. J. Biol. Chem. 1991; 266: 14143-14146.

45 Bender M. A., Gelinas R. E., Miller A. D. A majority of mice show long-term expression of a human b-globin gene after retrovirus transfer into hematopoietic stem cells. Mol. Cell. Biol. 1989; 9: 1426-1434.

46 Bodine D. M., Karlsson S., Nienhuis A. W. Combination of interleukins 3 and 6 preserves stem cell function in culture and enhances retrovirus-mediated gene transfer into hematopoietic stem cells. Proc. Natl. Acad. Sci. U.S.A. 1989; 86: 8897-8901.

47 Grosveld F., van Assendelft G. B., Greaves D. R., Kollias G. Position-independent, high-level expression of the human beta-globin gene in transgenic mice. Cell 1987; 51: 975-985.

48 Sadelain M., Wang C. H., Antoniou M., Grosveld F., Mulligan R. C. Generation of a high-titer retroviral vector capable of expressing high levels of the human beta-globin gene. Proc. Natl. Acad. Sci. U.S.A. 1995; 92: 6728-6732.

49 Ferrari G., Salvatori G., Rossi C., Cossu G., Mavilio F. A retroviral vector containing a muscle-specific enhancer drives gene expression only in differentiated muscle fibers. Hum. Gene Ther. 1995; 6: 733-742.

50 Vile R. G., Hart I. R. In vitro and in vivo targeting of gene expression to melanoma cells. Cancer Res. 1993; 53: 962-967.

51 Bauer T. R., Osborne W. R. A., Kwok W. W., Hickstein D. D. Expression from leukocyte integrin promoters in retroviral vectors. Hum. Gene Ther. 1994; 5: 709-716.

TABLE 1. Vectors for gene transfer

Vector	MoMLV	AdV	HSV	AAV	Non viral systems
Insert capacity	7-10 kb	7-10 kb	> 20 kb	5 kb	unrestricted
Titer (10^n IFU/ml)[a]	6-7	11	8	> 9	not applicable
Integration	yes	no	no	?	no/rare
Long term expression	yes	transient	poor	?	no
In vivo delivery	poor	yes	yes	yes	transient
Transduction of post-mitotic cells	no	yes	yes	yes	yes
Viral protein expression	no	yes	yes	yes	no

Abbreviations: MoMLV, Moloney murine leukemia virus; AdV, Adenovirus; HSV, Herpes Simplex
virus; AAV, Adeno-associated virus.

[a]IFU/ml, infectious units per ml.

Gene Therapy for Cancer

Susan A. Zullo*, Natasha J. Caplen, and R. Michael Blaese
Clinical Gene Therapy Branch, National Human Genome Research Institute, National Institutes of Health, Bethesda, MD, 20892, USA.

Keywords. Cancer gene therapy, immunotherapy, suicide genes, bystander effect

1 Introduction

Cancer is a complex acquired disease that affects millions of individuals each year. Underlying the pathogenesis of cancer are a variety of molecular genetic abnormalities which can be inherited or environmentally induced and result in unregulated cell proliferation. Conventional treatment strategies used to treat cancer: surgery, chemotherapy, and radiation have been only partially successful and new treatment options are critically needed. Gene therapy is currently being explored experimentally as an alternative or addition to established treatment options for malignant melanoma, leukemia, glioma, and others. The aim of this therapy is the introduction of a gene or genes into cells to provide a new set of permanent or temporary instructions for those cells resulting in the indirect elimination or direct killing of tumor cells. Indirect approaches to the treatment of cancer by gene therapy include the augmentation of the immune system; direct approaches include restoration of the normal function of a mutated tumor suppressor gene or expression of a tumoricidal gene. This review will describes these and other strategies.

2 Indirect Treatment Strategies

2.1 Immunotherapy and tumor vaccination

Cancer cells are often poorly immunogenic and can effectively elude the immune system. Immunotherapy is an indirect gene therapy strategy which aims to enhance or program the body's immune system to attack tumor cells on a systemic level. This approach is of particular interest because it could potentially eliminate every malignant cell including metastases within the body and respond to neoplastic recurrences. Currently, there are three principal approaches to immunotherapy using genetic manipulation: 1) *ex vivo* modification of autologous infiltrating T-lymphocytes, 2) *ex vivo* modification of autologous tumor cells, and 3) *in situ* transient expression of immuno-stimulatory genes

2.1.1 *Ex vivo* Modification of T-Lymphocytes

The systemic administration of cytokines and other anti-tumor agents has shown some promise as a means of enhancing cytotoxic T-lymphocyte (CTL) mediated destruction or necrosis of tumors, however, the dose required to evoke a response often induces significant adverse side-effects. T lymphocytes have the natural capacity to home in on tumor tissue, thus, it has been hoped that this could be exploited to facilitate cytokine delivery directly to a tumor mass. By using genetically modified tumor-infiltrating T-

NATO ASI Series, Vol. H 105
Gene Therapy
Edited by Kleanthis G. Xanthopoulos

lymphocytes (TILs) it has been proposed that the local secretion of cytokine would have the dual advantage of enhancing CTL anti-tumor activity and reducing unwanted systemic side effects. The feasibility of this approach was first assessed using a marker gene (Rosenberg *et al.*, 1990; Merrouche *et al.*, 1995); subsequently a clinical trial using retroviral modified TILs carrying tumor necrosis factor (TNF) was initiated in 1991 (Rosenberg 1990). Unfortunately, as a result of poor transduction efficiency and low cytokine expression, this approach has proven difficult to validate and thus there has been a shift to modification of tumor cells themselves (Culver and Blaese 1994).

2.1.2 *Ex vivo* Modification of Autologous Tumor Cells

The advantage of using autologous tumor cells expressing an exogenous immunostimulatory gene is the possibility of inducing a systemic immune response that will both destroy tumor cells and vaccinate the patient against recurrence of the tumor. A variety of transgenes have been proposed to mediate such a response when used to modify autologous tumor cells including those for interleukin (IL)-2, IL-4, IL-6, granulocyte-macrophage colony-stimulating factor (GM-CSF) and immune costimulatory molecules such as B7.1. Patient tumor cells are removed and theoretically could be modified with a gene for one or any combination of these types of molecules. The altered cells are then expanded in culture and returned to the same patient to elicit a specific and robust immune response modulated by the expressed gene product(s). Cytokine secretion should lead to the activation of the immune response in the area surrounding the injected tumors cells by induction of cytotoxic T lymphocytes, while presentation of tumor-associated antigens should lead to the attack of other neoplastic cells throughout the body.

Animal studies with IL-1, IL-2, IL-4, IL-6, IL-7, IL-10, IL-12, IL-13, IL-15, interferon-alpha (IFNα), interferon-gamma (IFNγ), TNFα, and GM-CSF all result in inhibition of tumor formation with a majority of the interleukins producing long-lasting immunity to subsequent tumor challenge (Blankenstein *et al.*, 1989; Tepper *et al.*, 1989; Watanabe *et al.*, 1989; Fearon *et al.*, 1990; Douvdevani *et al.*, 1992; Porgador *et al.*, 1992; Dranoff *et al.*, 1993; Ferrantini *et al.*, 1993; Hock *et al.*, 1993; Miller *et al.*, 1993; Richter *et al.*, 1993; Tahara *et al.*, 1994; Munger *et al.*, 1995). Recently a phase I clinical trial was reported upon in which autologous melanoma tumor cells were modified using a retrovirus carrying the IFNγ gene; these cells were then used to vaccinate, using repeated immunizations, a total of 13 patients. Over half the patients showed a humoral immune response against melanoma cells and two patients showed clinical tumor regression (Abdel-Wahab *et al.*, 1997).

Similarly, retroviral delivery of the accessory molecule B7.1 to tumor cells stimulates the interaction of antigen presenting cells and immune effector cells. This results in diminished tumorigenicity and protection from subsequent challenges as the tumor cells function as professional antigen presenting cells (Chen *et al.*, 1992; Townsend and Allison 1993). Furthermore, recently discovered tumor-specific antigens such as Mart-1 in malignant melanoma and other tumor types could be used (Kawakami *et al.*, 1994). The immune system of the individual would become primed for presentation of the same antigen on tumor cells that could appear at a later time. Finally, a related approach making use of the immunomodulatory effects of cytokines is the construction of antibody vaccines for B-cell activation. Tumor-specific idiotype DNA sequences can be fused to sequences encoding cytokines including GM-CSF and IL-2 which are then introduced as a protein directly back into the patient or transfected into autologous tumor cells. These cells would then function as a vaccine against the type of cancer specific to the idiotype sequence (Tao and Levy 1993; Chen *et al.*, 1994).

2.1.3 Transient Expression of Immuno-stimulatory Genes.

The first attempt to genetically modify tumors *in situ* involved the direct injection of an allogene encoding the histocompatability antigen HLA-B7 complexed with a cationic liposome. The transient expression of the foreign antigen HLA-B7 on the cell surface of the tumor, in this case melanoma lesions, should induce an immune reaction against the altered tumor cell. Several patients have now been treated using this approach and the results of two clinical trials have been published (Nabel *et al.*, 1993; Nabel *et al.*, 1996). In both studies, evidence for vector transfer and transgene expression was seen. In addition, at least one patient in the first trial and two in the second showed evidence of a systemic immune response against tumor cells. Further, one of the two patients from the second trial received treatment with tumor-infiltrating lymphocytes derived from gene-modified tumor tissue. Cells were administered on two separate occasions in combination with subsequent multiple infusions of IL-2. Partial tumor regression was observed after the first infusion and complete regression after the second, which has persisted for more than 21 months following the initial treatment. This observation is particularly interesting as it may in the long term high light a need, as with conventional approaches to treatments of cancers, to combine different gene therapy strategies to obtain completely remission of disease.

2.2 Chemoprotection

Gene therapy has also been proposed as a means of protecting hematopoietic cells from the toxic effects of chemotherapy. This strategy makes use of the genes encoding the multi-drug resistance (MDR) proteins which pump chemotherapy agents such as colchicine, doxorubicin, and vinblastine from cells (Ueda *et al.*, 1987) leaving them resistant to treatment with these agents. Clinically it is envisaged that extracted bone marrow would be transfected *ex vivo* using a retroviral vector containing an MDR gene; the manipulated cells would then be reinfused into the patient so as to repopulate the hematopoietic system. The gene-modified cells should then be more tolerant to the toxic therapeutic agents used to kill the cancer cells. The altered cells would also be tagged with a selectable marker. Animal models and human studies have been investigated to test the efficacy of this system. Retroviral-mediated transfer of the *mdr-1* gene into mouse bone marrow cells by several groups results in both chemoprotection and selective marking of the transduced population [Boesen, 1993 #20; (Gottesmann *et al.*, 1994). Several trials using this strategy have been proposed (Deisseroth *et al.*, 1994; Hesdorffer *et al.*, 1994); preliminary data from one of these trials has shown successful re-engraftment of hematopoietic cells transduced with a retrovirus carrying the *MDR-1* gene (Hannania *et al.*, 1996). Unfortunately, the possibility of imparting this resistance to an early or unrecognized cancer cell within the bone marrow must be recognized as a distinct disadvantage to this approach.

3 Direct Treatments

3.1 Introduction

Direct gene therapy treatment strategies for cancer capitalize on the current understanding of crucial pathways and vital genes within the course of carcinogenesis. One strategy involves the direct genetic modification of tumor cells with the aim of correcting an activated proto-oncogene or replacing a missing or mutant tumor suppressor gene. Proto-oncogenes function in the regulation of cell proliferation and differentiation, while tumor suppressor genes are involved in cell cycle arrest and induction of apoptosis (programmed cell death). Both types of genes have been found to be genetically altered in many types of cancer. Antisense oligonucleotides and ribozymes are being assessed as a means of correcting activated proto-oncogenes, such as *K-ras* and *bcr-abl*, and the introduction of genes, such as *p53* and *Rb*, are being used to correct mutations in tumor suppressor genes. Another approach to cancer gene therapy is the direct alteration of tumor cells so as to provoke cell death. In this strategy, sensitivity to a non-toxic pro-drug is conferred to the tumor by transfer of a so called "suicide gene". Administration of a pro-drug to a modified cell results in the generation of a cytotoxic anabolite and cell death. Enzyme systems such as herpes simplex virus thymidine kinase (HSV-tk) in conjunction with the drug Ganciclovir (GCV) and cytosine deaminase in combination with 5-fluorocytosine have already progressed to clinical trial for several cancers including glioma and malignant melanoma.

3.2 Genetic Manipulation

3.2.1 Proto-oncogene Blockage

Antisense oligonucleotide blockers and ribozymes can be used to down regulate or inactive undesirable gene expression of activated oncogenes such as the *bcr-abl* fusion gene and *k-ras*. Introduction of exogenous single-stranded oligonucleotides usually in the form of DNA bind to their corresponding RNA sequences to inhibit transcription and translation from these oncogenes. An initial clinical trial making use of this antisense technology involved the *bcr-abl* oncogene fusion found in chronic myeloid leukemia. Blast cells derived from patients with the disease and treated with DNA fragments targeted to the fusion region showed decreased colony formation, while normal cells displayed a survival advantage (Szczylik *et al.*, 1991). Human lung carcinoma cells have been successfully treated with a recombinant plasmid to inhibit *k-ras* expression resulting in a three-fold down-regulation of the gene and inhibition of tumorigenicity (Mukhopadhyay *et al.*, 1991; Zhang *et al.*, 1993). Subsequent *in vivo* studies in a human lung cancer model resulted in marked tumor reduction (Georges *et al.*, 1993). A clinical protocol for treatment of non-small cell lung cancer using a retroviral vector containing an antisense *k-ras* expression cassette has been proposed and is likely to be initiated shortly (Roth 1996b). Another oncogene *BCL-2* is often over-expressed in non-Hodgkin lymphoma leading to reduced apoptosis and increased tumorigenesis. The use of antisense oligonucleotides (in the absence of a gene transfer agent) infused directly into tumors has shown specific down regulation of *BCL-2* mRNA, and thus increased apoptosis, first in animal models (Cotter *et al.*, 1994; Kitada *et al.*, 1994), and more recently in human subjects. With respect to the latter clinical study, evidence of reduced *BCL-2* mRNA and protein was detected in some treated patients and there was also encouraging evidence of clinical changes (Webb *et al.*, 1997).

An alternative methodology to regulate the mRNA transcripts from oncogenes is to use ribozymes. These molecules function by combining antisense and catalytic properties allowing for sequence-specific cleavage of host RNA. Ribozymes targeted to the *bcr-abl* fusion gene previously mentioned for its role in leukemia can be transferred to tumor cells, resulting in a slowed cancer progression (Lange *et al.*, 1993).

3.2.2 Tumor Suppressor Gene Replacement

Introduction of tumor suppressor genes to compensate for absent or mutated genes is a second form of direct genetic manipulation as a treatment for cancer. The most commonly mutated tumor suppressor gene, found in over 50% of human tumors, is p53 (Levine *et al.*, 1994). Delivery of the intact gene has been shown to restore the function of the wild-type tumor suppressor protein leading to cell-cycle arrest; consequently repair of damaged DNA or apoptosis can occur (Friedmann 1992). Both retroviral and adenoviral mediated transfer of the *p53* gene has been shown to successfully modulate tumor formation in animal models (Cai *et al.*, 1993; Clayman *et al.*, 1995). As a result of these, and other similar studies, a clinical trial using *in vivo* retroviral mediated delivery of the *p53* gene has been performed in patients with lung cancer (Roth 1996c). Tumor regression was seen in some patients, while others showed no change in the size of the treated tumor. Interestingly, apoptosis was detected more frequently in cells obtained from treated tumors than seen in sample taken prior to gene transfer suggesting that this may be the primary mechanism resulting tumor cell death (Roth *et al.*, 1996). The same group has proposed a trial using adenoviral mediated delivery of the *p53* gene (Roth 1996a) and other trials include treatment of head and neck squamous cell carcinoma and liver cancer and planned (Clayman 1995; Venook and Warren 1995).

The critical limitation to these direct gene transfer approaches is the apparent need to deliver an expressing tumor suppressor gene or anti-oncogene to **each** and **every** tumor cell. Even with 95% gene transfer efficiency the cancer could re-establish from untransduced residual cells.

3.1 Suicide Genes

Gene therapy for solid tumors using suicide genes is another direct approach for cancer treatment. Suicide genes were first developed as a safety feature for gene transfer using retroviral vectors. Since retroviruses integrate randomly within the host cell genome, the potential for insertional mutagenesis exists. Addition of a suicide gene within the retroviral vector would allow control of such a malignancy. Similarly, suicide genes may be used to control tumorous cells caused by other mutations. In this method a microbial gene encoding a benign gene product which confers sensitivity to a relatively non-toxic drug, such as an antibiotic, is utilized.

The most common microbial gene used to induce sensitivity to a pro-drug is the herpes simplex virus thymidine kinase (HSV-*tk*) gene in combination with Ganciclovir (GCV) (Moolten 1986). The HSV-*tk* gene is transfected into localized tumors *in vivo* usually using a viral gene transfer vector. The transduced cells and those nearby are killed by treatment with GCV, an anti-viral agent used in treatment of herpetic infections (Faulds and Heal 1990). Cell killing is mediated by the phosphorylation of GCV by the herpes simplex thymidine kinase gene product to form GCV-monophosphate (Elion 1980; McKnight 1980). GCV-monophosphate is then converted by cellular kinase to GCV-triphosphate which acts as a DNA chain terminator, inhibiting DNA polymerase activity during DNA replication (Cheng *et al.*, 1983; Field *et al.*, 1983).

Gene therapy for malignant brain tumors using this approach has been tested in a clinical trial. Retroviral vector producer cells expressing HSV-tk were implanted intra-tumorally and followed by intravenous administration of GCV. Localized tumor regression was seen initially in several patients but the tumors eventually continued to progress in most. However, one patient with glioblastoma multiforme has been in remission for four and half years following this treatment (Oldfield *et al.*, 1993; Culver *et al.*, 1994). Several other trials assessing the HSV-tk/GCV system are on-going or will be initiated shortly, using both retroviral producer cells and adenoviral vectors, in a variety of tumors including brain tumors, ovarian cancer and melanoma (Eck *et al.*, 1996; Klatzmann 1996; Link *et al.*, 1996; Stockhammer *et al.*, 1997). In addition, it may be possible to make suicide gene therapy more effective by enhancing differential transcription of the enzyme to specific types of tumors using tissue-specific promoters. For example, the *tyrosinase* promoter can direct the expression of HSV-*tk* gene in melanoma cells specifically (Vile *et al.*, 1994).

Other suicide genes such as cytosine deaminase (CD) are also being utilized. In this scenario, cells transduced with the *CD* convert the anti-fungal agent 5-fluorocytosine to 5-fluorouracil, a toxic compound often used in traditional chemotherapy (Mullen *et al.*, 1992; Crystal *et al.*, 1997). A critical feature of the suicide gene strategy has been the observation that only a small percentage of transduced cells is required for complete regression of the tumor mass. The phenomenon by which unmodified neighboring cells are killed is referred to as the 'bystander effect' (Culver *et al.*, 1992; Bi *et al.*, 1993; Pitts 1994; Ishii-Morita *et al.*, 1997).

3.1.1 The Bystander Effect

The existence of the bystander effect makes direct gene transfer of the suicide gene HSV-*tk* a feasible treatment option since only a fraction of the tumor mass needs to be modified within a localized tumor mass for GCV to exert its antineoplastic effect. Early experiments by Moolten and colleagues showed that transfer of the HSV-*tk* gene to cancer cells renders them sensitive to GCV cytotoxicity *in vitro*. Further studies have shown that transplanted tumors engineered to express the HSV-*tk* gene could be eliminated by GCV administration (Moolten and Wells 1990). Interestingly, it was noted *in vitro* that when mixtures of HSV-*tk* containing tumor cells and non-HSV-*tk* containing cells were mixed together, there was no diminished cytotoxicity by GCV. Certain tumors showed complete regression *in vivo* when as little as 1-2% of cells were modified by HSV-*tk* gene transfer (Culver *et al.*, 1992). The bystander effect seems to be mediated by the transfer of the toxic nucleotide analogue GCV-phosphate via "metabolic co-operation" involving cellular gap junctions, see figure 1 (Bi *et al.*, 1993; Pitts 1994) (Ishii, Touraine, and Blaese unpublished data). Therefore, confluency of the cells *in vitro* and cell-to-cell contact *in vivo* along with the prevalence of gap junctions are critical for determining whether a tumor is bystander sensitive or resistant (Ishii-Morita *et al.*, 1997). These findings do support the hopeful possibilities for cancer gene therapy using this strategy since only a small fraction of the affected neoplastic cells will need to be modified for successful eradication of the entire mass.

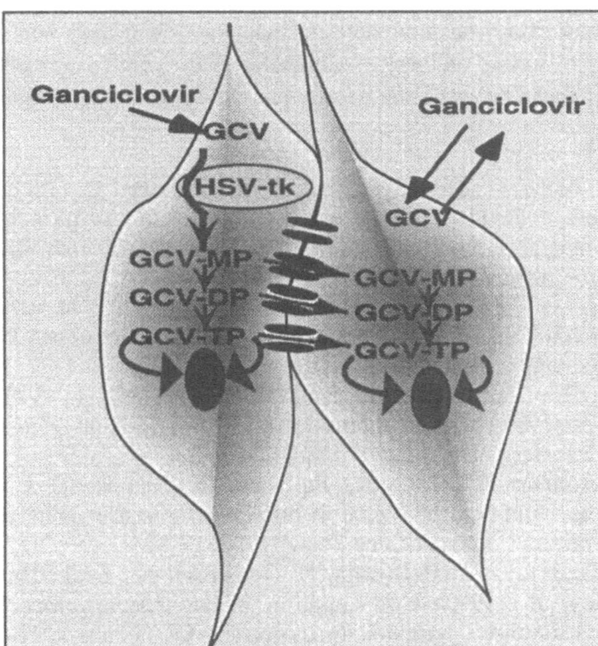

Figure 1. The Bystander Effect
A cell modified with the herpes simplex virus thymidine kinase (HSV-*tk*) gene is treated with a prodrug, Ganciclovir (GCV). GCV is then activated by the HSV-*tk* gene product to become monophosphorylated. Additional phosphorylation events by cellular kinases lead to production of GCV-triphosphate, a cytotoxic compound, which can transfer to adjacent cells through gap junctions. Cell death in a cell-cycle-dependent manner results in both gene-modified and naive cells through this bystander mechanism.

4 Future Direction

Unfortunately, no gene transfer system exists which allows for specific targeting for metastatic cancer or which can guarantee gene transfer to all cells of a localized tumor. Although hopeful for use in the potential treatment and cure of genetic and epigenetic disorders, gene therapy still has many limitations. A primary experimental difficulty impeding the success of gene transfer is the lack of an effective *in situ* gene delivery method which would allow for delivery to solid organs and tissues and accessibility to target cells. The use of replication competent viral vector delivery systems including adenoviruses and vaccinia viruses, or those with low immunoreactivity to allow for repeat administration will need to be further explored. A more complete understanding of the pathogenesis of cancer will certainly elucidate the role of other critical genes and pathways that may become the target for gene therapy in the future. The type of tumor including its location, specific antigens, and susceptibility to bystander effect will also be critical in determining the best combination of traditional and experimental strategies to use in treating this devastating disease. In theory, conventional treatments could first be implemented followed by direct gene strategies to restore tumor suppressor gene function, antisense or ribozyme approaches to inactivate oncogenes and growth factors, and tumor specific toxin delivery. Indirectly, the immune system could be simultaneously enhanced for systemic host defense to the invading tumor cells.

5 References

Abdel-Wahab, Z., Weltz, C., Hester, D., Pickett, N., Vervaert, C., Barber, J.R., Jolly, D., and Seigler, H.F. (1997) A phase I clinical trial of immunotherapy with interferon-gamma gene-modified autologous melanoma cells: monitoring the humoral immune response. Cancer **80**, 401-412.

Bi, W.L., Parysek, L., Warnick, R., and Stambrook, P.J. (1993) *In vitro* evidence that metabolic cooperation is responsible for the bystander effect observed with HSV tk retroviral gene therapy. Human Gene Therapy **4**, 725-731.

Blankenstein, T., Qin, Z., Uberla, K., Muller, W., Rosen, H., Volk, H.-D., and Diamantstein, T. (1989) Tumor suppression after tumor cell-targeted tumor necrosis factor-α gene transfer. J Exp Med **173**, 1047-1057.

Cai, D., Mukhopadhyay, T., Lui, T., Fujiwara, T., and Roth, J. (1993) Stable expression of the wild-type p53 gene in human lung cancer cells after retrovirus-mediated gene transfer. Human Gene Therapy **4**, 617-624.

Chen, L., Ashe, S., Brady, W., Hellstrom, I., Hellstrom, K., Ledbetter, J., Mcgowan, P., and Linsley, P. (1992) Costimulation of antitumor immunity by the B7 counter-receptor for the T-lymphocyte molecules CD28 and CTLA-4. Cell **71**, 1093-1102.

Chen, T., Tao, M., and Levy, R. (1994) Idiotype-cytokine fusion proteins as cancer vaccines. Relative efficacy of IL-2, IL-4, and granulocyte-macrophage colony-stimulating factor. J. Immunol. **153**, 4775-4787.

Cheng, Y.C., Grill, S.P., Dutschman, G.E., Nakayama, K., and Bastow, K.F. (1983) Metabolism of 9-(1,3-dihydroxy-2-propoxymethyl) guanine, a new anti-herpes virus compound in herpes simplex virus infected cells. J. Biol. Chem. **258**, 12460-12464.

Clayman, G. (1995) Clinical protocol for modification of tumor suppressor gene expression in head and neck squamous cell carcinoma (HNSCC) with an adenovirus vector. Human Gene Therapy **6**, 1082-1086.

Clayman, G., Elnagger, A., Roth, J., Zhang, W., Goepfert, H., Taylor, D., and Liu, T. (1995) *In vivo* molecular therapy with p53 adenovirus for microscopic residual head and neck squamous carcinoma. Cancer Research **55**, 1-6.

Cotter, F.E., Johnson, P., Hall, P., Pocock, C., AlMahdi, N., Cowell, J.K., and Morgan, G. (1994) Antisense oligonucleotides supress B-cell lymphoma growth in a SCID-hu mouse model. Oncogene **9**, 3049-3055.

Crystal, R.G., Hirschowitz, E., Liberman, M., Daly, J., Kazam, E., Henschke, G., Yankelevitz, D., Kemeny, N., Silverstein, R., Ohwada, A., Russi, T., Mastrangeli, A., Sanders, A., Cooke, J., and Harvey, B.G. (1997) Phase I study of direct administration of a replication deficient adenovirus vector containing the *E. coli* cytosine deaminase gene to metastatic colon carcinoma of the liver in association with oral administration of the pro-drug 5-flurocytosine. Human Gene Therapy **8**, 985-1001.

Culver, K.W. and Blaese, R.M. (1994) Gene therapy for cancer. Trends in Genetics **10**, 174-178.

Culver, K.W., Ram, Z., Walbridge, S., Ishii, H., Oldfield, E.H., and Blaese, R.M. (1992) *In vivo* gene transfer with retroviral vector producer cells for treatment of experimental brain tumors. Science **256**, 1550-1552.

Culver, K.W., VanGilder, J., Link, C.J., Carlstrom, T., Buroker, T., Yuh, W., Koch, K., Schabold, K., Doornbas, S., Wetjen, B., and Blaese, R.M. (1994) Gene therapy for the treatment of malignant brain tumors with *in vivo* tumor transduction with the herpes simplex thymidine kinase gene/Ganciclovir system. Human Gene Therapy **5**, 343-379.

Deisseroth, A.B., Kavangh, J., and Champlin, R. (1994) Use of safety-modified retroviruses to introduce chemotherapy resistance sequences into normal hematopoietic cells for chemoprotection during the therapy of ovarian cancer: a pilot trial. Human Gene Therapy **5**, 1507-1522.

Douvdevani, A., Huleihel, M., Zoller, M., Segal, S., and Apte, R. (1992) Reduced tumorigenicity of fibrosarcomas which constitutively generate Il-1-alpha either spontaneously or following IL-1-alpha gene transfer. Int J Cancer **51**, 822-830.

Dranoff, G., Jaffee, E., Lazenby, A., Golumbek, P., Levitsky, H., Brose, K., JAckson, V., Hamada, H., Pardoll, D., and Mulligan, R. (1993) Vaccination with irradiated tumor cells engineered to secrete murine granulocyte macrophage colony-stimulating factor stimulates potent, specific, and long-lasting anti-tumor immunity. Proc Natl Acad Sci USA **90**, 3539-3543.

Eck, S.L., Alavi, J.B., Alavi, A., Davis, A., Hackney, D., Judy, K., Mollman, J., Philpis, P.C., Wheeldon, E.B., and Wilson, J.M. (1996) Treatment of advanced CNS malignancies with recombinant adenovirus H5.010RSVTK: a phase I trial. Human Gene Therapy **7**, 1465-1482.

Elion, G.B. (1980) The chemotherapeutic exploitation of virus-specified enzymes. Adv. Enzyme Regul. **18**, 53-66.

Faulds, D. and Heal, R. (1990) Ganciclovir. A review of its antiviral activity, pharmacokenetics properties and therapeutic efficacy in cytomegalovirus infection. Drugs **39**, 5997-6331.

Fearon, E., Pardoll, D., Itaya, T., Golumbek, P., Levitsky, H., Simons, J., Karasuyama, H., Vogelstein, B., and Frost, P. (1990) Interleukin-2 production by tumor cells bypasses T-helper function in the generation of an antitumor response. Cell **60**, 397-403.

Ferrantini, M., Proietti, E., Santodonato, L., Gabriele, L., Peretti, M., Plavec, I., Meyer, F., Kaido, T., Gresser, I., and Belardelli, F. (1993) α1-interferon gene transfer into metastatic Friend leukemia cells abrogated tumorigenicity in immunocompetent mice: antitumor therapy by means of interferon-producing cells. Cancer Res **53**, 1107-1112.

Field, A.K., Davies, M.E., DeWitt, C., Perry, H.C., Liou, R., Germershausen, J., Karkas, J.D., Ashton, W.T., Johnston, D.B.R., and Tolman, R.L. (1983) 9-{[2-hydroxy-1-9hyroxymethyl) ethoxy]-methyl} guanine: a selective inhibitor of herpes group virus replication. Proc. Natl. Acad. Sci. USA **80**, 4139-4143.

Friedmann, T. (1992) Gene therapy of cancer through restoration of tumor suppressor functions. Cancer **70**, 1810-1817.

Georges, R.N., Mukhopadhyay, T., Zhang, Y., Yen, N., and Roth, J.A. (1993) Prevention of orthotopic human lung cancer growth by intratracheal instillation of a retroviral antisense K-ras construct. Cancer Research **53**, 1743-1746.

Gottesmann, M., Germann, U., Aksentijevich, I., Sugimoto, Y., Cardarelli, C., and Pastan, I. (1994) Gene transfer of drug resistance genes, implications for cancer therapy. Ann NY Acad Sci **716**, 26-38.

Hannania, E.G., Giles, R.E., Kavangh, J., Ellerson, D., Zu, Z., Wang, T., Su, Y., Kudelka, A., Rahman, Z., Holmes, F., Horbagyi, G., Claxton, D., Bachier, C., Thall, P., Cheng, S., Hester, J., Ostrove, J.M., Bird, R.E., Chang, A., Korbling, M., Seong, D., Cotes, R., Holzmayer, T., Mechetner, E., Heimfeld, S., Berenson, R., Burtness, B., Edwards, C., Bast, R., Andreeff, M., Champlin, R., and Deisseroth, A.B. (1996) Results of MDR-1 vector modification trial indicate that granulocyte/macrophage colony-forming unit cells do not contribute to posttransplant hematopoietic recovery following intensive systemic therapy. Proc. Natl. Acad. Sci. USA **93**, 15346-15351.

Hesdorffer, C., Antman, K., Bank, A., Fetell, M., Mears, G., and Begg, M. (1994) Human MDR gene transfer in patients with advanced cancer. Human Gene Therapy **5**, 1151-1160.

Hock, H., Dorsch, M., Kunzerdurf, U., Überle, K., Qin, Z., Diamantstein, T., and Blankenstein, T. (1993) Mechanisms of rejection induced by tumor cell-targeted gene transfer of interleukin-2, interleukin-4, interleukin-7, tumor necrosis factor, or interferon-gamma. Proc Natl Acad Sci USA **90**, 2774-2778.

Ishii-Morita, H., Agbaria, R., Mullen, C., Hirano, H., Koeplin, D., Ram, Z., Oldfield, E., Johns, D., and Blaese, R. (1997) Mechanism of 'bystander effect' killing in the herpes simplex thymidine kinase gene therapy model of cancer treatment. Gene Therapy **4**, 244-251.

Kawakami, Y., Eliyahu, S., Delgado, C., Robbins, P., Rivoltini, L., Topalian, S., Miki, T., and Rosenberg, S. (1994) Cloning of the gene for a shared human melanoma antigen recognized by autologous T-cells infiltrating into tumor. Proc Natl Acad Sci USA **91**, 3515-3519.

Kitada, S., Takayama, S., deReil, K., Tanaka, S., and Reed, J. (1994) Reversal of chemoresistance of lymphoma cells by antisense-mediated reduction of bcl-2 gene expression. Antisense Res. Devel. **4**, 71-79.

Klatzmann, D. (1996) Gene therapy for metastatic malignant melanoma: evaluation of tolerance to intratumoral injection of cells producing recombinant retroviruses carrying the herpes simplex virus type 1 thymidine kinase gene, to be followed by Ganciclovir administration. Human Gene Therapy **7**, 255-267.

Lange, W., Cantin, E., Finke, J., and Dolken, G. (1993) *In vitro* and *in vivo* effects of synthetic ribozymes targeted against *bcr/abl* mRNA. Leukemia **7**, 1786-1794.

Levine, A., Perry, M., Chang, A., Silver, A., Dittmer, D., Wu, M., and Welsh, D. (1994) The role of the p53 tumor-suppressor gene in tumorigenesis. British Journal of Cancer **69**, 409-416.

Link, C.J., Moorman, D., Seregina, T., Levy, J.P., and Schabold, K.J. (1996) A phase I trial of *in vivo* gene therapy with the herpes simplex thymidine kinase/Ganciclovir system for the treatment of refractory or recurrent ovarian cancer. Human Gene Therapy **7**, 1161-1179.

McKnight, S.L. (1980) The nucleotide sequence and transcript map of the herpes simplex thymidine kinase gene. Nucl. Acids Res. **8**, 5949-5964.

Merrouche, Y., Negrier, S., Bain, C., Combaret, V., Mercatello, A., Coronel, B., Moskovtchenko, J.F., Tolstoshev, P., Moen, R., Philip, T., and Favrot, M.C. (1995) Clinical application of retroviral gene-transfer in oncology: results of a French study with tumor-infiltrating lymphocytes transduced with the gene of resistance to neomycin. J. Clin. Oncol. **13**, 410-418.

Miller, A., McBride, W., Dubinett, S., Dougherty, G., Thacker, J., Shau, H., Kohn, D., Moen, R., Walker, M., Chiu, R., Schuck, B., Rosenblatt, J., Huang, M., Dhanani, S., Rhoades, K., and Economou, J. (1993) Transduction of human melanoma cell lines with the human interleukin-7 gene using retroviral-mediated gene transfer - comparison of immunologic properties with interleukin-2. Blood **82**, 3686-3694.

Moolten, F.L. (1986) Tumor chemosensitivity conferred by inserting herpes thymidine kinase genes: paradigm for a prospective cancer control strategy. Cancer Research **46**, 5276-5281.

Moolten, F.L. and Wells, J.M. (1990) Curability of tumors bearing herpes thymidine kinase gene transferred by retroviral vectors. J Natl Cancer Inst **82**, 297-300.

Mukhopadhyay, T., Tainsky, M., Cavender, A.C., and Roth, J.A. (1991) Specific inhibition of K-ras expression and tumorigenicity of lung cancer cells by antisense RNA. Cancer Research **51**, 1744-1748.

Mullen, C., Kilstrup, M., and Blaese, R. (1992) Transfer of the bacterial gene for cytosine deaminase to mammalian cells confers lethal sensitivity to 5-fluorocytosine; a negative selection system. Proc Natl Acad Sci USA **89**, 33-37.

Munger, W., DeJoy, S., Jeyaseelan, R., Torley, L., Grabstein, K., Eisenmann, J., Paxton, R., Cox, T., Wick, M., and Kerwar, S. (1995) Studies evaluating the antitumor activity and toxicity of interleukin-15, a new T cell growth factor: comparison with interleukin-2. Cell Immunol **165**, 289-293.

Nabel, G.J., Gordon, D., Bishop, D.K., Nickoloff, B.J., Yang, Z.-Y., Aruga, A., Cameron, M.J., Nabel, E.G., and Chang, A.E. (1996) Immune response in human melanoma after transfer of an allogenic class I major histocompatibility complex gene with DNA-liposome complexes. Proc. Natl. Acad. Sci. USA **93**, 15388-15393.

Nabel, G.J., Nabel, E.G., Yang, Z.-Y., Fox, B.A., Plutz, G.E., Gao, X., Huang, L., Shu, S., Gordon, D., and Chang, A.E. (1993) Direct gene transfer with DNA-liposome complexes in melanoma: expression, biologic activity, and lack of toxicity in humans. Proc. Natl. Acad. Sci. USA **90**, 11307-11311.

Oldfield, E.H., Ram, Z., Culver, K.W., Blaese, R.M., DeVroom, H.L., and Anderson, W.F. (1993) Gene therapy for the treatment of brain tumors using intra-tumoral transduction with the thymidine kinase gene and intravenous Ganciclovir. Human Gene Therapy **4**, 39-69.

Pitts, J.D. (1994) Cancer gene therapy: a bystander effect using the gap junctional pathway. Molecular Carcinogenesis **11**, 127-30.

Porgador, A., Tzehoval, E., Katz, A., Vadai, E., Revel, M., Feldman, M., and Eisenbach, L. (1992) Interleukin 6 gene transfection into Lewis lung carcinoma tumor cells suppresses the malignant phenotype and confers immunotherapeutic competence against parental metastatic cells. Cancer Res **52**, 3679-3686.

Richter, G., Krasagakes, S.K., Hein, G., Huls, C., Schmitt, E., Diamantstein, T., and Blankenstein, T. (1993) Interleukin 10 transfection into Chinese hamster ovary cells prevents tumor growth and macrophage infiltration. Cancer Res **53**, 4134-4137.

Rosenberg, S. (1990) Gene therapy of patients with advanced cancer using tumor infiltrating lymphocytes transduced with the gene coding for tumor necrosis factor. Human Gene Therapy **1**, 441-480.

Rosenberg, S.A., Aebersold, P., Cornetta, K., Kasid, A., Morgan, R., Moen, R., Karson, E.M., Lotze, M.T., Yang, J.C., Tapalian, S.L., Merino, M.J., Culver, K., Miller, A.D., Blaese, R.M., and Anderson, W.F. (1990) Gene transfer into humans - immunotherapy of patients with advanced melanoma, using tumor - infiltrating lymphocytes modified by retroviral gene transduction. N. Eng. Jour. Med. **323**, 570-578.

Roth, J., Nguyen, D., Lawrence, D.D., Kemp, B.L., Carrasco, C.H., Ferson, D.Z., Hong, W.K., Komaki, R., Lee, J.J., Nesbitt, J.C., Pisters, K.M., Putnam, J.B., Schea, R., Shin, D.M., Walsh, G.L., Dolormente, M.M., Han, C.L., Martin, F.D., Yen, N., Xu, K., Stephens, L.C., McDonnell, T.J., Mukhopadhyay, T., and Cai, D. (1996) Retrovirus-mediated wild-type p53 gene transfer to tumors of patients with lung cancer. Nature Medicine **9**, 985-991.

Roth, J.A. (1996a) Clinical protocol for the modifciation of tumor suppressor gene expression and induction of apoptosis in NSCLC with an Ad vector expressing wild-type p53 and cisplatin. Human Gene Therapy **7**, 1013-1030.

Roth, J.A. (1996b) Modification of mutant K-ras gene expression in non-small cell lung cancer (NSCLC). Human Gene Therapy **7**, 875-889.

Roth, J.A. (1996c) Modification of tumor suppressor gene expression in non-small cell lung cancer (NSCLC) with a retroviral vector expressing wildtype (normal) p53. Human Gene Therapy **7**, 861-874.

Stockhammer, G., Brotchi, J., Leblanc, R., Berstein, M., Schackert, G., Weber, F., Ostertag, C., Mulder, N.H., Mellstedt, H., Seiler, R., Yonekawa, Y., Twerdy, K., Kostron, H., Witte, O.D., Lambermont, M., Velu, T., Laneuville, P., Villemure, J.G., Rutka, J.T., Warnke, P., Laseur, M., Mooij, J.J., Boethius, J., Marianii, L., Meyer, M., Brändli, C., Frei, K., Künu, D., and Gianella-Borradori, A. (1997) Gene therapy for glioblastoma multiform: *in vivo* tumor transduction with the herpes simplex thymidine kinase gene followed by Ganciclovir. J. Mol. Med. **75**, 300-304.

Szczylik, C., Skorski, T., Nicolaides, N.C., Manzella, L., L;, L.M., Venturelli, D., AM;, A.M.G., and Calabretta, B. (1991) Selective inhibition of leukemia cell proliferation by BCR-ABL antisense oligodeoxynucleotides. Science **253**, 562-565.

Tahara, H., Zeh, H., Storkus, W., Pappo, I., Watkins, S., Gubler, U., Wolf, S., Robbins, P., and Lotze, M. (1994) Fibroblasts genetically engineered to secrete interleukin-12 can suppress tumor growth and induce antitumor immunity to a murine melanoma *in vivo*. Cancer Res **54**, 182-189.

Tao, M. and Levy, R. (1993) Idiotype/granulocyte-macrophage colony-stimulating factor fusion protein as a vaccine for B-cell lymphoma. Nature **362**, 755-758.

Tepper, R., Pattengale, P., and Leder, P. (1989) Murine interleukin-4 displays potent anti-tumor activity *in vivo*. Cell **57**, 503-512.

Townsend, S. and Allison, J. (1993) Tumor rejection after direct costimulation of CD8+ T-Cells by B7 -transfected melanoma cells. Science **259**, 368-370.
Ueda, K., Cardarelli, C., Gottesman, M., and Pastan, I. (1987) Expression of a full length cDNA for the human "MDR 1" gene confers resistance to colchicine, doxorubicin, and vinblastin. Proc Natl Acad Sci USA **84**, 3004-3008.
Venook, A. and Warren, R. (1995) Gene therapy of primary and metastatic malignant tumors of the liver using ACN53 via hepatic artery infusion: a phase I study. Human Gene Therapy **6**, 1086-1090.
Vile, R., Miller, N., Chernajovsky, Y., and Hart, I. (1994) A comparison of the properties of different retroviral vectors containing the murine tyrosinase promoter to achieve transcriptionally targeted expression of the HSVtk or IL-2 genes. Gene Therapy **1**, 307-316.
Watanabe, Y., Kuribayashi, K., Miyatake, S., Nishihara, K., Nakayama, E., Taniyama, T., and Sakata, T. (1989) Exogenous expression of mouse interferon-γ cDNA in mouse C1300 neuroblastoma cells results in reduced tumorigenicity by augmented anti-tumor immunity. Proc Natl Acad Sci USA **86**, 9456-9460.
Webb, A., Cunningham, D., Cotter, F., Clarke, P.A., diStefano, F., Ross, P., Corbo, M., and Dziewanowska, Z. (1997) BCl-2 antisense therapy in patients with non-Hodgkin lymphoma. Lancet **349**, 1137-1141.
Zhang, Y., Mukhopadhyay, T., Donehower, L.A., Georges, R.N., and Roth, J. (1993) Retroviral vector-mediated transduction of K-ras antisense RNA into human lung cancer cells inhibits expression of the malignant phenotype. Human Gene Therapy **4**, 451-460.

* In partial fulfillment of the Ph.D. requirements in the Graduate Genetics Program at The George Washington University.

Feasibility Study Intended for In Vivo Retroviral Mediated Gene Transfer of Bladder Urothelium

N. Dumey[1], M. Masset[1] P. Devauchelle[2], M. Marty[1], and O. Cohen-Haguenauer[1].

[1] Laboratoire Transfert Génétique et Oncologie Moléculaire, Institut d'Hématologie, Hopital Saint-Louis, 75475 Paris Cedex 10.

[2] Centre de Cancérologie Vétérinaire, Ecole Nationale Vétérinaire d'Alfort, 94704 Maisons Alfort, France.

1. Introduction

Current strategies aiming at gene therapy are often focused on *ex-vivo* manipulation of cells that can be selected for positive gene uptake and which will later be re implanted (Culver et al., 1990; Salvetti et al., 1995). The demands of therapeutic interventions in the field of cancer mostly translate into a requirement to address tumoural cells *in vivo*. *In vivo* gene transfer strategies, although conceptually attractive, still need to overcome many technical problems. Difficult access to the target cells and intrinsic impossibility for primary control of gene transfer efficiency represent major limitations. In addition, safety issues have to be considered relating both to diffusion of vectors to other organs or tissues than those of interest and to the consequences of expression of foreign genes in normal tissues. In animal models, introduction and expression of chosen genes in a sufficiently high fraction of tumours cells have been shown to inhibit tumour growth, induce tumour regression, or prevent invasion (Dougherty et al., 1996).

Superficial bladder tumours often relapse and invasion is frequently observed after endoscopic resection (Heney, 1992; Herr, 1991). In order to prolong remission, endovesical instillation of either cytotoxic agents (Tolley et al., 1996) or non specific immunostimulators (Herr 1989a, 1989b) have been performed. Thus far, no conventional treatment has shown to achieve cure. Therefore additional therapeutics means are required. Instillation of therapeutic coumpounds into the bladder have already proved safe. The efficiency of the procedure can be easily and repeatedly monitored by simple endoscopy.

We have explored the feasibility of *in vivo* transduction of bladder urothelium by means of retrovirus-mediated gene transfer in a canine animal model, since spontaneous carcinomas are often observed in domestic urban dogs, mimicking human pathology. Murine retroviruses are known to integrate the genome of dividing cells preferentially (Miller et al., 1990). We have thus investigated the potential for selective transduction of tumoural cells (Culver et al., 1992; Hurford et al., 1995).

A retroviral vector encoding both the ß-galactosidase and Neomycin resistance genes has been used. The ability of this vector to transduce urothelial cells *in vitro* was analysed, as well as its half life and resistance to urine both *in vitro* and *in vivo*.

NATO ASI Series, Vol. H 105
Gene Therapy
Edited by Kleanthis G. Xanthopoulos

2.Transduction of human cell lines derived from bladder transitional carcinoma

Titration of supernatant was performed on five independent human bladder cell lines in parallel to NIH3T3 cells used as reference controls. All five cell lines proved accessible to retroviral transduction. The cell lines 647V (primary bladder tumour), RT112 and SD148 (both well-differentiated tumours) showed levels of transduction equivalent to NIH3T3 cells, with titers of 10^5 cfu/ml. J82S (from metastases) and T24 (anaplastic) cell lines were respectively one and two order of magnitude less sensitive to transduction as compared to NIH3T3 cells.

It seems unlikely that these variations would be due to cycling time since T24 cell line, the least susceptible of all, presents a cycling time of 19h (Bubenik et al., 1973), which is close to NIH3T3 cells. These differences in transduction may be due to different origins or malignancy grade of the carcinoma cell lines (Hastings et al., 1981).

3. Infections of primary cultures of dog bladder urothelium.

Primary cultures were obtained from biopsies of a normal Beagle, and from normal and tumoural specimens of a Cocker spaniel and a Bruno du Jura. Fresh biopsies were cut in small fragments and placed in Petri dishes or flasks. After 7 days of culture, monolayers of urothelial and fibroblastic cells were growing from the explants and cultures were infected for 3 hours with the recombinant retrovirus. After X-Gal staining, foci of LacZ positive cells could be observed in all infected cultures. Infections were localised predominantly at the edges of the foci, i.e. amid growing areas. Fibroblasts growing in the flasks were transduced as well. Cultures originating from normal or tumoural specimens were equally transduced. Untransduced primary cultures used as control stained negative.

A retrovirus vector harbouring the murine 4070A envelope with amphotropic host range thus also proved capable of transducing canine bladder epithelial cells in primary cultures. Penetration of the murine defective retrovirus was thus evidenced in urothelial cells of different origins, indicating that these cells types all express the amphotropic receptor.

4. Analysis of retrovirus particles kinetics

4.1 *In vitro* half-life of retrovirus particles and resistance to urine.

In order to estimate the half life of the recombinant retrovirus, five independent batches of supernatant were filtered and incubated at 37°C. Retrovirus supernatant titers were measured at timed intervals. For each supernatant, a curve of the titer versus time was obtained (fig 1). The half life of viral particles was deduced from a linear regression curve fit equation (calculated with Cricket Graph III software, Computer Associates International). Mean half life obtained from the five supernatants was 5h30 +/-2h30 (sd).

Potential inactivation of the retroviral vector by urine components was then investigated. Titers of supernatants mixed with an equal volume of human filtered urines and kept at 37°C were measured at timed intervals. During the first 6 to 9 hours (fig 2), decay in titer remained similar to the control (supernatant mixed with PBS). After 10 hours, the supernatant mixed with urine showed a 2 times shorter half-life (2h30) than the control (5h). Heat inactivation of urine at 56°C during 30 minutes prior mixing did not result in a modified half-life.

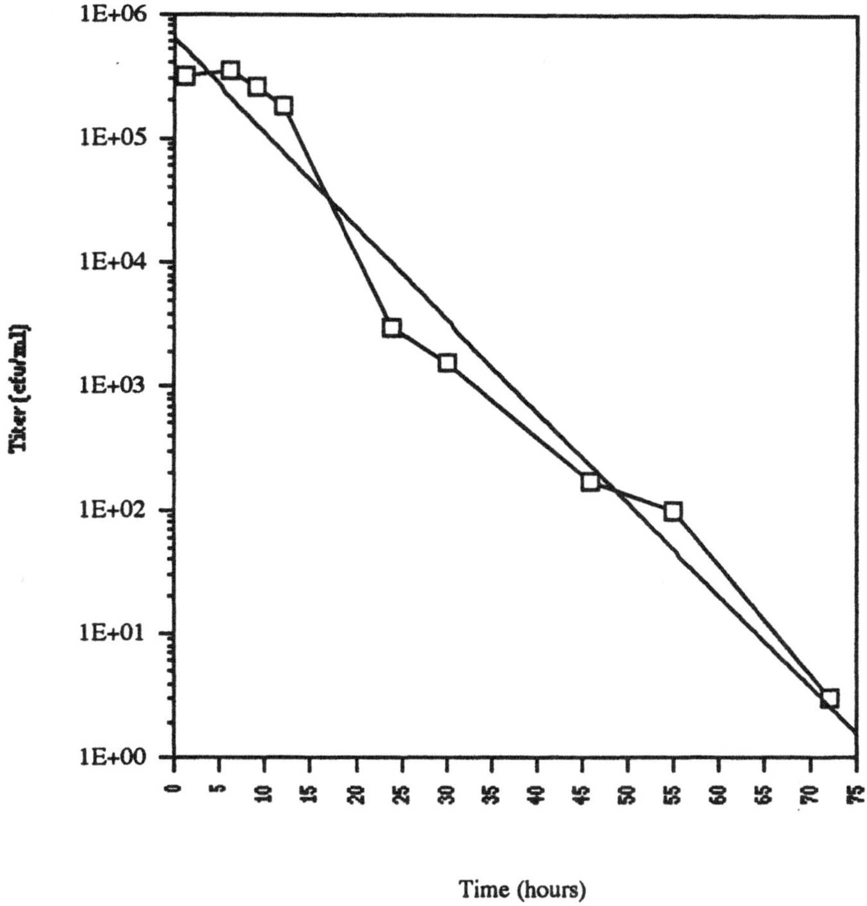

Figure 1 : Representative time related decay in titer of one batch of retroviral vector supernatant.
Viral supernatant was filtered and kept at 37°C. NIH3T3 cells were infected using serial dilution of this supernatant at timed intervals, indicated in hours. Titers are indicated on a logarithmic scale (10^1 cfu/ml is quoted as 1E+01) and data represent the product of the number of foci of blue stained cells by the dilution factor. The half life of infectious retrovirus particles was determined from the equation of a linear regression curve fit calculated using Cricket Graph software.

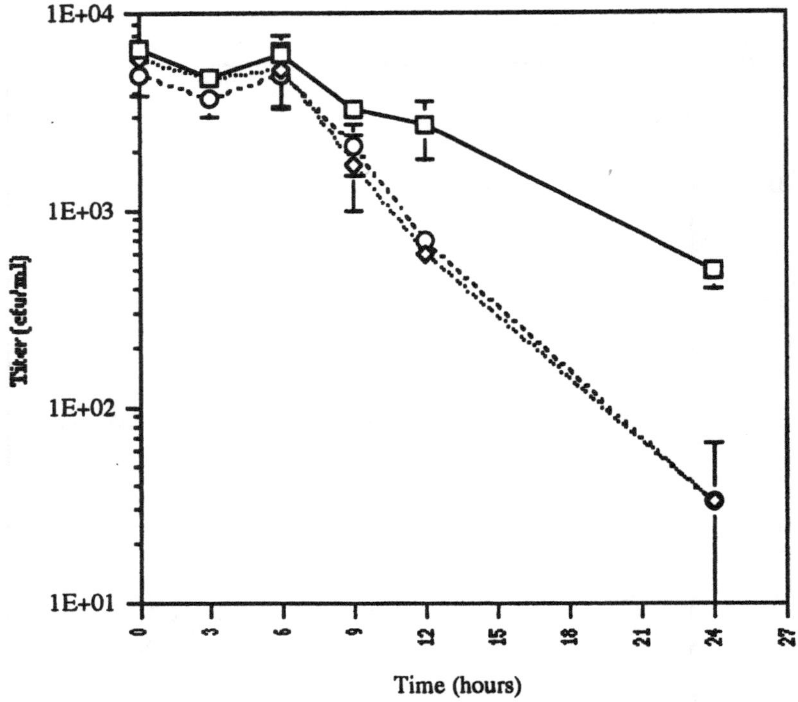

Figure 2 : Effect of urines on *in vitro* survival of infectious retroviral particles. Retroviral supernatant was either mixed with an equal volume of PBS (-□-), urine (-◊-) or heat inactivated urine (-O-), kept at 37°C, and repeatedly titered at timed intervals, indicated in hours. Data have been calculated from three separate experiments (mean ± SD).

4.2 Evaluation of adequacy between duration of infusion period and infectivity of virus particles.

In order to investigate to which extent a prolonged incubation of virus with target cells will increase infection rate, NIH3T3 cells were incubated with retroviral supernatant for various periods of time ranging from 1 to 12 hours. Two days after infection, all cultures were confluent and cells were fixed, stained, and scored for ß-galactosidase activity in microscopic fields. As shown in figure 3, the increase in the number of transduced cells was logarithmic, with a mean of 200 infected cells per mm^2 after 1 hour incubation, 350 infected cells/mm^2 following 6 hours and tended to a plateau close to 400 infected cells/mm^2 when infections were performed for 6 to 12 hours.

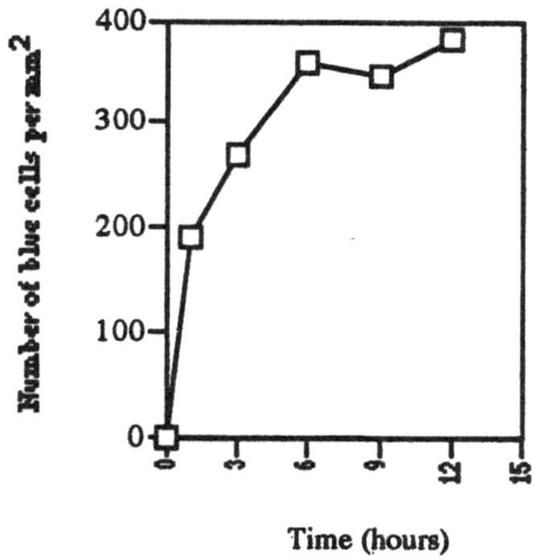

Figure 3 : Time-related increase in the number of transduced NIH3T3 cells. NIH3T3 cells were exposed to a viral supernatant during indicated periods of time (hours). Cells were then left to grow another 48 hours and X-Gal stained. Blue cells were scored under a microscope in 1mm^2 areas. Data represent a mean of 5 different areas.

The half life of this vector, produced by Psi-CRIP packaging cells, was of 5h30 on average. This results are in accordance with other reports on defective retroviral particles produced by Psi-CRIP (Chuck et al., 1996), or by other ecotropic or amphotropic producer cell lines (Kotani et al., 1994; Layne et al., 1989; Levin et al., 1976; Paul et al., 1993; Sanes et al., 1986). While many biological products are known to reduce or inhibit retroviral infection, even at low concentrations (Kimura et al., 1996; Russell et al., 1995; Takeuchi, et al. 1994), no deleterious effect of urine on viral titer could be observed *in vitro*, within the first 6 to 9 hours of incubation. These results are compatible with *in vivo* infusion procedure, physiologically limited to a few hours. This period appeared to be sufficient to allow optimal transduction, as *in vitro* incubation of viral particles with target cells beyond 6 hours did not yield significant increase in the overall number of infected cells.

4.3 Monitoring vector infectivity during instillation in normal dog bladder.

Domestic dogs were recruited at the Centre de Radiotherapie-Scanner (Maisons-Alfort, France), after informed consent of the owners. They were kept on water diet conditions to limit urine production during infusion.

Bladder infusions were performed via two lines balloon catheters (de Foley catheter). One line was dedicated to balloon inflation, to prevent leaking, while the other port was used for infusion and withdrawal of fluids.

Prior to infusion of viral supernatant, bladders were emptied from urine and washed with PBS. Viral supernatant (50 to 70 ml) was then instilled allowing repletion of the organ but avoiding high pressure infusion. Bladder emptying and washing were done in a safe and controlled manner 3 to 6 hours later.

Retroviral supernatant was obtained from confluent cultures of producer cell lines, filtered and supplemented with polybrene. *In vivo* infusion of supernatant was performed into the bladder of a normal (non-tumoural) female Beagle. Three millilitre samples were collected from the bladder at timed intervals (2h, 3h, 6h), and titered. As a control, viral supernatant was kept in acellular medium at 37°C.

While supernatant kept *in vitro* showed a 6.5 hours half life, the titer of supernatant infused *in vivo* showed a slightly faster decrease (fig 4). Six hours after infusion, the bladder was emptied and washed with PBS. All fluids were retrievable without any leaking. Catheter was left in place and further production of urine was recovered. No infectious particles could be detected the next day (seventeen hours later).

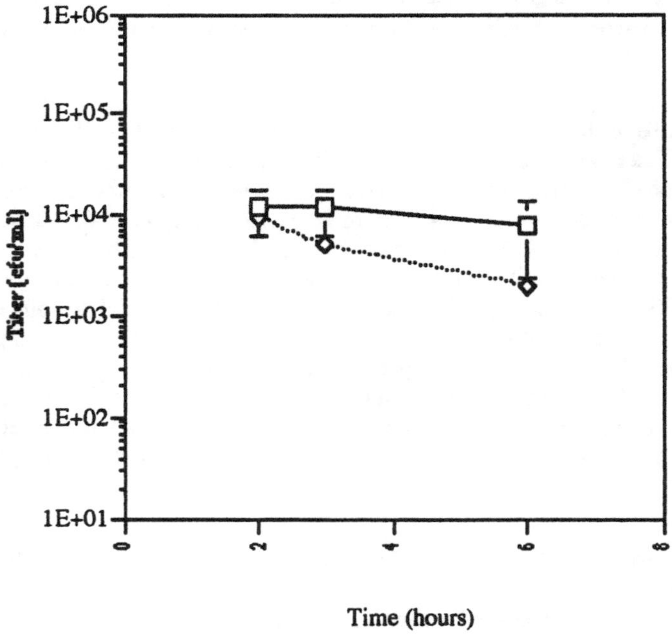

Figure 4 : Survival of retroviral particles during *in vivo* instillation.
Supernatant was infused intravesically in dog bladder. At indicated times (hours), samples were retrieved and titered on NIH3T3 cells (-O-), and compared to uninfused supernatant kept at 37°C (-□-) as control. Data represent a mean (± SD) of three independent measures.

5. *In vivo* transduction of normal and tumoural urothelium ; single instillation.

Infusions were performed on two normal female Beagles (as described above) and on a tumoural male Greyhound. In order to allow integration of the provirus, and sufficient expression of the ß-galactosidase transgene, surgical biopsies were taken 5 and 8 days post infusion.

Transduction was investigated following *in vivo* infusion in normal and tumoural dogs by several approaches. Expression of the marker gene was revealed by X-Gal staining of primary cultures of bladder biopsies from infused normal dogs, where clusters of epithelial cells harbouring blue nuclei were observed. No transduced fibroblast could be detected within the cultures (Dumey et al, submitted).

No transduction was revealed on cryosections of specimens from the normal animals or from sound tissue from the tumour-bearing dog, but some *in vivo* transduced tumours contained X-Gal stained cells. The cells expressing the ß-galactosidase enzyme were detected in regions where the tumour appeared to be actively developing, in the absence of necrosis. Staining could be observed in all layers of the urothelium but not in the stroma. Transduction of tumour cells was further confirmed by PCR. Positive signals were obtained on DNA extracted from tumours that were X-Gal positive on cryosections. No signal could be detected from sound bladder specimens of either tumoural or sound animals (Dumey et al, submitted).

6. Repeated instillations in normal dogs results in rare but long-lasting transduction

As previously described for *in vivo* gene transfer experiments (Barka et al., 1996; Cardoso et al., 1993), we observed a low level of transduction in normal urothelium after infusion of viral supernatant. Nevertheless, since we could demonstrate a low level of transduction after infusion of viral supernatant into the bladder we chose to investigate stable expression of marker genes (Naviaux et al, 1992).

Following repeated infusions (four times over 40 hours) that could be easily performed, surgical biopsies were taken at either 8 or 60 days. In both instances, primary cultures of urothelium could evidence foci of blue-stained epithelial cells.

7. Repeated instillations in tumoural dogs results in quantitatively siginificant and preferential transduction of actively growing malignant cells

In order to improve transduction in a tumoural dog (Bruno du Jura), 1st/ concentrated viral supernatant was used following precipitation of viral particles intended to enhance virus to cell contacts (Morling et al., 1995) and 2nd/ repeated infusions over 40 hours were performed. Cryosections were analysed from blocks taken at 15 days after transduction. No blue stained cell was observed in non-tumoural areas. Over 20% tumoural cells showed blue staining. Transduction was further confirmed by PCR performed from cryosections of both normal and/or tumoural tissue. A positive signal was present only in those areas where malignant cells would also stain X-Gal positive (Dumey et al, submitted).

We have demonstrated for the first time to our knowledge, that transduction of urothelial cells is achievable by direct *in vivo* infusion of retroviral supernatant into the bladder. The observed resistance of the virus to urine during a 6 hours limited procedure made amphotropic MuLV derived vectors good candidates to achieve gene transfer and mediate long-lasting expression of genes of interest in malignant bladder urothelium. Indeed, in vivo gene transfer into tumour-bearing bladders translates into preferential transduction of actively dividing malignant urothelial cells. Further studies focus on increasing expression in using epithelial-cells specific promotors to drive genes with therapeutic potential. These include various mechanisms such as anti-invasion, conditional suicide (Culvetr et al, 1992) or inhibition of signal transduction. On the other hand, potential to generate systemic immunity (Dilloo et al., 1996; Zier et al., 1995) following immmunostimulatory rationale remains questionable.

Acknowledgements

We thanks N.Ferry and J.M. Heard for providing us with the LLZ producer clone.We are grateful to F. Delisle and J. Gibouin for their expert contribution in handling animals. N.D. is a recipient of a fellowship from the Ministère de la Recherche et de l'Enseignement (MRE) and from the Association pour la Recherche contre le Cancer (ARC). This work was supported by grants from the MRE, the Comité départemental de Paris de la Ligue contre le Cancer and the ARC.

References

Barka, T. and Van der Noen, H.M. (1996). Retrovirus-mediated gene transfer into salivary glands in vivo. Hum Gene Ther. **7**, 613-618.

Bass, C., Cabrera, G., Elgavish, A., Robert, B., Siegal, G.P., Anderson, S.C., Maneval, D.C. and Curiel, D.T. (1995). Recombinant adenovirus-mediated gene transfer to genitourinary epithelium *in vitro* and *in vivo*. Cancer Gene Ther. **2**, 97-104.

Bubenik, J., Baresova, M., Viklicky, V., Jakoubkova, J., Sainerova, H. and Donner, J. (1973). Established cell line of urinary bladder carcinoma (T24) containing tumour-specific antigen. Int J Cancer. **11**, 765-773.

Cardoso, J.E., Branchereau, S., Jeyaraj, P.R., Houssin, D., Danos, O. and Heard, J.M. (1993). In situ retrovirus-mediated gene transfer into dog liver. Hum Gene Ther. **4**, 411-418.

Chuck, A.S., Clarke, M.F. and Palsson, B.O. (1996). Retroviral infection is limited by Brownian motion. Hum Gene Ther. **7**, 1527-1534.

Culver, K.W., Morgan, R.A., Osborne, W.R., Lee, R.T., Lenschow, D., Able, C., Cornetta, K., Anderson, W.F. and Blaese, R.M. (1990). *In vivo* expression and survival of gene-modified T lymphocytes in rhesus monkeys. Hum Gene Ther. **1**, 399-410.

Culver, K.W., Ram, Z., Wallbridge, S., Ishii, H., Oldfield, E.H. and Blaese, R.M. (1992). *In vivo* gene transfer with retroviral vector-producer cells for treatment of experimental brain tumors [see comments]. Science. **256**, 1550-1552.

Danos, O. and Mulligan, R.C. (1988). Safe and efficient generation of recombinant retroviruses with amphotropic and ecotropic host ranges. Proc Natl Acad Sci U S A. **85**, 6460-6464.

Dilloo, D., Bacon, K., Holden, W., Zhong, W., Burdach, S., Zlotnik, A. and Brenner, M. (1996). Combined chemokine and cytokine gene transfer enhances antitumor immunity. Nat Med. **2**, 1090-1095.

Dougherty, G.J., Chaplin, D., Dougherty, S.T., Chiu, R.K. and McBride, W.H. (1996). *In vivo* gene therapy of cancer. tumor target. **2**, 106-115.

Dumey N, Masset M, Devauchelle P, Cussenot O, Cotard JP, Le Duc A, Marty M and Cohen-Haguneuaer O. *In vivo* retroviral-mediated gene transfer into bladder urothelium results in preferential transduction of tumour cells. *submitted*

Hastings, R.J. and Franks, L.M. (1981). Chromosome pattern, growth in agar and tumorigenicity in nude mice of four human bladder carcinoma cell lines. Int J Cancer. **27**, 15-21.

Heney, N.M. (1992). Natural history of superficial bladder cancer. Prognostic features and long-term disease course. Urol Clin North Am. **19**, 429-433.

Herr, H.W. (1989a). BCG therapy of superficial bladder tumors. Prog Clin Biol Res. **303**, 369-374.

Herr, H.W. (1991). Transurethral resection and intravesical therapy of superficial bladder tumors. Urol Clin North Am. **18**, 525-528.

Herr, H.W., Badalament, R.A., Amato, D.A., Laudone, V.P., Fair, W.R. and Whitmore, W.F., Jr. (1989b). Superficial bladder cancer treated with bacillus Calmette-Guerin: a multivariate analysis of factors affecting tumor progression. J Urol. **141**, 22-29.

Hurford, R.K., Jr., Dranoff, G., Mulligan, R.C. and Tepper, R.I. (1995). Gene therapy of metastatic cancer by *in vivo* retroviral gene targeting. Nat Genet. **10**, 430-435.

Kimura, H., Sakamoto, T., Cardillo, J.A., Spee, C., Hinton, D.R., Gordon, E.M., Anderson, W.F. and Ryan, S.J. (1996). Retrovirus-medicated suicide gene transduction in the vitreous cavity of the eye: Feasibility in prevention of proliferate vitreoretinopathy. Hum Gene Ther. **7**, 799-808.

Kotani, H., Newton, P.B.r., Zhang, S., Chiang, Y.L., Otto, E., Weaver, L., Blaese, R.M., Anderson, W.F. and McGarrity, G.J. (1994). Improved methods of retroviral vector transduction and production for gene therapy. Hum Gene Ther. **5**, 19-28.

Layne, S.P., Spouge, J.L. and Dembo, M. (1989). Quantifying the infectivity of human immunodeficiency virus. Proc Natl Acad Sci U S A. **86**, 4644-4648.

Lee, S.S., Eisenlohr, L.C., McCue, P.A., Mastrangelo, M.J. and Lattime, E.C. (1994). Intravesical gene therapy: *in vivo* gene transfer using recombinant vaccinia virus vectors. Cancer Res. **54**, 3325-3328.

Levin, J.G. and Rosenak, M.J. (1976). Synthesis of murine leukemia virus proteins associated with virions assembled in actinomycin D-treated cells: evidence for persistence of viral messenger RNA. Proc Natl Acad Sci U S A. **73**, 1154-1158.

Mickel, F.S., Weidenbach, F., Swarovsky, B., LaForge, K.S. and Scheele, G.A. (1989). Structure of the canine pancreatic lipase gene. J Biol Chem. **264**, 12895-12901.

Miller, A.D. (1990). Retrovirus packaging cells. Hum Gene Ther. **1**, 5-14.

Miller, A.D. and Rosman, G.J. (1989). Improved retroviral vectors for gene transfer and expression. Biotechniques. **7**, 980-982, 984-986, 989-990.

Miller, D.G., Adam, M.A. and Miller, A.D. (1990). Gene transfer by retrovirus vectors occurs only in cells that are actively replicating at the time of infection [published erratum appears in Mol Cell Biol 1992 Jan;12(1):433]. Mol Cell Biol. **10**, 4239-4242.

Morling, F.J. and Russell, S.J. (1995). Enhanced transduction efficiency of retroviral vectors coprecipitated with calcium phosphate. Gene Ther. **2**, 504-508.

Morris, B.D., Jr., Drazan, K.E., Csete, M.E., Werthman, P.E., Van Bree, M.P., Rosenthal, J.T. and Shaked, A. (1994). Adenoviral-mediated gene transfer to bladder *in vivo*. J Urol. **152**, 506-509.

Naviaux, R.K. and Verma, I.M. (1992). Retroviral vectors for persistent expression *in vivo*. Curr Opin Biotechnol. **3**, 540-547.

Paul, R.W., Morris, D., Hess, B.W., Dunn, J. and Overell, R.W. (1993). Increased viral titer through concentration of viral harvests from retroviral packaging lines. Hum Gene Ther. **4**, 609-615.

Russell, D.W., Berger, M.S. and Miller, A.D. (1995). The effects of human serum and cerebrospinal fluid on retroviral vectors and packaging cell lines. Hum Gene Ther. **6**, 635-641.

Salvetti, A., Moullier, P., Cornet, V., Brooks, D., Hopwood, J.J., Danos, O. and Heard, J.M. (1995). *In vivo* delivery of human alpha-L-iduronidase in mice implanted with neo-organs. Hum Gene Ther. **6**, 1153-1159.

Sanes, J.R., Rubenstein, J.L. and Nicolas, J.F. (1986). Use of a recombinant retrovirus to study post-implantation cell lineage in mouse embryos. Embo J. **5**, 3133-3142.

Sorge, J., Wright, D., Erdman, V.D. and Cutting, A.E. (1984). Amphotropic retrovirus vector system for human cell gene transfer. Mol Cell Biol. **4**, 1730-1737.

Tolley, D.A., Parmar, M.K., Grigor, K.M., Lallemand, G., Benyon, L.L., Fellows, J., Freedman, L.S., Grigor, K.M., Hall, R.R., Hargreave, T.B., Munson, K., Newling, D.W., Richards, B., Robinson, M.R., Rose, M.B., Smith, P.H., Williams, J.L. and Whelan, P. (1996). The effect of intravesical mitomycin C on recurrence of newly diagnosed superficial bladder cancer: a further report with 7 years of follow up. J Urol. **155**, 1233-1238.

Williams, R.D. (1980). Human urologic cancer cell lines. Invest Urol. **17**, 359-363.

Zier, K.S. and Gansbacher, B. (1995). The impact of gene therapy on T cell function in cancer. Hum Gene Ther. **6**, 1259-1264.

A Gene Therapy Approach for the Treatment of ALS

Patrick Aebischer and Diego Braguglia
Gene Therapy Center and Division of Surgical Research
CHUV, Lausanne University Medical School
1011 Lausanne, Switzerland
e-mail: Patrick.Aebischer@chuv.hospvd.ch

1 Abstract

Amyotrophic lateral sclerosis is a fatal neurodegenerative disease leading to paralysis and death within 3 to 5 years. The mechanisms and causes of the disease remain unknown and to date no efficient therapeutic approaches are able to significantly alter the course of the disease. Two forms of ALS, a sporadic form accounting for 90% of the cases and a familial form for 10% of the ALS individuals have been identified. Approximately 20% of the familial cases correlates with a mutation in the SOD1 gene.

Neurotrophic factors hold promise for the treatment of neurodegenerative diseases. Neuroprotective effects of trophic factors in the central nervous system (CNS) and in various animal models of neurodegeneration has led to the development of strategies for the treatment of neurodegenerative disorders. The presence of the blood brain barrier remains, however, a major concern for the delivery of neurotrophic factors in the CNS. A technique involving the intrathecal implantation of polymer encapsulated cell-lines genetically engineered to release neurotrophic factors provides a means to continuously deliver neurotrophic factors directly within the CNS, avoiding the numerous side effects observed following their systemic administration.

2 Neurotrophic factors, encapsulated cells and ALS

Among the motor neuron diseases (MND), amyotrophic lateral sclerosis (ALS) is the most common. ALS is age-related and typically appears in the middle of adult life, afflicts the motoneuron system and leads to paralysis and death within 3 to 5 years (Williams and Windebank, 1991). Neuropathologically, ALS is characterized by the degeneration and loss of the large anterior horn cells of the spinal cord and lower cranial motor nuclei of the brainstem, atrophy of striated muscles as consequence of denervation, and degeneration of the upper motoneurons (i.e. Beltz cells of the motor cortex). At the cellular level, ALS induced alterations include a reduction in synaptophysin expression in anterior horns, which parallels the severity of the neuronal loss (Kato et al., 1987), the accumulation of neurofilaments in the proximal segment of anterior hom cells axon (considered as one of the early cytoplasmic change in ALS) (Carpenter, 1968; Sobue et al., 1990) and the presence of ubiquitin-labelled skein-like structures in anterior horn cells (Carpenter, 1968; Leigh et al., 1988; Sobue et al., 1990; Leigh et al., 1991; Suenaga et al., 1993). In CNS cells, the Golgi apparatus plays a key role in the fast axonal transport. ALS patients displayed a fragmented Golgi apparatus in 30% of their motoneurons (Mourelatos et al., 1990; Gonatas et al., 1992; Mourelatos et al., 1994).

NATO ASI Series, Vol. H 105
Gene Therapy
Edited by Kleanthis G. Xanthopoulos

2.1 Causes of ALS and other motor neuron diseases

Three different aspects of ALS must be considered: the mechanism leading to cell death, the selective and time-delayed motoneuron degeneration. Several hypotheses have been proposed. The first one relies on an atypical polioviral infection, potentially explaining motoneuron selectivity. This is unlikely since no polioviruses have been detected in ALS cerebrospinal fluids and brain samples. A second hypothesis includes a role of metallotoxins in motoneuron pathology. To date no convincing results support this hypothesis. The observation that many patients suffering from ALS possess antibodies directed against their own distal motor neuron termini suggested the possibility of ALS being an autoimmune disease. However an autoimmune reaction against voltage-dependent Ca^{2+} channels could not be demonstrated as the cause of ALS. These patients also do not benefit from an immunosuppressive therapy (Brown, 1995). A more likely hypothesis points out the toxicity caused by neurotransmitters such as glutamate, which level is abnormally elevated in sporadic ALS patients and which transport is impaired in synaptosomal preparation from the ALS brain (Rothstein et al., 1990; 1992). In the recent past, genetic investigation has identified two proteins that are defective in some ALS patients: the superoxide dismutase (SOD) and neurofilaments heavy chains (Brown, 1995).

The current belief is that motoneuron degeneration and death observed in ALS result from a complex interaction between oxidative stress, excitotoxic stimulation and altered function of mitochondria and neurofilaments.

2.1.1. The excitotoxic hypothesis

Neurotransmitter mediated toxicity may play an important role in the pathogenesis of ALS. Normally glutamate is released in a Ca^{2+} dependent process in depolarized presynaptic termini. Glutamate can activate specific motoneurons receptors [NMDA (N-methyl-D-aspartate) and non-NMDA], which control ion entry. The glutamate is then cleared from the synaptic cleft and extracellular environment by specific transporters, localized to surrounding astrocytes and neuronal elements. Since elevated glutamate levels can be neurotoxic, an efficient maintenance of low extracellular glutamate concentration is necessary. Several lines of evidence suggest that an inefficient glutamate transport leads to an excessive accumulation of this neurotransmitter and thus to neurotoxicity. Oligo antisense knock out of the astroglial specific GLAST and GLT-1 glutamate transporters suggest key roles for these transporters in maintaining a low extracellular glutamate concentration (Rothstein, 1996). In ALS, a defective glutamate transport is implicated with the observed chronic loss of motoneurons (Rothstein et al., 1992; Shaw et al., 1994).

Glutamate exerts its toxic activity in two ways: via a series of intracellular cascades, activated by an excess calcium influx, which in turn leads to an activation of oxidant-generating enzymatic pathways, including the formation of nitric oxide, peroxynitrite, hydroxyl radicals and superoxide anions (Choi, 1988). The glutamate excitotoxicity can also be exerted by the so called weak excitotoxic pathway (Albin and Greenamyre, 1992) with normal concentration of glutamate becoming neurotoxic on metabolically compromised neurons.

2.1.2 Superoxide dismutase and ALS

Superoxide dismutase (SOD) dysfunction has been proposed to be linked to motoneuron degeneration observed in ALS. Although sporadic and familial forms of ALS are clinically indistinguishable, the average onset and the duration of the disease vary considerably among these two groups (Tandan and Bradley, 1985; Siddique, 1991). Mutations in the SOD1 gene (Deng et al., 1993; Rosen et al., 1993), located on the chromosome 21 (21q21) (Siddique et al., 1991), have been found in about 20% of familial ALS individuals. The genetic linkage for the other 80% cases has not yet been established.

In humans, three isoforms (encoded by different genes) of superoxide dismutase protein have been identified: SOD1, also known as Cu-Zn-SOD1 or cytosolic SOD, SOD2 mitochondrial or Mn-SOD, and SOD3, or extracellular SOD (Ec-SOD) (Groner et al., 1986). The SOD1 isoform is ubiquitously expressed and is particularly abundant in neurons (Pardo et al., 1995). Beside its main enzymatic activity, superoxide dismutase displays also a marginal peroxidase activity (Hodgson and Fridovich, 1975; Symonyan and Nalbandyan, 1972). Hydrogen peroxide, the product of the dismutase activity, acts as an inhibitor of SOD1 peroxidase activity (Symonyan and Nalbandyan, 1972).

A transgenic mice model for FALS

Transgenic mice overexpressing the mutated forms of human SOD1 represent the most relevant animal model for FALS (Gurney et al., 1994; Ripps et al., 1995; Wong et al., 1995). These mice develop ALS related clinical symptoms, like weakness and paralysis and die at variable ages depending on the transgenic mice line. For one mutation, Gly93Ala the onset and age of death seems to be dependent on the number of copies integrated in the mouse genome (Dal Canto and Gurney, 1995). Mice overexpressing human wild-type SOD1 do not show any clinical symptoms for at least 2 years. Pathological and histological studies confirmed that motoneurons degeneration is the cause of the paralysis observed in SOD1 transgenic mice (Dal Canto and Gurney, 1995).

2.1.3 ALS and neurofilaments

One of the common hallmark in several motoneuron degenerative diseases, including ALS is the aberrant accumulation of neurofilaments (NF) in the affected motoneurons. The hypothesis of potential involvements of NF in these diseases received credit and experimental support from transgenic mice overexpressing mice NF-L or human NF-H (Cote et al., 1993; Xu et al., 1993). These animals develop loss of kinetic activity, muscular atrophy and paralysis. These alterations are similar to those observed in sporadic and familial ALS (Hirano et al., 1984a; 1984b, 1988). However, the NF-L and NF-H mice models are not a very satisfying model for ALS since paralysis arising from neuronal failure and derived muscular atrophy is not associated with a significant motoneuron death even at the terminal stage of the disease. By expressing mutated form of NF in mice, a more suggestive model for ALS-like diseases was created. These animals developed early abnormal gait and weakness in both upper and lower limbs. Similarly to ALS, large neurofilaments-rich axons are lost, while smaller axons with less neurofilaments are spared (Kawamura et al., 1981).

2.2 Rationale for using neurotrophic factors in ALS

Several lines of evidence, including transgenic mice and *in vitro* and *in vivo* experiments, suggest that neurotrophic factors (NTF) play a critical role in supporting survival and differentiation of various neuronal populations. Motoneurons respond to a variety of NTF's, namely CNTF, LIF, IGF-1, BDNF, NT3, NT4/5, GDNF and Neurturin. Transgenic animals with targeted disruption of NTF genes and their receptors (Emfors et al., 1994a;

1994b; Frim et al., 1994; Jones et al., 1994; Klein et al., 1993; Masu et al., 1993; Smeyne et al., 1994; Tojo et al., 1995) have demonstrated and emphasized the NTF's physiological role in the development and survival of motoneurons. *In vitro,* survival of rat embryonic motoneurons is increased by a variety of NTF, including CNTF, BDNF, NT3, NT4/5 and GDNF. In addition these factors have further been reported *in vivo* to reduce axotomy induced motoneurons death (Magal et al., 1993; Sendtner et al., 1990; Wong et al., 1993; Zurn et al., 1994; Zurn and Werren, 1994). Animal models for neurodegenerative diseases also provided additional evidence on the central role of NTF in the biology of motoneurons (Ikeda et al., 1995a; 1995b; Mitsumoto et al., 1994; Sagot et al., 1995; Sendtner et al., 1992).

3 Encapsulation technology

The presence of the blood brain barrier represents a major hurdle for the delivery of neurotrophic factors to the CNS. A technique involving the intratechal implantation of polymer encapsulated cell-lines genetically engineered to release neurotrophic factors provides a means to continuously deliver directly within the CNS.

Several relevant advantages of this technology are the characterization of the used clonal cell-lines for parameters including neurotrophic factor secretion and stability, *in vitro* and *in vivo* cell-survival, cell banking and the absence of adventitious agents (including bacteria, yeasts, mycoplasma, viruses, retroviruses). The chosen cell line has to be exempt from transforming (tumorigenicity) activities and toxicity, as assessed on small (mice and rats) and large (monkeys and sheep) animals. The availability of both master cell bank (MCB) and working cell bank (WCB) will guarantee homogeneity and reproducibility in the treatment, ensuring that all patients will be implanted with an identical clonal cell line.

In vitro and *in vivo* experiments strongly suggest that administration of a combination of factors have additive and in some combinations, synergistic effects on the survival of motoneurons. A second generation of genetically engineered cell-lines allows the delivery of a combination of neurotrophic factors. The third generation of engineered cell-lines is under investigation and would provide *in vivo* modulation of gene expression. This system is based on transcriptional regulation driven by small, stable molecules (tetracycline and its derivatives), which diffuse across the blood brain barrier (Gossen et al., 1995). This system has already been reported in the generation of conditionally immortalized cell lines (Ewald et al., 1996) and in the control of the level of expression of Epo (Bohl et al., 1997).

The encapsulation technique allows the transplantation of xenogeneic cells in the absence of a pharmacological host immunosuppression. The immunoisolation is obtained by surrounding the cells by a selective semipermeable membrane barrier with a controlled pore size, allowing the inward diffusion of nutrients and the outward diffusion of cell secreted bioactive factor(s) (Chick et al., 1975; 1977). The semipermeable membrane serves as a critical immunoisolation barrier. Cells from the host immune system, immunoglobulins, as well as complement factors are excluded from the graft. The semisolid gel or matrix in which the cells are embedded allowing the encapsulation of dissociated cells in a favorable biochemical and mechanical environment dramatically influence cell viability, i.e. by inhibiting cell clustering or gravimetric settling (Zielinski and Aebischer, 1994). Identification of the best matrix, membrane composition and membrane filling parameters depends widely on cell phenotype and growth characteristics and has to be adapted empirically for each cell line.

Transplantation of encapsulated cells provides a powerful approach for the continuous and local delivery of recombinant proteins. A shut off mechanism is easily obtained by the retrieval of the device at the first appearance of adverse side effects and a safety feedback control is incorporated to ensure selected rejection of the implanted cells in the unlikely event of capsule breakage. The herpes simplex virus tymidine kinase (HSV-TK) gene which catabolises ganciclovir into a cell toxic chemical only in cells expressing TK provides a molecular mechanism to eliminate the transplanted cells. The high immunogenicity of xenogeneic cells becomes an advantage; unencapsulated cross-species cell lines are efficiently rejected even in immunoprivileged sites (Aebischer and Lysaght, 1995).

The encapsulated cell technology confers significant advantages in the treatment of neurological diseases. The presence of the highly selective permeable blood brain barrier (BBB) makes the delivery of systemically administered trophic factors to the CNS a challenging task. Intravenous administration of neurotrophic factors, even at low amounts, leads often to undesirable side effects due to the widespread distribution of their receptors outside the CNS. In addition neurotrophic factors are quite unstable and their bioactivity decrease rapidly (Tan et al., 1996). Due to the impermeability of the blood brain barrier, the only mechanism for a systemically delivered factor to reach the CNS is by retrograde transport from peripheral fibers.

A continuous delivery of neurotrophic factors by means of the immunoisolation technology was applied to various animal model for motoneurons degeneration including the axotomy-induced cell death and the pmn-pmn mice. In addition, neurotrophic factors were successfully delivered by the same technology in the unilateral medial forebrain bundle (MFB) axotomy (Tseng et al., 1997), a model for Parkinson's disease, in a quinolinic acid (QA) rodent and primate models for Huntington's disease (Emerich et al., 1996; 1997) and to the lateral ventricle of fimbria-fornix-lesioned rats, a model harboring homologies to the Alzheimer's disease (Hoffman et al., 1993; Hoffman et al., 1990).

Clinical trial using encapsulated genetically modified cells

A phase I clinical trial based on the enrollment of 12 ALS patients has demonstrated the feasibility and the safety of the encapsulated cell technology for the intrathecal constant delivery of hCNTF (Aebischer et al., 1996). Patients were implanted with polymer capsules containing genetically engineered baby hamster kidney (BHK) cells to produce hCNTF (between 0.5 and I µg/day). Nanogram levels of CNTF were detected in patients' cerebrospinal fluid (CSF) up to 60 weeks post-transplantation, whereas no CNTF was detectable prior to implantation. The intrathecal delivery of CNTF was not associated with any of the limiting adverse side effects. All the implanted devices were intact and secreted CNTF at the explant. No evidence of tissue or cell adherence to the external surface was observed confirming the excellent biocompatibility of the implant. The small number of patients and the short observation periods do not allow for the moment assessment of a potential slowing of the neurodegenerative process. Diffusion studies indicate that the bioactive secreted factor can diffuse within the CSF. Simultaneous CSF lumbar and cervical levels of CNTF were measured in one patient. The cervical concentration was 14.6% of the lumbar level of CNTF, demonstrating that all the spinal axis is exposed to the trophic factor after lumbar placement of the implant (Aebischer et al., 1996).

In vitro and *in vivo* studies suggest that the combination of factors may have an additive or in some cases even a synergistic effect (Zurn et al., 1996). The possibility of mixing two different clonal cell lines secreting each a factor may encounter difficulties due to the unavoidable difference in term of growth between the two populations. The approach undertaken to release two factors can be achieved by different methods: the first straight forward possibility is the implantation of two devices, one for each factor. The second deals with the generation of genetically modified cell lines able to produce a combination of factors. The effect of the combined factors administration was tested on purified spinal cord cultures (Zurn et al., 1996). In all the combinations so far tested (CNTF-GDNF, CNTF-NT4/5, CT1-NT4/5), a marked synergistic effect on ChAT activity in purified spinal cord cultures was demonstrated. These results suggest that an enhanced effect on motoneurons survival and rescue after injury could be achieved by the application of a combination of factors.

4 Perspectives

The encapsulation of xenogeneic modified cell lines has the potential of overcoming many of the problems associated with systemic delivery and local administration of recombinant proteins. This technique minimizes the need for repeated injections and the effects of poor protein stability. Although encapsulated cells were placed in the intrathecal region (Aebischer et al., 1996), their clinical applications may not be restricted to the CNS. Encapsulated cell therapy may constitute a relevant advantage in diseases (i.e. diabetes, anemia, obesity) requiring a constant continuous delivery of pharmacological levels of bioactive recombinant proteins. Future generation of implants may allow the release of recombinant proteins (i.e. insulin and erytropoietin) upon response to physiological stimuli, i.e. glucose concentration for insulin and hypoxia for erythropoietin. In addition, the availability of a shut-off mechanism represents a relevant advantage, allowing the interruption of the recombinant factor release by retrieval of the devices at the first apparition of side effects.

Further generation of genetically engineered cell lines may allow the simultaneous administration of combination of growth factors. In addition, the artificial *in vivo* modulation of transgene expression, based on the diffusion of small molecules, may allow the identification of the ideal therapeutic window for each factor.

5 References

Aebischer, P., Schluep, M., Deglon, N., Joseph, J. M., Hirt, L., Heyd, B., Goddard, M., Hammang, J. P., Zurn, A. D., Kato, A. C., Regli, F., and Baetge, E. E. (1996). Intrathecal delivery of CNTF using encapsulated genetically modified xenogeneic cells in amyotrophic lateral sclerosis patients. Nature Medicine 2, 696-699.

Aebischer, P., and Lysaght, M.J. (1995). Immunoisolation and cellular xenotransplantation. Xeno 3, 43-48.

Albin, R. L., and Greenamyre, J. T. (1992). Alternative excitotoxic hypotheses. Neurology 42, 733-738.

Bohl, D., Naffakh, N., and Heard, J. M. 1997. Long-term control of erytropoietin secretion by doxycycline in mice transplanted with engineered primary myoblasts. Nature Medicine 3, 299-305.

Brown, R. H., Jr. (1995). Amyotrophic lateral sclerosis: recent insights from genetics and transgenic mice. Cell 80, 687-692.

Carpenter, S. (1968). Proximal axonal enlargement in motor neuron disease. Neurology 18, 841-851.

Chick, W. L., Like, A. A., and Lauris, V. (1975). Beta cell culture on synthetic capillaries: an artificial endocrine pancreas. Science 187, 847-849.

Chick, W. L., Perna, J. J., Lauris, V., Low, D., Galletti, P. M., Panol, G., Whittemore, A. D., Like, A. A., Colton, C. K., and Lysaght, M. J. (1977). Artificial pancreas using living beta cells: effects on glucose homeostasis in diabetic rats. Science 197, 780-782.

Choi, D. W. (1988). Glutamate neurotoxicity and diseases of the nervous system. Neuron 1, 623-634.

Cote, F., Collard, J. F., and Julien, J. P. (1993). Progressive neuronopathy in transgenic mice expressing the human neurofilament heavy gene: a mouse model of amyotropic lateral sclerosis. Cell 73, 35-46

Dal Canto, M. C., and Gurney, M. E. (1995). Neuropathological changes in two lines of mice carrying a transgene for mutant human Cu,Zn SOD, and in mice overexpressing wild type human SOD: a model of familiar amyotrophic lateral sclerosis (FALS). Brain Research 676, 2540.

Deng, H. X., Hentati, A., Tainer, J. A., lqbal, Z., Cayabyab, A., Hung, W. Y., Getzoff, E. D., Hu, P., Herzfeldt, B., Roos, R. P., and et al. (1993). Amyotrophic lateral sclerosis and structural defects in Cu,Zn superoxide dismutase. Science 261, 1047-1051.

Emerich, D.F., Linder, M.D., Winn, S.R., Chen, E.Y., Frydel, B.R., and Kordower, J. H. (1996). Implants of encapsulated human CNTF-producing fibroblasts prevent behavioral deficits and striatal degeneration in a rodent model for Huntigton's disease. Journal of neuroscience 16, 5168-5181.
Emerich, D.F., Winn, S. R., Hantraye, P. M., Peschanski, M., Chen, E.Y., Chu, Y., McDermott, P., Baetge, E. E., and Kordower, J. H. (1996). Protective effect of encapsulated cells producing neurotrophic factor CNTF in a monkey model of Huntinton's disease. Nature 386, 395-399.
Emfors, P., Lee, K. F., and Jaenisch, R. (1994a). Mice lacking brain-derived neurotrophic factor develop with sensory deficits. Nature 368, 147-150.
Ernfors, P., Lee, K. F., Kucera, J., and Jaenisch, R. (1994b). Lack of neurotrophin-3 leads to deficiencies in the peripheral nervous system and loss of limb proprioceptive afferents. Cell 77, 503-512.
Ewald, E., Li, M., Efrat, S., Auer, G., Wall, R. J., Furth, P.A., and Hennighausen, L. (1996). Time-sensitive reversal of hyperplasia in transgenic mice expressing SV40 T Antigen. Science 273, 1384-1386.
Frim, D. M., Uhler, T. A., Galpern, W. R., Beal, M. F., Breakefield, X. O., and Isacson, 0. (1994). Implanted fibroblasts genetically engineered to produce brain-derived neurotrophic factor prevent 1-methyl-4-phenylpyridinium toxicity to dopaminergic neurons in the rat. Proceedings of the National Academy of Sciences of the United States of America 91, 5104-5108.
Gonatas, N. K., Stieber, A., Mourelatos, Z., Chen, Y., Gonatas, J. O., Appel, S. H., Hays, A. P., Hickey, W. F., and Hauw, J. J. (1992). Fragmentation of the Golgi apparatus of motoneurons in amyotrophic lateral sclerosis. American Journal of Pathology 140, 731-737.
Goner, Y., Geiman-Hurvitz, J., Dafri, N., et al. The human Cu/Zn superoxide dismutase gene family: architecture and expression of the chromosome 21-ecoded functional gene and its processed peudogenes. In: Rotilis, G., ed. Superoxide and superoxide dismutase in chemistry, biology and medicine. Amsterdam: Elsevier Science Publisher, Biochemical Division, 1986, 247-255.
Gossen, M., Freundlieb, S., Bender, G., Müller, G., Hillen, W., and Bujard H. (1995). Transcriptional activation by tetracyclines in mammalian cells. Science 268, 1766-1769.
Gurney, M. E., Pu, H., Chiu, A. Y., Dal Canto, M. C., Polchow, C. Y., Alexander, D. D., Caliendo, J., Hentati, A., Kwon, Y. W., Deng, H. X., and et al. (1994). Motor neuron degeneration in mice that express a human Cu,Zn superoxide dismutase. Science 264, 17721725.
Hirano, A., Donnenfeld, H., Sasaki, S., and Nakano, I. (1984a). Fine structural observations of neurofilamentous changes in amyotrophic lateral sclerosis. Journal of Neuropathology & Experimental Neurology 43, 461-470.
Hirano, A., Nakano, I., Kurland, L. T., Mulder, D. W., Holley, P. W., and Saccomanno, G. (1984b). Fine structural study of neurofibrillary changes in a family with amyotrophic lateral sclerosis. Journal of Neuropathology & Experimental Neurology 43, 471-480.
Hirano, A. Color atlas of pathology of the nervous system. 2nd ed. New York: Igakushoin 1988, 99.
Hodgson, E. K., and Fridovich, 1. (1975). The interaction of bovine erythrocyte superoxide dismutase with hydrogen peroxide: inactivation of the enzyme. Biochemistry 14, 5294-5299.
Hoffman, D., Wahlberg, L., and Aebischer, P. (1990). NGF released from a polymer matrix prevents loss of ChAT expression in basal forebrain neurons following a fimbria-fornix lesion. Experimental Neurology 110, 39-44.
Hoffman, D., Breakefield, X. O., Short, M. P., and Aebischer, P. (1993). Trans-plantation of a polymer-encapsulated cell line genetically engineered to release NGF. Experimental Neurology 122, 100-106.
Ikeda, K., Klinkosz, B., Greene, T., Cedarbaum, J. M., Wong, V., Lindsay, R. M., and Mitsumoto, H. (1995a). Effects of brain-derived neurotrophic factor on motor dysfunction in wobbler mouse motor neuron disease. Annals of Neurology 37, 505-511.
Ikeda, K., Wong, V., Holmlund, T. H., Greene, T., Cedarbaum, J. M., Lindsay, R. M., and Mitsumoto, H. (1995b). Histometric effects of ciliary neurotrophic factor in wobbler mouse motor neuron disease. Annals of Neurology 37, 47-54.
Jones, K. R., Farinas, I., Backus, C., and Reichardt, L. F. (1994). Targeted disruption of the BDNF gene perturbs brain and sensory neuron development but not motor neuron development. Cell 76, 989-999.
Kato, T., Hirano, A., and Donnenfeld, H. (1987). A Golgi study of the large anterior horn cells of the lumbar cords in normal spinal cords and in amyotrophic lateral sclerosis. Acta Neuropathologica 75, 34-40.
Kawamura, Y., Dyck, P. J., Shimono, M., Okazaki, H., Tateishi, J., and Doi, H. (1981). Morphometric comparison of the vulnerability of peripheral motor and sensory neurons in amyotrophic lateral sclerosis. Journal of Neuropathology & Experimental Neurology 40, 667-675.
Klein, R., Smeyne, R. J., Wurst, W., Long, L. K., Auerbach, B. A., Joyner, A. L., and Barbacid, M. (1993). Targeted disruption of the trkB neurotrophin receptor gene results in nervous system lesions and neonatal death. Cell 75, 113-122.
Leigh, P. N., Anderton, B. H., Dodson, A., Gallo, J. M., Swash, M., and Power, D. M. (1988). Ubiquitin deposits in anterior horn cells in motor neurone disease. Neuroscience Letters 93, 197-203.
Leigh, P. N., Whitwell, H., Garofalo, O., Buller, J., Swash, M., Martin, J. E., Gallo, J. M., Weller, R. O., and Anderton, B. H. (1991). Ubiquitin-immunoreactive intraneuronal inclusions in amyotrophic lateral sclerosis. Morphology, distribution, and specificity. Brain 114, 775-788.
Magal, E., Louis, J. C., Oudega, M., and Varon, S. (1993). CNTF promotes the survival of neonatal rat corticospinal neurons *in vitro*. Neuroreport 4, 779-782.
Masu, Y., Wolf, E., Holtmann, B., Sendtner, M., Brem, G., and Thoenen, H. (1993). Disruption of the CNTF gene results in motor neuron degeneration. Nature 365, 27-32.
Mitsumoto, H., Ikeda, K., Klinkosz, B., Cedarbaum, J. M., Wong, V., and Lindsay, R. M. (1994). Arrest of motor neuron disease in wobbler mice cotreated with CNTF and BDNF. Science 265, 1107-1110.

Mourelatos, Z., Adler, H., Hirano, A., Donnenfeld, H., Gonatas, J. O., and Gonatas, N. K. (1990). Fragmentation of the Golgi apparatus of motoneurons in amyotrophic lateral sclerosis revealed by organelle-specific antibodies. Proceedings of the National Academy of Sciences of the United States of America 87, 4393-4395.

Mourelatos, Z., Hirano, A., Rosenquist, A. C., and Gonatas, N. K. (1994). Fragmentation of the Golgi apparatus of motoneurons in amyotrophic lateral sclerosis (ALS). Clinical studies in ALS of Guam and experimental studies in deafferented neurons and in beta,beta'-iminodipropionitrile axonopathy. American Journal of Pathology 144, 1288-1300.

Pardo, C. A., Xu, Z., Borchelt, D. R., Price, D. L., Sisodia, S. S., and Cleveland, D. W. (1995). Superoxide dismutase is an abundant component in cell bodies, dendrites, and axons of motoneurons and in a subset of other neurons. Proceedings of the National Academy of Sciences of the United States of America 92, 954-958.

Ripps, M. E., Huntley, G. W., Hof, P. R., Morrison, J. H., and Gordon, J. W. (1995). Transgenic mice expressing an altered murine superoxide dismutase gene provide an animal model of amyotrophic lateral sclerosis. Proceedings of the National Academy of Sciences of the United States of America 92, 689-693.

Rosen, D. R., Siddique, T., Patterson, D., Figlewicz, D. A., Sapp, P., Hentati, A., Donaldson, D., Goto, J., O'Regan, J. P., Deng, H. X., and et al. (1993). Mutations in Cu/Zn superoxide dismutase gene are associated with familial amyotrophic lateral sclerosis. Nature 362, 59-62.

Rothstein, J. D. (1996). Excitotoxicity hypothesis. Neurology 47, 19-25.

Rothstein, J. D., Martin, L. J., and Kuncl, R. W. (1992). Decreased glutamate transport by the brain and spinal cord in amyotrophic lateral sclerosis. New England Journal of Medicine 326, 1464-1468.

Rothstein, J. D., Tsai, G., Kuncl, R. W., Clawson, L., Cornblath, D. R., Drachman, D. B., Pestronk, A., Stauch, B. L., and Coyle, J. T. (1990). Abnormal excitatory amino acid metabolism in amyotrophic lateral sclerosis. Annals of Neurology 28, 18-25.

Sagot, Y., Tan, S. A., Baetge, E., Schmalbruch, H., Kato, A. C., and Aebischer, P. (1995). Polymer encapsulated cell lines genetically engineered to release ciliary neurotrophic factor can slow down progressive motor neuronopathy in the mouse. European Journal of Neuroscience 7, 1313-1322.

Sendtner, M., Carroll, P., Holtmann, B., Hughes, R. A., and Thoenen, H. (1994). Ciliary neurotrophic factor. Journal of Neurobiology 25, 1436-1453.

Sendtner, M., Holtmann, B., Kolbeck, R., Thoenen, H., and Barde, Y. A. (1992). Brain-derived neurotrophic factor prevents the death of motoneurons in newborn rats after nerve section. Nature 360, 757-759.

Sendtner, M., Kreutzberg, G. W., and Thoenen, H. (1990). Ciliary neurotrophic factor prevents the degeneration of motoneurons after axotomy. Nature 345, 440-441.

Sendtner, M., Schmalbruch, H., Stockli, K. A., Carroll, P., Kreutzberg, G. W., and Thoenen, H. (1992). Ciliary neurotrophic factor prevents degeneration of motoneurons in mouse mutant progressive motor neuronopathy. Nature 358, 502-504.

Shaw, P. J., Chinnery, R. M., and Ince, P. G. (1994). [3H]D-aspartate binding sites in the normal human spinal cord and changes in motor neuron disease: a quantitative autoradiographic study. Brain Research 655, 195-201.

Siddique, T. (1991). Molecular genetics of familial amyotrophic lateral sclerosis. Advances in Neurology 56, 227-231.

Siddique, T., Figlewiez, D. A., Pericak-Vance, M. A., Haines, J. L., Rouleau, G., Jeffers, A. J., Sapp, P., Hung, W. Y., Bebout, J., McKenna-Yasek, D., and et al. (1991). Linkage of a gene causing familial amyotrophic lateral sclerosis to chromosome 21 and evidence of genetic-locus heterogeneity. New England Journal of Medicine 324, 1381-1384.

Smeyne, R. J., Klein, R., Schnapp, A., Long, L. K., Bryant, S., Lewin, A., Lira, S. A., and Barbacid, M. (1994). Severe sensory and sympathetic neuropathies in mice carrying a disrupted Trk/NGF receptor gene. Nature 368, 246-249.

Sobue, G., Hashizume, Y., Yasuda, T., Mukai, E., Kumagai, T., Mitsuma, T., and Trojanowski, J. Q. (1990). Phosphorylated high molecular weight neurofilament protein in lower motoneurons in amyotrophic lateral sclerosis and other neurodegenerative diseases involving ventral horn cells. Acta Neuropathologica 79, 402-408.

Suenaga, T., Matsushima, H., Nakamura, S., Akiguchi, I., and Kimura, J. (1993). Ubiquitin-immunoreactive inclusions in anterior horn cells and hypoglossal neurons in a case with Joseph's disease. Acta Neuropathologica 85, 341-344.

Symonyan, M. A., and Nalbandyan, R. M. (1972). Interaction of hydrogen peroxide with superoxide dismutase from erythrocytes. FEBS Letters 28, 22-24.

Tan, S. A., Deglon, N., Zurn, A. D., Baetge, E. E., Bamber, B., Kato, A. C., and Aebischer, P. (1996). Rescue of motoneurons from axotomy-induced cell death by polymer encapsulated cells genetically engineered to release CNTF. Cell Transplantation 5, 577-587.

Tandan, R., and Bradley, W. G. (1985). Amyotrophic lateral sclerosis: Part 1. Clinical features, pathology, and ethical issues in management. Annals of Neurology 18, 271-280.

Tojo, H., Kaisho, Y., Nakata, M., Matsuoka, K., Kitagawa, M., Abe, T., Takami, K., Yamamoto, M., Shino, A., Igarashi, K., and et al. (1995). Targeted disruption of the neurotrophin-3 gene with lacZ induces loss of trkC-positive neurons in sensory ganglia but not in spinal cords. Brain Research 669, 163-175.

Tseng, J.L., Beatge, E. E., Zurn, A. D., and Aebischer, P, (1997). GDNF reduces drug-induced rotational behavior after medial forebrain bundle transaction by a mechanism not involving striatal dopamine. Journal of Neuroscience 17, 325-333.

Williams, D. B., and Windebank, A. J. (1991). Motor neuron disease (amyotrophic lateral sclerosis). Mayo Clinic Proceedings 66, 54-82.

Wong, P. C., Pardo, C. A., Borchelt, D. R., Lee, M. K., Copeland, N. G., Jenkins, N. A., Sisodia, S. S., Cleveland, D. W., and Price, D. L. (1995). An adverse property of a familial ALS-linked SOD1 mutation causes motor neuron disease characterized by vacuolar degeneration of mitochondria. Neuron 14, 1105-1116.

Wong, V., Arriaga, R., Ip, N. Y., and Lindsay, R. M. (1993). The neurotrophins BDNF, NT-3 and NT-4/5, but not NGF, up-regulate the cholinergic phenotype of developing motoneurons. European Journal of Neuroscience 5, 466-474.

Xu, Z., Cork, L. C., Griffin, J. W., and Cleveland, D. W. (1993). Increased expression of neurofilament subunit NF-L produces morphological alterations that resemble the pathology of human motor neuron disease. Cell 73, 23-33.

Zielinski, B. A., and Aebischer, P. (1994). Chitosan as a matrix for mammalian cell encapsulation. Biomaterials 15, 1049-1056.

Zurn, A. D., and Werren, F. (1994). Development of CNS cholinergic neurons *in vitro:* selective effects of CNTF and LIF on neurons from mesencephalic cranial motor nuclei. Developmental Biology 163, 309-315.

Zurn, A. D., Baetge, E. E., Hammang, J. P., Tan, S. A., and Aebischer, P. (1994). Glial cell line-derived neurotrophic factor (GDNF), a new neurotrophic factor for motoneurones. Neuroreport 6, 113-118.

Zurn, A. D., Winkel, L., Menoud, A., Djabali, K., and Aebischer, P. (1996). Combined effects of GDNF, BDNF, and CNTF on motoneuron differentiation *in vitro.* Journal of Neuroscience Research 44, 133-141.

Modular Fusion Proteins for Receptor-mediated Gene Delivery

Christoph Uherek[1], Jesús Fominaya[2] and Winfried Wels[1]

1 Institute for Experimental Cancer Research, Tumor Biology Center, Breisacher Str. 117, D-79106 Freiburg, Germany

2 Department of Immunology and Oncology, Centro Nacional de Biotecnologfa/CSIC, Universidad Autónoma de Madrid, Campus Cantoblanco, Madrid E-28049, Spain

Keywords. Non-viral gene delivery, fusion protein, *Pseudomonas* exotoxin A, diphtheria toxin, GAL4, ErbB2

1 Self-assembling Systems for Targeted Gene Delivery

The development of vectors for target-cell specific gene delivery is a major goal of gene therapeutic strategies. Thereby non-viral gene delivery vectors are gaining increasing interest [1]. Progress has been made in the understanding how individual activities of viruses can be mimicked and methodologies have been developed which allow to combine different functions required for gene transfer into an artificial complex. An attractive approach for the design of such modular self-assembling systems for gene delivery is based on fusion proteins engineered to incorporate in a single polypeptide chain several cooperating functions [2-4]. Each of these domains can account for a distinct activity required for DNA-binding, cell recognition and intracellular delivery. Such fusion proteins in contrast to similar chemical conjugates can be produced in suitable expression systems in their final form and the resulting products are generally homogeneous in their composition.

2 Bacterial Toxins as a Building Plan for Multifunctional Proteins

Bacterial toxins such as *Pseudomonas* exotoxin A (ETA) and diphtheria toxin (DT) bind specifically to receptors on the surface of target cells and, after internalization via receptor-mediated endocytosis, utilize the acidification of intracellular vesicles as a signal to activate an endosome escape function. This results in the translocation of an enzymatically active toxin fragment from the endosome to the cytosol where it inactivates the cellular protein synthesis machinery. Thereby distinct toxin domains provide the different functions required [5, 6].

This natural design of bacterial toxins as multidomain proteins has been exploited for different purposes. By replacing the original cell recognition domain with tumor-specific ligands, recombinant toxins have been derived which are being evaluated as potent anti-cancer therapeutics [7]. In addition, the potential of these bacterial proteins for the intracellular delivery of heterologous protein domains has been demonstrated [8]. We have extended these approaches and based on the molecular organization of ETA we have constructed a modular fusion protein which facilitates target-cell specific

NATO ASI Series, Vol. H 105
Gene Therapy
Edited by Kleanthis G. Xanthopoulos

gene delivery [3]. Thereby the toxin's natural cell binding domain was replaced by a tumor specific recognition domain. Likewise, the enzymatic domain was exchanged with a DNA-binding function which enables interaction with plasmid DNA. The toxin's internal translocation domain was retained. The resulting chimeric molecule consists of three functionally different domains arranged in a general structure which still resembles that of the natural toxin. However, the recombinant protein was nontoxic, exhibited an altered target-cell tropism, and functioned as a DNA carrier.

3 The 5EG Fusion Protein as a Prototype of a Modular DNA Carrier

The 5EG fusion protein which is schematically shown in Figure 1A serves as a prototype of this type of gene delivery vehicle. It carries at the N-terminus a single chain (sc) Fv domain derived from the ErbB2 specific monoclonal antibody FRP5. The cell recognition domain is followed by residues 252 to 366 of ETA which harbor the toxin's translocation domain and facilitate endosome escape. At the C-terminus the chimeric 5EG protein carries a DNA-binding domain comprising amino acids 2 to 147 of the yeast transcription factor GAL4. This domain mediates the formation of 5EG - DNA complexes. GAL4 binds as a dimer in a sequence specific manner to a palindromic sequence of 17 base pairs [9]. Therefore this recognition motif has to be included into plasmid DNA to allow complex formation.

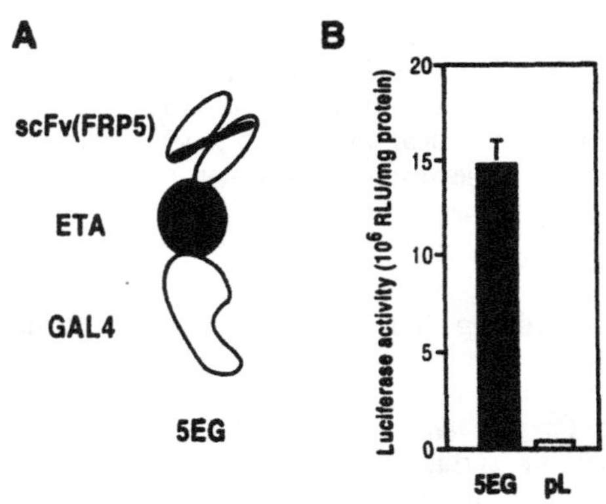

Figure 1. (A) Schematic representation of the modular DNA carrier protein 5EG. The bacterially expressed molecule contains at the N-terminus the ErbB2 specific scFv(FRP5) antibody domain connected to the ETA translocation domain and the DNA-binding domain of the yeast transcription factor GAL4. (B) Gene transfer mediated by the carrier protein 5EG. ErbB2 expressing COS-1 cells were treated with a complex containing the pSV2G4LUC luciferase reporter plasmid, 5EG fusion protein, and poly-L-lysine as described [3] (5EG). A similar complex lacking the 5EG fusion protein was used as a negative control (pL). Luciferase activity in cellular lysates was analyzed 48 h later and is expressed in relative light units per mg of total protein.

Recombinant 5EG fusion protein was produced in *E. coli* and purified from bacterial lysates. Both, the EfbB2 specific cell recognition domain and the GAL4 DNA-binding domain of the molecule were functional. Simultaneous binding of 5EG to a GAL4 specific oligonucleotide and to ErbB2 expressing SKBR3 human breast carcinoma cells could be observed. Complexes of the 5EG multidomain protein with pSV2G4LUC plasmid DNA carrying two GAL4 specific recognition sequences and a luciferase reporter gene under the control of the SV40 early promoter were prepared. DNA is a polyanion and requires neutralization of the negative charge as well as condensation to decrease particle size and allow insertion into endocytic vesicles [10]. Therefore the protein-DNA complexes were further incubated with poly-L-lysine, a polycation which binds to and condenses DNA, before they were applied to cells in tissue culture. As shown in Figure 1B the protein-DNA complexes were able to transfect ErbB2 expressing cells in a cell specific manner. Thereby the luciferase activity directly correlated with the amount of the fusion protein in the complex and was strictly dependent on the presence of the target receptor on the cell surface [3].

Efficient gene transfer mediated by the 5EG carrier protein also required the presence of a functional ETA translocation domain. The use of a fusion protein lacking this function resulted in drastically decreased reporter gene expression. Furthermore, a variety of agents which block endosomal acidification such as the carboxylic ionophores monensin or nigericin, and the inhibitor of vacuolar H+-ATPases bafilomycin Al, decreased the transfection efficiency of 5EG containing complexes [11]. This shows that the 5EG fusion protein is dependent on a low pH environment for full activity and suggests that the bacterial translocation domain in the chimeric DNA carrier protein functions in a fashion very similar to the wildtype toxin.

4 Exploiting the Modular Concept of DNA Carrier Proteins

Structural domains which are able to fold independently from other regions in the parental protein are likely to retain their functionality as isolated domains and are good candidates for the use in chimeric fusion proteins. The example of the 5EG molecule shows that such heterologous protein domains of mammalian, bacterial and yeast origin can be assembled into a functional chimeric protein which thereby acquires a novel biological activity. Due to this modular concept similar carrier proteins with modified activities, e.g. different target cell specificity, could be obtained by replacing individual protein domains. This might assist in the development of gene transfer strategies optimized for distinct target cell populations and is currently being investigated.

By exchanging the N-terminal single chain antibody domain of the 5EG prototype protein with a natural peptide ligand for the EGF receptor, a carrier protein was derived which facilitates gene transfer specifically into EGF receptor expressing cells. Replacement of the GAL4 DNA-binding domain of 5EG with a synthetic sequence of 28 lysine residues resulted in a protein which formed complexes with plasmid DNA via non-specific interaction of the lysine tail and facilitated efficient gene transfer thereby displaying the same target cell tropism as 5EG containing complexes.

Finally, the ETA endosome escape function was replaced by the translocation domain of diphtheria toxin. In wildtype DT in contrast to ETA the enzymatically active domain is located at the N-terminus. Based on this different molecular organization a chimeric fusion protein termed GD5 was constructed which carries the GAL4 DNA-binding domain at the N-terminus followed by the DT translocation domain and the heterologous scFv cell recognition domain at the C-terminus. Recently the DT translocation domain chemically coupled to poly-L-lysine has been used in *trans* to facilitate endosome escape of ligand-poly-L-lysine conjugate - DNA complexes [12]. Here the GD5 fusion protein links the DT translocation domain to cell targeting and DNA-binding functions in *cis* which upon complexation of the fusion protein with plasmid DNA resulted in efficient gene transfer.

These results show that the modular concept of DNA carrier proteins allows to easily adopt novel or improved functions. This might aid in the further development of artificial virus-like particles suitable for a broad range of applications in gene therapy.

References

1. Cooper, M. J. Noninfectious gene transfer and expression systems for cancer gene therapy. Semin. Oncol. 23: 172-187, 1996.

2. Chen, S. Y., Zani, C., Khouri, Y., and Marasco, W. A. Design of a genetic immunotoxin to eliminate toxin immunogenicity. Gene Ther. 2: 116-123, 1995.

3. Fominaya, J. and Wels, W. Target cell-specific DNA transfer mediated by a chimeric multidomain protein. Novel non-viral gene delivery system. J. Biol. Chem. *271:* 10560-10568, 1996.

4. Paul, R. W., Weisser, K. E., Loomis, A., Sloane, D. L., LaFoe, D., Atkinson, E. M., and Overell, R. W. Gene transfer using a novel fusion protein, GAL4/invasin. Hum. Gene Ther. 8: 1253-1262, 1997.

5. Hwang, J., Fitzgerald, D. J., Adhya, S., and Pastan, I. Functional domains of Pseudomonas exotoxin identified by deletion analysis of the gene expressed in E. coli. Cell. 48: 129-136, 1987.

6. Choe, S., Bennett, M. J., Fujii, G., Curmi, P. M., Kantardjieff, K. A., Collier, R. J., and Eisenberg, D. The crystal structure of diphtheria toxin. Nature. 357: 216-222, 1992.

7. Wels, W., Groner, B., and Hynes, N. E. Intervention in receptor tyrosine kinase-mediated pathways: recombinant antibody fusion proteins targeted to ErbB2. Curr. Top. Microbiol. Immunol. 213: 113-128, 1996.

8. Novoa, I., Cotten, M., and Carrasco, L. Hybrid proteins between Pseudomonas aeruginosa exotoxin A and poliovirus 2Apro cleave p220 in HeLa cells. J. Virol. 70: 3319-3324, 1996.

9. Carey, M., Kakidani, H., Leatherwood, J., Mostashari, F., and Ptashne, M. An aminoterminal fragment of GAL4 binds DNA as a dimer. J. Mol. Biol. 209: 423-432, 1989.

10. Wagner, E., Cotten, M., Foisner, R., and Birnstiel, M. L. Transferrin-polycation-DNA complexes: the effect of polycations on the structure of the complex and DNA delivery to cells. Proc. Natl. Acad. Sci. U S A. 88: 4255-4259, 1991.

11. Wels, W. and Fominaya, J. Peptides and fusion proteins as modular DNA carriers. *In*: L. W. Seymour, A. V. Kabanov, and P. L. Felgner (eds.), Self-assembling complexes for gene delivery: from chemistry to clinical trial. John Wiley & Sons Ltd., New York: 1998, in press.

12. Fisher, K. J. and Wilson, J. M. The transmembrane domain of diphtheria toxin improves molecular conjugate gene transfer. Biochem. J. *321:* 49-58, 1997.

Intracellular Combinatorial Chemistry with Peptides in Selection of Caspase-like Inhibitors

S. Michael Rothenberg, Department of Molecular Pharmacology, Program in Cancer Biology, Joan Fisher, Department of Molecular Pharmacology, David Zapol, Department of Molecular Pharmacology, Immunology Program, Stanford University School of Medicine, Stanford, CA 94305, David Anderson, Rigel Pharmaceuticals, Inc., Sunnyvale, CA, Yasumichi Hitoshi, Department of Molecular Pharmacology, Philip Achacoso, Department of Molecular Pharmacology and Garry P. Nolan, Department of Molecular Pharmacology, and Dept. of Microbiology and Immunology, e-mail: gnolan@cmgm.stanford.edu

Abstract

Fas is a cell surface receptor that can transmit signals for programmed cell death. Using a retroviral expression system, we have demonstrated that a short peptide derived from the cleavage site in a cellular target of a pro-apoptotic cysteine protease can be expressed within intact cells with sufficient activity to inhibit Fas-mediated apoptosis. *In vitro* analysis demonstrates that this retrovirally-expressed peptide is as potent as 150uM levels of the chemically synthesized peptide. Furthermore, using retroviral peptide library-based functional cloning we identified variants of this peptide with apparent anti-apoptotic activity. This approach is likely to lead to the identification of peptide variants with activity against a variety of signaling processes, both normal and pathological.

Introduction

Apoptosis is an important homeostatic mechanism that maintains cell number, positioning and differentiation. One of the best characterized processes for regulating apoptosis is initiated by the cell surface receptor Fas. Clustering of the Fas cytoplasmic domain by binding of Fas ligand or crosslinking antibody to the extracellular domain is capable of activating the interleukin-1b converting enzyme family of cysteine proteases (Caspases) - the proteolytic executioners of apoptosis (Enari et al., 1995)(Enari et al., 1996)(Los et al., 1995)(Tewari and Dixit, 1995). Recent studies implicate Caspase-8 (MACH/FLICE/Mch5) as a link between the Fas receptor and downstream Caspases via its association with FADD/MORT1, which itself associates with the Fas death domain (Boldin et al., 1996)(Fernandes-Alnemri et al., 1996)(Muzio et al., 1996). Other Caspases activated during Fas-mediated apoptosis include Caspase 1 (ICE) and Caspase 3 (Cpp32/Yama/Apopain)(Enari et al., 1996). The cleavage of a number of cellular substrates of the Caspases, including structural proteins, signaling molecules and DNA repair enzymes, coincides with the onset of apoptosis (Nicholson et al., 1995)(Na et al., 1996)(Tewari et al., 1995)(McConnell et al., 1997)(Zhivotovsky et al., 1997).

One of these substrates, Poly (ADP-ribose) polymerase (PARP), is proteolytically cleaved at the onset of apoptosis by Cpp32 (Caspase 3/Yama/Apopain)(Nicholson et al., 1995). The tetrapeptide aldehyde Ac-DEVD-CHO, corresponding to the cleavage site in PARP, residues 213-216, was shown to inhibit both the activity of CPP32 *in vitro* and the ability of apoptotic cell extracts to cause DNA fragmentation of isolated nuclei.

NATO ASI Series, Vol. H 105
Gene Therapy
Edited by Kleanthis G. Xanthopoulos

(Nicholson et al., 1995)(Enari et al., 1996). The tetrapeptide aldehyde Ac-YVAD-CHO, corresponding to the cleavage site in IL-1b, a substrate for Caspase 1/ICE, has a similar, albeit weaker, anti-apoptotic effect (Thornberry et al., 1992)(Thornberry et al., 1994)(Enari et al., 1996). A difficulty in using such chemically synthesized peptides to modify apoptosis in intact cells is their poor entry into cells, short half-life and requirement for N- and C-terminal modification for effect (Thornberry, 1994).

We have devised a system for introducing short peptides into, and expressing within, living cells using retroviruses. The approach results in continuous expression of sufficiently high levels of peptide to modify intracellular signaling cascades. Here we demonstrate that a retrovirally-expressed, ribosomally-synthesized peptide corresponding to the Cpp32 cleavage site in PARP can, without any chemical modifications, protect cells from apoptosis mediated by Fas. Furthermore, we have selected variants of this peptide from a large, randomized library using a genetic screen within living mammalian cells. The results demonstrate the power of retroviral expression of short peptides for altering cellular phenotypes. Furthermore, the ability to select peptides of novel sequence with activity against apoptosis provides a new tool for isolating novel signaling proteins and for developing pharmaceutical leads capable of modulating pathological signaling processes.

Materials and Methods

Cell Lines: Phoenix-E producer cells expressing retroviral Gag, Pol and ecotropic envelope proteins were grown, and transfections performed, as per Pear et al (Pear et al., 1993). Briefly, cells were passaged into 6 cm tissue culture plates at 1.5 million per plate 18 hours prior to transfection. Transfections of vector or library DNA were accomplished by calcium phosphate precipitation. Cell media was adjusted to 25 uM chloroquine just prior to transfection. After application of precipitate, fresh media was added to the cells at 8 and again at 24 hours after transfection, at which time cells were transferred to 32°C for an additional 48 hours. Jurkat T cells expressing the mouse ecotropic retroviral (Baker et al., 1992) were resuspended at 5×10^5 cells per ml viral supernatant and spin infections were carried out in 24 well plates at 2500 rpm for 1.5 hours at 32°C, in the presence of 5uM polybrene. The plates were kept at 32°C for 10-14 hours, at which time cells were resuspended in fresh RPMI/10% FCS and transferred to 37°C for an additional 48 hours. Transduction efficiencies were estimated by independent transfections with ß-galactosidase retroviral vectors and FACS-Gal analysis of transduced cells (Nolan et al., 1988).

Vectors and Peptide Libraries: Synthetic oligonucleotide inserts encoding Caspase inhibitor peptides or peptide libraries were prepared as described previously (Mattheakis et al., 1994). Briefly, peptide or library oligonucleotides prepared by standard degenerate oligonucleotide synthesis were converted to double-stranded DNA inserts by primed DNA polymerization, digested with Bst XI and subcloned into the retroviral vector pMSCV/PC. Ligations were transformed by electroporation into *E. Coli* and the DNA recovered after overnight plating. For library preparation, a small aliquot of the combined electroporated samples was plated to determine library size. The remaining sample was inoculated directly into LB broth, incubated overnight and the library DNA prepared by Maxiprep (Qiagen). Insert sequence of the randomized region was NNKNNKNNKGAKNNK, where N encodes A,C,G or T and K encodes G or T.

Anti-Fas Selection: Jurkat T cells transduced with peptide or library vectors were selected with IgM anti-Fas antibody Ch-11 (Kamiya Corp.). Briefly, cells were resuspended in RPMI/2.5% FCS plus 50ng/ml CH-11 and distributed into 24 well plates at 10^5 cells per well. After 5 days at 37°C, 1 ml RPMI/20% FCS was added to each well and the cells were placed at 37°C for an additional two - three weeks. Wells containing live cells were identified by visual inspection for nutrient depletion of the media and the presence of cell colonies. XTT assay of triplicate samples of each well were performed to confirm outgrowth of resistant cells.

Cpp 32 Enzyme Assays: Cell extracts were prepared and enzyme assays were carried out as per Lazebnik et al. (Lazebnik et al., 1993) Briefly, apoptotic extracts cells were prepared by repeated freeze/thawing and then were incubated with the chromogenic CPP32 substrate DEVD-pNA or the fluorogenic substrate DEVD-AMC (Clontech) in the presence or absence of increasing concentrations of chemically-synthesized CINP peptide. Substrate cleavage was measured in a plate reader or spectrofluorimeter for release of free chromogen/fluorophore.

LC/MS Analysis: After incubation of Fas-activated cellular extracts with chemically-synthesized CINP peptide, the extracts were passed over an HPLC column, desalted and loaded directly into a mass spectrometer. Individual major peaks in the size ranges expected for the full length peptide and potential cleavage products were scanned for peaks corresponding to the expected molecular weight values.

PCR Rescue and Library Analysis: Poly-A purified mRNA was prepared from surviving cells and RT-PCR carried out using the Titan RT-PCR kit (Boehringer-Mannheim) and primers flanking the peptide-encoding insert. Primers used were 5'GATCCTCCCTTTATCCAG3' and 5'CAGGTGGGGTCTTTCATTCC3'. Amplified samples were purified, digested with Bst XI and subcloned into the pMSCV/PC retroviral vector as described for peptide library generation. DNA was prepared from individual colonies and the sequence of each insert determined by cycle sequencing with the same upstream primer used for PCR rescue.

Statistical Analysis and Alignment of Sequences: Library sequences before and after selection were analyzed to determine the statistical significance of amino acid frequency differences. Separate comparisons for the completely randomized positions and the D/E position were carried out. The null hypothesis for each test was that there was no bias, so the observed frequency to the expected frequency of each amino acid given the nucleotide degeneracy was compared. The significance of the deviation was noted at the highest level of confidence given the results of a G-test (Microsoft Excel).

Results

Expression of Caspase inhibitor peptides can rescue T cells from Fas-mediated apoptosis in long-term assays.

To determine whether a peptide could inhibit apoptosis when expressed in cells with retroviruses, the retroviral vector pMSCV/PC/CINP was constructed (Figure 1). This construct encodes residues 209-218 of PARP, including the scissile bond D216-G217 downstream of a Kozak translation consensus sequence (Kozak, 1986) and should result in translation of the free peptide, without any N- or C-terminal chemically synthetic modifications, when expressed in cells. Jurkat T cells were transduced with virus prepared with this construct, a b-galactosidase (b-gal) construct or several nonspecific peptide constructs. Each transduced cell population was divided into 10 wells of a 24 well plate

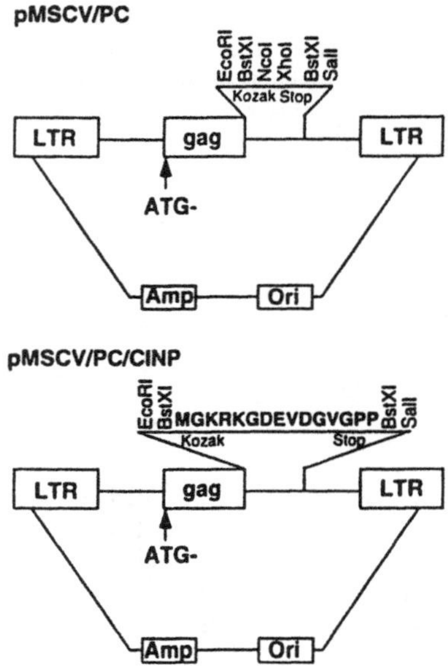

Figure 1
Schematic structures of pMSCV/PC and pMSCV/PC/CINP vectors. The pMSCV/PC vector (*Upper*) encodes a Kozak translation initiation site, stop codon and peptide stabilizing residues within two nonidentical BstXI sites. These sites were used to create the pBabe/PC/CINP vector (*Lower*), which encodes the potential Caspase inhibitor peptide.

at 10^5 cells per well and stimulated with CH-11, a monoclonal antibody against the Fas receptor. After two weeks, cell survival in each well was assayed in triplicate using XTT (Roehm et al., 1991). As shown in Figure 2, the two Jurkat T cell populations independently transduced with the pMSCV/PC/CINP construct showed significantly increased survival by XTT assay compared to T cells transduced with b-gal or any of the nonspecific peptide constructs.

Mechanism of peptide action: pseudosubstrate inhibition.

The DEVD peptide encoded by the pMSCV/PC/CINP construct was able to protect cells transduced with it from Fas-induced apoptosis, even though it contains the scissile D216-G217 bond in PARP actually cleaved by CPP32 and lacks the C-terminal aldehyde required for covalent inactivation of the enzyme's catalytic cysteine (Thornberry et al., 1994). To gain some insight into the expressed peptide's mechanism of protection, we compared residual Cpp32 enzyme activity against the added chromogenic substrate DEVD-AMC in apoptotic cell extracts which had been mixed with the chemically synthesized CINP peptide to the activity in extracts of apoptotic cells that had been transduced with the retroviral vector encoding the same peptide. The amount of residual CPP 32 enzyme activity in transduced cells was equivalent to the residual activity in untransduced extracts containing 150uM chemically synthesized peptide (data not shown). We conclude that a PARP cleavage site peptide without any chemical modifications can serve as a pseudosubstrate inhibitor of Cpp32, decreasing the amount of endogenous substrate cleavage enough to protect cells from apoptosis if present at sufficiently high concentration, even if the peptide itself is cleaved. However, we doubt that the actual concentration of the peptide in cells approaches 150uM. We are currently exploring the possibility that the peptide acts in some other manner (see discussion) to give this apparent inhibitory concentration.

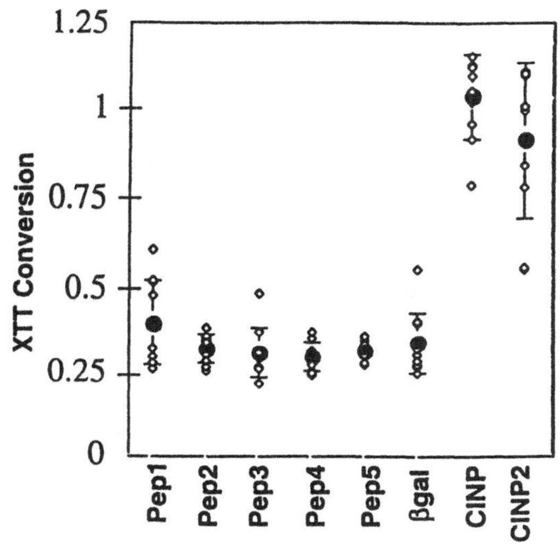

Figure 2
Retrovirally-expressed CINP peptide protects cells from Fas-induced apoptosis. Jurkat T cells were transduced with peptide control virus (Pep 1-5), b-gal virus or virus encoding the CINP peptide (CINP and CINP2). Cells were distributed into 24-well plates and treated with anti-Fas monoclonal CH-11 as described in Methods. Cell outgrowth after antibody treatment was monitored by XTT assay. Data shown are the results of 10 independent cell samples, with XTT measurement for each data point done in duplicate.

To determine whether the CINP peptide itself was capable of being cleaved by activated cell extracts, the chemically synthesized CINP peptide was mixed with Fas-activated apoptotic cell extracts, and LC/MS analysis carried out on the cleavage products. As shown in Figure 3, prominent peaks corresponding to the molecular weights of the products expected from cleavage by Cpp32 were found (peaks 2 and 4a), as well as two peaks corresponding to products formed by the cleavage of peptide bonds not recognized by cysteine proteases (peaks 3 and 4b). The presence of such additional cleavage products is not surprising, given that extracts of cells could contain a variety of endo- and exopeptidases in addition to cysteine proteases. These results suggest that the CINP peptide is capable of being cleaved by activated Cpp32 and support the hypothesis that overexpression of a peptide substrate of Cpp32 can decreases the amount of endogenous substrate cleavage sufficiently to protect cells in which Cpp32 has been activated from apoptosis.

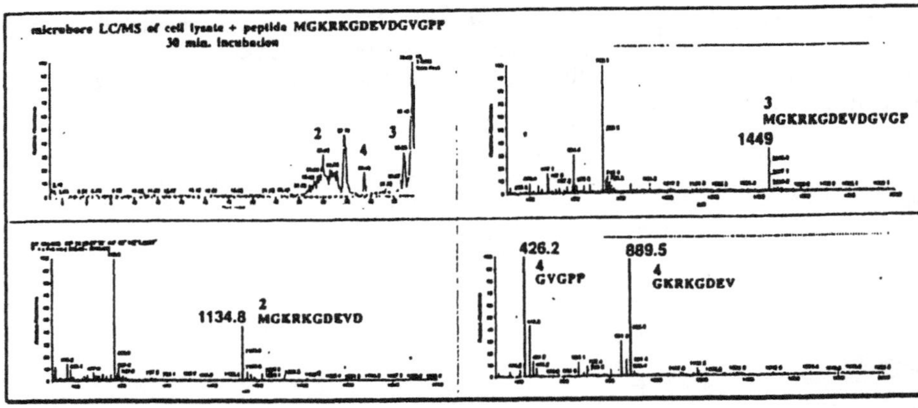

Figure 3
Mass spectrometry analysis of peptide activity in Fas-activated cellular extracts. Peptide was incubated in cellular extract from activated activated Jurkat T cells. Extracts were passed over HPLC, desalted, and run directly into the mass spectrometer. Individual major peaks in the range of sizes expected for the peptide and its cleavage products were scanned for peaks corresponding to expected value. Major peaks were found for the products in peaks 2 and 4. Peak 4 also contained a peptide product corresponding to an unknown proteolytic mechanism (GKRKGDEV).

Selection of peptides within T cells shows statistical variance from randomness.

To verify that rare retrovirus expression constructs could be phenotypically selected from a large library using the Phoenix system, co-transfections were carried out with a b-gal retrovirus vector and serial molar dilutions of a neomycin-resistance *(neo)* retroviral vector (Figure 4). After transduction of NIH 3T3 cells and selection in 1 mg/ml G418, neomycin-resistant colonies were detected down to 1 neo construct in 10^7 β-gal constructs. In such

βgal	1	1	1	1	1	1	1
Neo	1->10^{-4}	10^{-5}	10^{-6}	10^{-7}	10^{-8}	10^{-9}	0
Colonies	Confluent	60	20	31	10	0	0

Figure 4
The Phoenix retroviral packaging system can select for rare phenotypes. Phoenix producer cells were co-transfected with a b-galactosidase retroviral construct (bgal) and serial molar dilutions of a neomycin resistance gene-encoding retroviral construct (Neo). 48 hours later, the viral supernatant was used to transduce populations of NIH 3T3 cells, which were selected in 1mg/ml G418. The number of G418-resistant cell colonies after two weeks of selection is shown.

experiments, the b-gal constructs represent the total library and the *neo* constructs the rare members of the library capable of conferring a selectable phenotype on the target cells. Thus, the Phoenix library system can detect rare events against a high background of undesirable events, with a sufficiently strong selection.

We next determined whether the Caspase inhibitor peptide described above or variants of it could be identified by a genetic selection in mammalian cells for peptide inhibitors of Fas-mediated apoptosis. A retroviral construct library was generated in which the nucleotide positions corresponding to amino acids P4,P3,P2 and P-1 were completely randomized, while the degeneracy in nucleotides corresponding to position P1 was limited to permit the generation of only Aspartate or Glutamate at that position in the library (Figure 5). It is known that an Aspartate at the P1 position is strongly preferred for substrate cleavage by all known Caspases, while position P4 is important for substrate recognition, with ICE-like proteases having a preference for hydrophobic residues like Tyrosine at P4 and CED-3/CPP32-like proteases having a preference for Aspartate (Rano et al., 1997)(Rano et al., 1997). Among known endogenous substrates for Caspases, positions P2 and P3 show the most variability. An analysis of the sequence of 25 clones from the unselected DNA

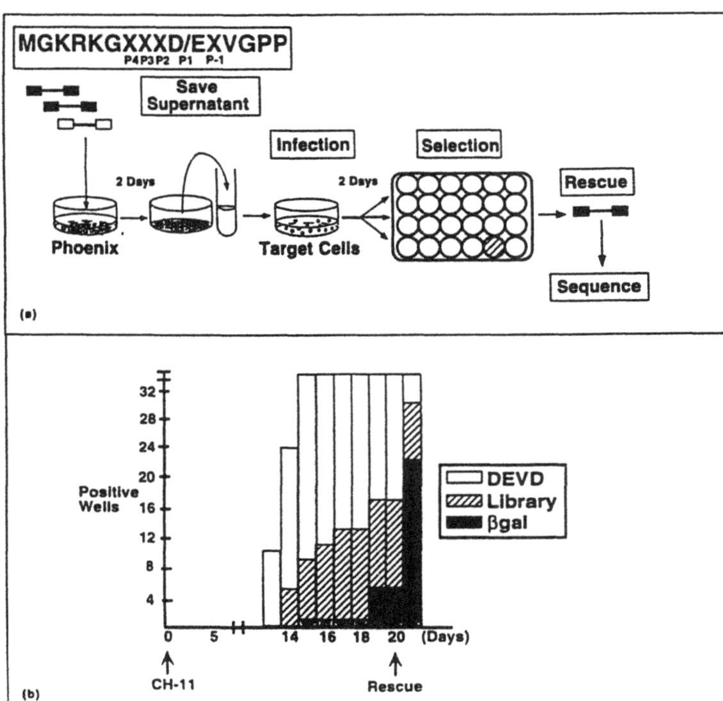

Figure 5
Genetic selection for variants of CINP that inhibit Fas-induced apoptosis. (a) Jurkat T cells transduced with peptide library virus (see Materials and Methods) were distributed into 24 well plates at 100,000 cells per well and treated with 50 ng/ml CH-11 monoclonal IgM antibody against the Fas receptor. (b) Wells containing live cells were identified by visual inspection and XTT assay.

construct library demonstrated no statistically significant variation from expected randomness (data not shown). The library was transfected into Phoenix Eco and Jurkat T cells cells expressing the ecotropic viral receptor (Baker et al., 1992) were transduced with the resulting virus. 10^7 cells were divided into 100 wells of a 24 well plate and selected in the presence of CH-11 antibody as described above (Figure 5a). After two weeks, wells containing surviving cells were examined by visual inspection for nutrient depletion in the media and the presence of cell colonies. There was a significant increase in positive wells of cells transduced with the library, as compared to cells transduced with a b-gal vector (17 for the library versus 6 for b-gal by day 22; Figure 5b). As expected, each well of cells that had been transduced with the pMSCV/PC/CINP construct contained live cells.

To identify the peptides that might be expressed in the surviving cells, RT-PCR was carried out on cells from each well (data not shown). After PCR amplification each library sample contained a predominant PCR band of the expected size, plus nonspecific bands; restriction

(a)

Amino Acid	potential codons	Observed no selection	Observed selection	Expected if Random
non P1				
F	1	4	5	3.1
L	3	14	8	9.4
S	3	7	9	9.4
Y	1	3	7	3.1
C	1	4	0	3.1
W	1	2	1	3.1
P	2	9	5	6.3
H	1	3	0	3.1
Q	1	2	2	3.1
R	3	12	8	9.4
I	1	3	4	3.1
M	1	4	7	3.1
T	2	4	13	6.3
N	1	5	5	3.1
K	1	2	1	3.1
A	2	8	11	6.3
D	1	3	5	3.1
E	1	1	3	3.1
G	2	2	2	6.3
V	2	4	13	6.3
STP	1	4	0	3.1
P1				
D	1	12	21	12.5
E	1	13	5	12.5

Figure 6
(a) Table of amino acid frequencies for selected peptides. Sequences were determined for individual unselected and *in vivo* selected peptides, and the frequencies of individual residues compared in the two groups. Table shows observed amino acid frequencies before and after selection as well as the frequencies expected from a purely random library. (b) Clustal W aligned peptides selected after Fas treatment. Sequences are presented as grouped by the PAM 250 weight matrix into most similar sequences.

FAS SELECTED PEPTIDES		
TASDA		SDRDI
RASDL		VLGDR
TPSDM		LSYDR
YPSDV		SYQDL
YASDV		MTPDP
YTSDA		MTIEA
YAVDE		WLLEF
YARDI		
RKTDA		RNTET
QTTDA		RETER
IVNDT		VVDEM
FGNDF		VVNEM
FPDDL		
FADDL		

(b)

digest with Bst XI generated the expected peptide-encoding insert plus the digested end products. Interestingly, none of 6 cell samples from the b-gal-transduced cells generated an RT-PCR product. Since the initial transduction efficiency was approximately 15% as measured by X-Gal of transduced cells, and since b- gal should not confer any survival advantage in the presence of Fas stimulation, the frequency of surviving cells actually transduced with a b-gal virus should approximate the initial transduction efficiency (15%). Therefore, the apparent lack of b-gal-transduced surviving cells among the negative control samples is not surprising.

For 10 of the 17 samples surviving from the library, the RT-PCR product was ligated into pMSCV/PC, electroporated into *E. Coli* and plated on LB/Amp plates overnight. The DNA prepared from 10 colonies for each sample was sequenced to identify the encoded peptide(s). An analysis of the types of residues in the selected versus the unselected library is shown in Figure 6a. A highly significant increase in the frequency of Asp in position P1 (21/26, or 83%, $p < 0.001$) was observed versus the frequency in the unselected library (50%), implying that cells expressing peptides with Asp in this position possess a survival advantage in the presence of Fas stimulation. Furthermore, the observed amino acid frequencies at the other positions differed significantly from the random distribution of residues expected if no selection had occurred.

A Pam 250 Weight Matrix of the recovered peptide sequences is shown in Figure 6b. Interestingly, several of the peptides recovered by the screen appear similar in sequence to actual or predicted cleavage sites in endogenous substrates of the Caspases (see discussion).

Discussion

The ability of peptides or peptide-like compounds to alter the physiology of cells is well documented from the examples of peptide-like antibiotics, anticancer agents and immune modulators (Moreno et al., 1995)(Katz and Donadio, 1995)(Gasson, 1995). However, it has never been reproducible demonstrated in a stably-expressing system that such peptides can confer significant long-term consequences. Furthermore, the full activity of such compounds often derives from post-translational modification of a ribosomally-synthesized pro-peptide form of the compound, for example cyclization or amino- or carboxy-terminal modification. Here, we demonstrate that a ribosomally synthesized, unmodified linear peptide can potently alter the physiology of mammalian cells when expressed in such cells with retroviruses. The retrovirally expressed CINP peptide, derived from the primary amino acid sequence of PARP, a cellular substrate of the prop-apoptotic cysteine protease Cpp32, is capable of protecting those cells from apoptosis induced by Fas (Figure 2). Consistent with this protective effect is the demonstration that apoptotic extracts of cells expressing this peptide contain signficantly reduced Cpp32 enzyme activity toward a chromogenic peptide substrate (data not shown).

Comparison of the residual Cpp32 activity in the extracts of apoptotic cells overexpressing the CINP peptide to extracts mixed with the chemically synthesized CINP peptide suggests that retroviral expression of the peptide can decrease Cpp32 activity in the extracts to the level of 150uM of chemically synthesized peptide added to such extracts (data not shown). Because cell contents are diluted during preparation of the extracts, the actual intracellular concentration would be predicted to be significantly higher. With retroviral expression, however, the peptide is present in the cell before pro-Cpp32 is cleaved to its activated form through induction by an apoptotic stimulus. Because this activation depends on cleavage of pro-Cpp32 at two Asp-Gly bonds contained within cleavage sites that fit the consensus for Cpp32 substrates, it is thought that Cpp32 might act on itself in a self-amplification reaction, or that the proenzyme is cleaved by other upstream cysteine proteases, for example ICE (Enari et al., 1996)(Faleiro et al., 1997)(Takahashi et al., 1997). Therefore, the reduction in Cpp32 enzyme activity observed in extracts of cells overexpressing the CINP peptide through retroviral transduction could be due to a decrease in the total amount of pro-Cpp32 cleaved to active Cpp32, in addition to direct inhibition of the activated enzyme. The activation state of Cpp32 could be determined by Western blot of cell extracts using antibodies that recognize both the pro- and activated form of the enzyme: decreased activation would be reflected in an increase in the higher molecular weight uncleaved pro form (and a decrease in the lower molecular weight cleavage products). Although we could detect cleaved PARP (a Cpp32 substrate) in extracts of Fas antibody-treated cells by Western blot with anti-PARP antibodies, we were unable to detect cleaved Cpp32 in the same extracts with a monoclonal antibody to the p12 subunit of Cpp32, which is present in both the pro- and activated forms of the enzyme (Y.H., unpublished results). A possible explanation is that too little pro-Cpp32 is cleaved to the activated enzyme to be detected by Western blot. A second approach to detecting cleaved Cpp32 is to mix the purified, ^{35}S-methionine-labelled enzyme with apoptotic cell extracts and then detect the cleaved products by PAGE. However, while this might permit us to determine whether retrovirally-expressed CINP peptide is capable of decreasing the cleavage of labeled proCpp32 by endogenous Cpp32 in the activated extracts, it does not permit us to determine whether the peptide can decrease the initial activation of Cpp32 in intact cells. We are currently attempting to label the activated Caspases in cell extracts with a biotinylated aldehyde inhibitor biotinDEVD-CHO, which might permit purification and qunatitation of the amounts of activated endogenous Caspases in cellular extracts using Streptavidin-conjugated secondary reagents (Faleiro et al., 1997).

The ability of the fifteen amino acid ribosomally-synthesized CINP peptide to protect cells from apoptosis suggested the possibility of using our retroviral peptide expression technology in a genetic selection for CINP variants which protect cells from apoptosis (Figure 5a). A retroviral construct library was designed in which selected residues in the CINP sequence were completely randomized while the key Aspartate residue at the P1 position was either maintained or mutated to Glutamate. We have previously shown that a Glutamate at P1 renders the retrovirally-expressed peptide inactive (S.M.R., unpublished data). The library was introduced into Jurkat T cells by the retroviral technique, and the cells were treated with the CH-11 monoclonal antibody against Fas to induce apoptosis. We observed an enrichment in wells containing living cell colonies in samples transduced with the library versus those transduced with a b-gal vector (figure 5b). The highly significant bias towards Aspartate in position P1 of the recovered peptide sequences is expected if Asp at P1 is critical to the activity of the peptide in protecting the cells from apoptosis. A comparison of the recovered peptide sequences to known or predicted cysteine protease cleavage sites shows a number of similarities (Figure 6). Of particular interest is a family of related sequences (YASDV,YTSDA,YAVDE,YARDI) which are quite similar to the cleavage site in interleukin-1b (YVADA). It has previously been shown that the tetrapeptide aldehyde YVAD-CHO is capable of inhibiting apoptosis in Jurkat T cells induced by Fas (Enari et al., 1995). Therefore, it is not surprising that our screen might have selected peptides with activity against other with required activity in inducing apoptosis. We are currently in the process of introducing the recovered peptides individually into cells to determine which can protect cells from Fas-induced apoptosis. We are also testing each peptide for its specific ability to inhibit the activity of the several known Caspases.

Not present in the recovered peptide sequences is a sequence encoding the CINP peptide itself, which should have been present in the original construct library at a frequency of 1 in 500,000. A possible explanation is that during library virus production a CINP-encoding sequence was not present. Ten million Jurkat T cells were actually used for transduction, which at a 30% transduction efficiently should have produced approximately three million transduced cells, or 1.5 times the theoretical total library construct size of two million at the nucleotide level. When considered with the wide heterogeneity in expression achieved with retroviral infection, it is therefore not unexpected that the CINP peptide sequence was not recovered from the surviving cells, either because it was never introduced into the cells before selection, or because it was introduced, but expressed at too low of a level to permit the transduced cell(s) to survive the anti-Fas selection. Another explanation is that numerous peptides in the library are capable of inhibiting apoptosis by acting at other points in the cascade or by more efficiently inhibiting other important Caspases. Since our peptides are conformationally flexible, it is possible that certain linear motifs assume more rigid structures capable of inhibiting distinct Caspases more efficiently. These and other possibilities will be tested.

Importantly, the technique described here offers the possibility that protein-protein interactions, often defined by linear peptide interfaces, might be disrupted or co-opted through expression of short peptides. Thus, a panoply of potential new gene therapeutics, derived from known protein interfaces, could be developed and delivered by viral or other expression techniques. The system has the advantage that short peptides might be largely non-immunogenic. As well, crucial determinants from proteins might be isolated for therapeutic effect while not introducing other protein determinants that provide unnecessary or pathological side effects.

We have demonstrated that a ribosomally synthesized linear peptide derived from a substrate for the pro-apoptotic cysteine protease Cpp32 can inhibit Fas-mediated apoptosis in Jurkat T cells when expressed in those cells with retroviruses. Furthermore, we have demonstrated that variants of this peptide can be selected from a large library. Our approach is broadly applicable to any physiological process which can be modeled in cell culture. We are currently in the process of screening retroviral peptide libraries for peptides with activity against signaling pathways involved in cancer, immune cell dysfunction and viral infection. Such peptides would represent ligands for isolating the proteins involved in these processes as well as potential lead compounds for pharmaceutical development.

Acknowledgements

The authors thank the members of the Nolan Lab and David Ferrick of Rigel Pharmaceuticals, Inc., for helpful suggestions. S.M.R. was supported by an NIH grant through the Program in Cancer Biology, Stanford University. G.P.N. is a Scholar of the Leukemia Society of America, a recipient of the Burrough's Wellcome New Investigator Award in Pharmacology and a Howard Hughes Young Investigator Award, and is supported by NIH grants AI 35304, AI 39646 and AR/AI 44565. Parts of this work were supported by a gift from Tularik, Inc.

References

Baker, B. W., Boettinger, D., Spooncer, E., and Norton, J. D. (1992). Efficient retroviral-mediated gene transfer into human B lymphoblastoid cells expressing mouse ecotropic viral receptor. Nucleic Acids Research 20, 5234-5238.

Boldin, M. P., Goncharov, T. M., Goltsev, Y. V., and Wallach, D. (1996). Involvement of MACH, a novel MORT1/FADD-interacting protease, in Fas/APO-1- and TNF receptor-induced cell death. Cell 85, 803-815.

Enari, M., Hase, A., and Nagata, S. (1995). Apoptosis by a cytosolic extract from Fas-activated cells. Embo J 14, 5201-5208.

Enari, M., Hug, H., and Nagata, S. (1995). Involvement of an ICE-like protease in Fas-mediated apoptosis. Nature 375, 78-81.

Enari, M., Talanian, R. V., Wong, W. W., and Nagata, S. (1996). Sequential activation of ICE-like and CPP32-like proteases during Fas- mediated apoptosis. Nature 380, 723-726.

Faleiro, L., Kobayashi, R., Fearnhead, H., and Lazebnik, Y. (1997). Multiple species of CPP32 and Mch2 are the major active caspases present in apoptotic cells. Embo J 16, 2271-2281.

Faleiro, L., Kobayashi, R., Fearnhead, H., and Lazebnik, Y. (1997). Multiple species of CPP32 and Mch2 are the major active caspases present in apoptotic cells. Embo J 16, 2271-2281.

Fernandes-Alnemri, T., Armstrong, R. C., Krebs, J., Srinivasula, S. M., Wang, L., Bullrich, F., Fritz, L. C., Trapani, J. A., Tomaselli, K. J., Litwack, G., and Alnemri, E. S. (1996). In vitro activation of CPP32 and Mch3 by Mch4, a novel human apoptotic. Proc Natl Acad Sci USA 93, 7464-7469.

Gasson, M. J. (1995). Lantibiotics. Biotechnology 28, 283-306.

Katz, L., and Donadio, S. (1995). Macrolides. Biotechnology 28, 385-420.

Kozak, M. (1986). Point mutations define a sequence flanking the AUG initiator codon that modulates translation by eukaryotic ribosomes. Cell 44, 283-292.

Lazebnik, Y. A., Cole, S., Cooke, C. A., Nelson, W. G., and Earnshaw, W. C. (1993). Nuclear events of apoptosis in vitro in cell-free mitotic extracts: a model system for analysis of the active phase of apoptosis. J Cell Biol 123, 7-22.

Los, M., Van de Craen, M., Penning, L. C., Schenk, H., Westendorp, M., Baeuerle, P. A., Droge, W., Krammer, P. H., Fiers, W., and Schulze-Osthoff, K. (1995). Requirement of an ICE/CED-3 protease for Fas/APO-1-mediated apoptosis. Nature 375, 81-83.

Mattheakis, L. C., Bhatt, R. R., and Dower, W. J. (1994). An in vitro polysome display system for identifying ligands from very large peptide libraries. Proc Natl Acad Sci USA 91, 9022-9026.

McConnell, K., Dynan, W., and Hardin, J. (1997). The DNA-dependent protein kinase catalytic subunit (p460) is cleaved during Fas-mediated apoptosis in Jurkat cells. J Immunol 158, 2083-2089.

Moreno, F., San Millan, J. L., Hernandez-Chico, C., and Kolter, R. (1995). Microcins. Biotechnology 28, 307-321.

Muzio, M., Chinnaiyan, A. M., Kischkel, F. C., O'Rourke, K., Shevchenko, A., Ni, J., Scaffidi, C., Bretz, J. D., Zhang, M., Gentz, R., Mann, M., Krammer, P. H., Peter, M. E., and Dixit, V. M. (1996). FLICE, a novel FADD-homologous ICE/CED-3-like protease, is recruited to the CD95 (Fas/APO-1) death--inducing signaling complex. Cell 85, 817-827.

Na, S., Chuang, T. H., Cunningham, A., Turi, T. G., Hanke, J. H., Bokoch, G. M., and Danley, D. E. (1996). D4-GDI, a substrate of CPP32, is proteolyzed during Fas-induced apoptosis. J. Biol. Chem. 271, 11209-11213.

Nicholson, D. W., Ali, A., Thornberry, N. A., Vaillancourt, J. P., Ding, C. K., Gallant, M., Gareau, Y., Griffin, P. R., Labelle, M., Lazebnik, Y. A., and et al. (1995). Identification and inhibition of the ICE/CED-3 protease necessary for mammalian apoptosis. Nature 376, 37-43.

Nicholson, D. W., Ali, A., Thornberry, N. A., Vaillancourt, J. P., Ding, C. K., Gallant, M., Gareau, Y., Griffin, P. R., Labelle, M., Lazebnik, Y. A., and et al. (1995). Identification and inhibition of the ICE/CED-3 protease necessary for mammalian apoptosis. Nature 376, 37-43.

Nolan, G. P., Fiering, S., Nicolas, J. F., and Herzenberg, L. A. (1988). Fluorescence-activated cell analysis and sorting of viable mammalian cells based on beta-D-galactosidase activity after transduction of Escherichia coli lacZ. Proc Natl Acad Sci USA 85, 2603-2607.

Pear, W. S., Nolan, G. P., Scott, M. L., and Baltimore, D. (1993). Production of high-titer helper-free retroviruses by transient transfection. Proc Natl Acad Sci USA 90, 8392-8396.

Rano, T. A., Timkey, T., Peterson, E. P., Rotonda, J., Nicholson, D. W., Becker, J. W., Chapman, K. T., and Thornberry, N. A. (1997). A combinatorial approach for determining protease specificities: application to interleukin-1beta converting enzyme (ICE). Chem Biol 4, 149-155.

Roehm, N. W., Rodgers, G. H., Hatfield, S. M., and Glasebrook, A. L. (1991). An improved colorimetric assay for cell proliferation and viability utilizing the tetrazolium salt XTT. J Immunol Methods 142, 257-265.

Takahashi, A., Hirata, H., Yonehara, S., Imai, Y., Lee, K. K., Moyer, R. W., Turner, P. C., Mesner, P. W., Okazaki, T., Sawai, H., Kishi, S., Yamamoto, K., Okuma, M., and Sasada, M. (1997). Affinity labeling displays the stepwise activation of ICE-related proteases by Fas, staurosporine, and CrmA-sensitive caspase-8. Oncogene 14, 2741-2752.

Tewari, M., Beidler, D. R., and Dixit, V. M. (1995). CrmA-inhibitable cleavage of the 70-kDa protein component of the U1 small nuclear ribonucleoprotein during Fas- and tumor necrosis factor- induced apoptosis. J Biol Chem 270, 18738-18741.

Tewari, M., and Dixit, V. M. (1995). Fas- and tumor necrosis factor-induced apoptosis is inhibited by the poxvirus crmA gene product. J Biol Chem 270, 3255-3260.

Thornberry, N. A. (1994). Interleukin-1 beta converting enzyme. Methods Enzymol 244, 615-631.

Thornberry, N. A., Bull, H. G., Calaycay, J. R., Chapman, K. T., Howard, A. D., Kostura, M. J., Miller, D. K., Molineaux, S. M., Weidner, J. R., Aunins, J., and et, a.l. (1992). A novel heterodimeric cysteine protease is required for interleukin-1 beta processing in monocytes. Nature 356, 768-774.

Thornberry, N. A., Peterson, E. P., Zhao, J. J., Howard, A. D., Griffin, P. R., and Chapman, K. T. (1994). Inactivation of interleukin-1 beta converting enzyme by peptide (acyloxy)methyl ketones. Biochemistry 33, 3934-3940.

Thornberry, N. A., Peterson, E. P., Zhao, J. J., Howard, A. D., Griffin, P. R., and Chapman, K. T. (1994). Inactivation of interleukin-1 beta converting enzyme by peptide (acyloxy)methyl ketones. Biochemistry 33, 3934-3940.

Zhivotovsky, B., Gahm, A., and Orrenius, S. (1997). Two different proteases are involved in the proteolysis of lamin during apoptosis. Biochem Biophys Res Commun 233, 96-101.

Lipid Gene Trasfer, a Story of Simplicity and Complexity

Natasha J. Caplen,
Clinical Gene Therapy Branch, National Human Genome Research Institute, National Institutes of Health, Bethesda, MD 20892, USA.

Keywords. Cationic liposomes, liposome-mediated gene transfer

1 Introduction

It is 10 years since the first description of the use of a positively charged lipid to deliver nucleic acids to mammalian cells (Felgner *et al.*, 1987). In essence the methodology of liposome-mediated transfer of nucleic acids is very simple. Cells are exposed to a complex of nucleic acid and lipid formed by simply mixing the two components together. The lipid in some way facilities both uptake of the nucleic acid by, and transport within a cell, usually enhancing transfer many fold over that seen if nucleic acid alone is used. However, this description is proving to be something of an over simplification of what is probably a much more complex process. This review will discuss the basic chemistry of the most widely used cationic liposomes, current models for DNA-liposome complexes, the chemical and biological factors that effect the efficiency of DNA transfer, and the proposed trafficking mechanisms that facilitate DNA transfer to the nucleus.

2. Liposomes

Liposomes are membranous lipid vesicles which enclose an aqueous volume. These spherical particles vary in size from 10s of nanometers to 10s of microns in diameter. The principal component is usually a phospholipid formed from a pair of fatty acids possessing a hydrophobic tail and a hydrophilic head group. In aqueous solutions the hydrophobic fatty acid tails self associate to exclude water, whilst the hydrophilic head group interacts with any aqueous liquid on the inside and the outside of the vesicle. The resultant lipid bilayer forms a continuous membrane which encloses a small volume of water. A single bilayer is unilamellar. Multilamellar liposomes can also be formed and consist of nested rings of bilayers arranged in a structure not dissimilar to the rings of an onion.

Phospholipids also form the largest structural component of biological membranes. Given this similarity a fusogenic process between cells and liposomes has long been envisaged as a possible mechanism for drug delivery. In recent years this has been particularly used by the beauty industry for the delivery of skin care products, but has also been used clinically for the delivery of drugs including Amphoteracin B in the treatment of systemic fungal infections. In these cases the active agent is enclosed in the internal volume of the liposome, which is usually made up of anionic or neutral lipids. Nucleic acids, particularly, DNA are usually in a molecular form which (unless

NATO ASI Series, Vol. H 105
Gene Therapy
Edited by Kleanthis G. Xanthopoulos

very condensed) is too large to be encapsulated in this way. Therefore to facilitate nucleic acid transfer cationic liposomes were developed.

3. Cationic liposomes

3.1 Background

Cationic liposomes by definition carry an overall positive charge. Nucleic acids, DNA, RNA and oligonucleotides, carry a negative charge as a result of the phosphate groups which make up the backbone of these molecules. The positively charged liposomes can thus interact with the negatively charged nucleic acid to form a complex. Cationic liposomes can be formed from double chain cationic lipids which form liposomes spontaneously, however, most positively charged liposomes manufactured for nucleic acid transfer contain a non-bilayer forming cationic lipid and a neutral "helper" lipid such as dioleyolphospatidylethanolamine (DOPE) which stabilizes the liposome.

3.2 Nomenclature

The technology of cationic lipid- and polymer-mediated gene transfer has rapidly developed in recent years leading to the use of a large number of often synonymous terms to describe the various components and composition of these self-assembling systems. In an attempt to overcome some of the more confusing aspects of this evolving terminology Felgner and colleagues have recently described a consensus nomenclature (Felgner *et al.*, 1997). Whilst this nomenclature has not been universally excepted as yet, wherever possible it is recommended that this be used. The main points are as follows: lipoplex refers to any cationic lipid-nucleic complex and polyplex refers to any cationic polymer-nucleic acid complex. Lipofection should be used to describe nucleic acid delivery by cationic lipids and polyfection, nucleic acid delivery mediated by cationic polymers. The nucleic acid can be DNA, RNA or oligonucleotide.

3.3 Structural aspects of cationic lipids

All cationic lipids contain four different functional domains: a positively charged head group, a spacer of varying length, a linker bond and a hydrophobic anchor. The head group of most known cationic lipids contain simple or multiple amine groups with different degrees of substitution. Cationic liposomes containing multivalent cationic lipids, such as DOSPA/DOPE (LipofectAMINE, Gibco BRL, Gaithersburg, Maryland, USA), and DOGS (Transfectam, Promega, Madison, Wisconsin, USA), usually show better transfection activities than those containing monovalent cationic lipids, such as DOTMA/DOPE (Lipofectin, Gibco BRL). This may be due to the fact that multivalent cationic lipids appear to form lipoplexes which are more compact than those formed between DNA and monovalent cationic lipids (Behr *et al.*, 1989; Zhou *et al.*, 1991). The spacer arm appears to be less critical, however, the linker bond does appear to be important in determining the chemical stability and the biodegradability of the cationic lipid and thus influences the toxicity profile of the liposome. Cationic lipids with ester bonds are more biodegradable but are less stable. Recently described novel cationic lipids have used a carbamate linker which appears to be biodegradable but more stable (Lee *et al.*, 1996). There are two forms of hydrophobic lipid anchor that have been used, anchors based on a cholesterol ring, such as DC-Chol (Gao and Huang 1991) and those based on a pair of aliphatic chains, such as DMRIE (Felgner *et al.*, 1994). Both show good transfection properties with both types being represented in those liposomes currently being investigated clinically. There have been several recent studies systematically assessing different combinations of these critical moieties and these should be referred to for further details (Felgner *et al.*, 1994; Lee *et al.*, 1996; Wheeler *et al.*, 1996).

Active cationic liposomes are usually small (100 to 400 nm), and unilamellar and are prepared by vortexing, sonication or microfluidization. In determining the usefulness of a particular cationic lipid for gene transfer several factors need to be considered in addition to its ability to mediate gene transfer at relatively high efficiency. These factors include: the stability of the lipid/liposome, the DNA binding and condensation capacity of the lipid, the consistency of the lipoplex formed, the stability of the lipoplex, the biodegradability and toxicity of the lipid(s), and a resistance to deactivation by serum and other *in vivo* components. Finally, the option to add additional synthetic molecules, such as antibodies, ligands, and endolytic peptides to either enhance specificity of uptake or DNA transfer may be a factor in the design of the lipid components.

3.4 DNA:lipid interactions

Cationic liposomes play 3 roles in the process of nucleic acid transfer; 1) complexation with, and condensation of the nucleic acid; 2) transfer across the cell membrane, and 3) trafficking of the nucleic acid from the cell membrane and perhaps into the nucleus. The head group of the cationic lipid is responsible for interactions between the liposomes and DNA, and between lipoplexes and cell membranes or other components of the cell. The nucleic acid transferred is normally DNA in the form of a plasmid, though RNA, oligonucelotides and DNA up to 150 kb in length have also been used. For the purposes of this review subsequent discussion will be concentrate on the transfer of plasmid DNA.

The interaction of the positively charged liposome with negatively charged DNA results in the spontaneous formation of a complex. Little is known about the interaction of the nucleic acid and the liposome, however, under optimal conditions the contact appears to be very tight protecting DNA from DNase digestion and chelation by ethidium bromide (Gershon *et al.*, 1993). The first proposed model of interaction was theoretical and pictured a "bead on string" structure with the DNA strand decorated with individual whole liposomes (Felgner and Rinegold 1989). However, electron microscopy studies have reported various structures including, string like structures with indications of the fusion of individual liposomes (Gershon *et al.*, 1993), tube like images possible depicting lipid bilayer covered DNA (Sternberg *et al.*, 1994) and entrapment of DNA within aggregate multilamellar structures (Gustafsson *et al.*, 1995). Most recently, using *in situ* optical microscopy and x-ray diffraction, lipoplexes were visualized of consisting of a higher ordered multilamellar structure with DNA sandwiched between cationic bilayers (Rädler *et al.*, 1997; Spector and Schnur 1997).

Whilst these models may be valid for the cationic liposomes under study and under the conditions that the particular lipoplexes were formed, there are many cationic liposomes now available incorporating lipids with a variety of chemical structures and with different amounts of neutral and charged motifs. Thus it is likely that different liposomes may interact with nucleic acids in different ways, particularly mono- versus multi-valent cationic lipids. It should also be noted that the lipoplexes formed have often been observed to be heterogeneous and dynamic and as yet no model has taken this fully into account.

3.5 Cell trafficking

Lipoplexes interact with mammalian cell membranes so facilitating delivery of the DNA to the cell and ultimately transgene expression, however, very little is known conclusively about this process. Endocytosis, phagocytosis, pinocytosis, and direct fusion with the cell membrane may all play a role in lipoplex cell entry depending on the cell type. The initial interaction between the lipoplex and the cell membrane is electrostatic as a result of the excess positive charge associated with the lipoplex and the net negative charge of the cell surface. At least *in vitro* contact time between the lipoplex and the cell surface is critical, usually 4 to 8 hours (Caplen *et al.*, 1995a; Zabner *et al.*, 1995). Whether this time is required for sufficient lipoplexes to settle on the cells to obtain a meaningful amount of DNA transfer and thus gene expression or because this amount of time is required for the active uptake of lipoplexes is unclear. Initially it was thought that lipoplexes entered directly into the cytoplasm as a result of fusion (Felgner *et al.*, 1987; Smith *et al.*, 1993). Alternatively, many groups have suggested that endocytosis is the predominate pathway of entry (Legendre and Szoka 1992; Zhou and Huang 1994; Wrobel and Collins 1995; Zabner *et al.*, 1995). However, the use of this pathway may only apply to small lipoplexes (<200 nm) and thus another report has suggested that phagocytosis is the predominate process of cellular uptake (Matsui *et al.*, 1997). It appears that DNA must be released from lipoplexes prior to entry into the nucleus as microinjection of lipoplexes directly into the nucleus of Xenopus oocytes results in very poor transgene expression in comparison with introduction of naked DNA (Zabner *et al.*, 1995). One suggestion is that DNA is released from the complex as a result of displacement by ionic molecules found in high concentrations in cells, for example, ATP, polypeptides, RNA, histone proteins, or anionic lipids (Xu and Szoka 1996). Finally it should be stressed that different cationic lipids may produce lipoplexes which are taken up by cells in different ways. Also, that different cell-types and even cells in various states of differentiation may use alternative pathways.

3.5 Formulation and DNA transfer efficiency

The need to optimize the formulation of a lipoplex can not be understated. A significant number of biochemical, biophysical and cellular variables can influence the final level of transgene expression mediated by a lipoplex by several orders of magnitude. The most critical, and thus most studied variables are, the ratio of the nucleic acid and cationic lipid, the concentration of these components, the total amount of nucleic acid or lipid, and the composition of the diluent in which the lipoplex is formed (Caplen *et al.*, 1995a; Fasbender *et al.*, 1995; Eastman *et al.*, 1997; Liu *et al.*, 1997).

The ratio of nucleic acid to cationic lipid can be expressed in several ways based on either a weight: weight formulation, a ratio based on the molarity of the two components or the ratio of the positive to negative charge. In the same statement addressing the issue of nomenclature, Felgner and colleagues have recommended that the ratio should in future be calculated and expressed as a charge ratio, as this best describes the electrostatic relationship between the nucleic acid and the cationic lipid (Felgner *et al.*, 1997). Charge ratio is calculated as the

$$\frac{\text{Positive charge equivalents of the cationic component}}{\text{Negative charge equivalents of the nucleic acid component.}}$$

Each cationic lipid will have a different positive charge equivalent and should take account of all protanable groups and the presence of any counterions. Each nucleotide monomer contains one negative charge and thus the negative charge equivalents can be calculated by using an average molecular weight per nucleotide monomer of 330, thus 1 μg DNA is equivalent to 3nmol of phosphate. Lipoplexes with an equal number of negative and positive charges will have a charge ratio of 1, a lipoplex with an excess positive charge will have a ratio greater than 1, and those lipoplexes with a net negative charge will have a charge ratio of less than 1.

The concentration of the DNA and the lipid appears to influence the interaction of the DNA and lipid and the degree of condensation rather than the interaction between the lipoplex and the cell. In contrast, the absolute amount of DNA or lipid appears to reflect optimization of the interaction of the lipoplex with the cell, probable reflecting that insufficient DNA results in reduced transgene expression and too much lipid results in cytotoxicity. The composition of the diluent is critical as this can modulate the available charge due to the presence of chelating agents, salts and the influence of pH. The lipoplex diluent and the media in which the lipofection is performed can also play a role as there is often a need to balance conditions which favor gene transfer with conditions which maintain cell viability.

Unfortunately the determined amount for each variable often does not translate between cell lines and more importantly from *in vitro* to *in vivo*. The degree to which these variables must be optimized will to some extent be dependent on the goal of the investigator. If lipofection is to be used to obtain transgene expression which is required, for example, to determine the relative efficiency of a promoter or the function of a new gene, or to establish a stable cell line using a dominate selectable marker, then only minimal optimization is probable required and often the use of standard conditions suggested by the manufacturer or previously published in the literature will be sufficient for this purpose. However, if the ultimate goal is to be related to gene therapy whether it be to determine the feasibility of using a new lipid, new plasmid construct or a novel target tissue or route of administration there must be due regard to the formulation used.

3.6 *In vitro* versus *in vivo*

Finally, the biggest problem of applying this technology to gene therapy has been that the cationic liposomes, the lipoplex formulations and the transfection conditions which give the best gene expression *in vitro* do not necessarily give the best results *in vivo*. In large part this should be expected as cell lines often exhibit atypical patterns of gene expression and differentiation. The physical chemistry of the cell surface, for example, the surface area, charge and charge density, may be substantially different *in vivo*. In addition, the availability of surface receptors and degree to which a particular cell-type facilitates up-take through endocytsosis or similar mechanisms may vary tremendously from that seen in cell lines and even primary cells. The physical contact time between the lipoplex and the cell surface could be substantially different *in vivo* as often the target tissue will have mechanisms designed to specifically remove particulate matter, for example, cilia in the respiratory and gastro-intestinal tracts. Also a significant proportion of lipoplexes could be subjected to elimination, degradation or trapped in non-target tissues depending on the route of administration and the stability of the lipoplex. In addition, the physiochemical properties of the administered complex can change with exposure to physiological salts, proteins, lipids, and carbohydrates which could result in aggregation, dissociation or reorganization of the lipoplexes.

Having said this significant progress has been made in obtaining reproducible *in vivo* lipofection at an efficiency which mediates meaningful levels of gene expression. *In vivo* routes of administration of lipoplexes have included direct intranasal and intratracheal application or nebulized delivery to the respiratory tract (Stribling *et al.*, 1992; Yoshimura *et al.*, 1992; Alton *et al.*, 1993; Hyde *et al.*, 1993; Canonico *et al.*, 1994; McLachlan *et al.*, 1995; Lee *et al.*, 1996). Other targets have included the gastro-intestinal tract (Alton *et al.*, 1993), brain (Roessler and Davidson 1994; Zhu *et al.*, 1996) and tumors (Plautz *et al.*, 1993; Xing *et al.*, 1997). Intra-arterial (Stephan *et al.*, 1996) and intra-venous administration have also been attempted. The efficacy of intravenous administration of lipoplexes has been somewhat controversial. It appears, however, that using well defined formulations intravenous injection is possible, with vascular endothelial cells, monocytes, and macrophages being the principal targets of transfection (Zhu *et al.*, 1993; Canonico *et al.*, 1994; Thierry *et al.*, 1995; Liu *et al.*, 1997; Smyth Templeton *et al.*, 1997). Importantly, whatever the route of administration, as yet, no DNA transfer has been detected in gonadal tissue following *in vivo* delivery of lipoplexes (Stewart *et al.*, 1992; McLachlan *et al.*, 1995).

5. Clinical application of cationic liposomes

As of the end of 1996 there are in excess of 20 clinical trials worldwide that have assessed or are assessing the application of lipoplexes to the treatment of disease (Marcel and Grausz 1997). At present four cationic liposomes have been, or are being assessed in clinical gene therapy trials, namely, DC-Chol:DOPE (Gao and Huang 1991), DMRIE (Vical Inc., San Diego, California USA) (Felgner *et al.*, 1994), DOTAP (Boehringer Mannheim, Mannheim, Germany) and GL67 (Genzyme Inc., Framingham, Mass., USA) (Lee *et al.*, 1996). Clinical trials have include studies in patients with both genetic and acquired diseases, principally cystic fibrosis (Caplen *et al.*, 1995b; Gill *et al.*, 1997; Porteous *et al.*, 1997)(see accompanying review) and melanoma (Nabel *et al.*, 1993; Nabel *et al.*, 1996). These trials have on the whole been successful. There have been no overt safety problems and in the majority of subjects there has been evidence of DNA transfer and in some cases there have been indications of efficacy. Research, however, must continue so as to improve the efficiency of cationic-liposome mediated DNA thus broadening it's potential clinical application.

5. References

Alton, E.W.F.W., Middleton, P.G., Caplen, N.J., Smith, S.N., Steel, D.M., Munkonge, F.M., Jeffery, P.K., Geddes, D.M., Hart, S.L., Williamson, R., Fasold, K.I., Miller, A.D., Dickinson, P., Stevenson, B.J., McLachlan, G., Dorin, J.R., and Porteous, D.J. (1993) Non-invasive liposome-mediated gene delivery can correct the ion transport defect in cystic fibrosis mutant mice. Nature Genetics **5**, 135-142.

Behr, J.P., Demeneix, B., Loeffler, J.P., and Mutul, J.P. (1989) Efficient gene transfer into mammalian primary endocrine cells with lipopolyamine-coated DNA. Proc. Natl. Acad. Sci. (USA) **86**, 6982-6986.

Canonico, A.E., Conary, J.T., Meyrick, B.O., and Brigham, K.L. (1994) Aerosol and intravenous transfection of human α1-antitrypsin gene to lungs of rabbits. Am. J. Respir. Cell Mol. Biol. **10**, 24-29.

Caplen, N.J., Kinrade, E., Sorgi, F., Gao, X., Gruenert, D., Geddes, D., Coutelle, C., Huang, L., Alton, E.W.F.W., and Williamson, R. (1995a) *In vitro* liposome-mediated DNA transfection of epithelial cell lines using the cationic liposome DC-Chol/DOPE. Gene Therapy **2**, 603-613.

Caplen, N.J., Alton, E.W.F.W., Middleton, P.G., Dorin, J.R., Stevenson, B.J., Gao, X., Durham, S., Jeffery, P.K., Hodson, M.E., Coutelle, C., Huang, L., Porteous, D.J., Williamson, R., and Geddes, D.M. (1995b) Liposome-mediated *CFTR* gene transfer to the nasal epithelium of patients with cystic fibrosis. Nature Medicine **1**, 39-46. Addendum 1 (3) 272.

Eastman, S.J., Tousignant, J.D., Lukason, M.J., Murray, H., Siegel, C.S., Constantino, P., Harris, D.J., Cheng, S.H., and Scheule, R.K. (1997) Optimization of formulations and conditions for the aerosol delivery of functionla cationic lipid:DNA complexes. Human Gene Therapy **8**, 313-322.

Fasbender, A.J., Zabner, J., and Welsh, M.J. (1995) Optimization of cationc lipid-mediated gene transfer to airway epithelia. Am. J. Physiol. **269** (Lung Cell. Mol. Physiol.), L45-L51.

Felgner, P.L., Gadek, T.R., Holm, M., Roman, R., Chan, H.W., Wenz, M., Northrop, J.P., Ringolg, G.M., and Danielsen, M. (1987) Lipofection: A highly efficient, lipid-mediated DNA-transfection procedure. Proc. Natl. Acad. Sci. (USA) **84**, 7413-7417.

Felgner, P.L. and Rinegold, G.M. (1989) Cationic liposome mediate transfection. Nature **337**, 387-388.

Felgner, J.H., Kumar, R., Sridhar, C.N., Wheeler, C.J., Tsai, Y.J., Border, R., Ramsey, P., Martin, M., and Felgner, P.L. (1994) Enhanced gene delivery and mechanism studies with a novel series of cationic lipid formulations. Jour. Biol. Chem. **269**, 2550-2561.

Felgner, P.L., Barenholz, Y., Behr, J.P., Cheng, S.H., Cullis, P., Huang, L., Jessee, J.A., Seymour, L., Szoka, F., Thierry, A.R., Wagner, E., and Wu, G. (1997) Nomenclature for synthetic gene delivery systems. Human Gene Therapy **8**, 511-512.

Gao, X. and Huang, L. (1991) A novel cationic liposome reagent for efficient transfection of mammalian cells. Biochem. Biophys. Res. Comm. **179**, 280-285.

Gershon, H., Ghirlando, R., Guttman, S.B., and Minsky, A. (1993) Mode of formation and structural features of DNA-cationic liposome complexes used for transfection. Biochemistry **32**, 7143-7151.

Gill, D., Southern, K.W., Mofford, K.A., Seddon, T., Huang, L., Sorgi, F., Thomson, A., MacVinish, L.J., Ratcliff, R., Bilton, D., Lane, D.J., Littlewood, J.M., Webb, A.K., Middleton, P.G., Colledge, W.H., Cuthbert, A.W., Evans, M.J., Higgins, C.F., and Hyde, S.C. (1997) A placebo-controlled study of liposome-mediated gene transfer to the nasal epithelium of patients with cystic fibrosis. Gene Therapy **4**, 199-209.

Gustafsson, J., Arvidson, G., Karlsson, G., and Almgren, M. (1995) Complexes between cationic liposomes and DNA visualized by cryo-TEM. Biochim. Biophys. Acta **1235**, 305-312.

Hyde, S., Gill, D., Higgins, C., Treizie, A., MacVinish, L.J., Cuthburt, A., Ratclife, R., Evans, M., and Colledge, W. (1993) Correction of the ion transport defect in cystic fibrosis transgenic mice by gene therapy. Nature **362**, 250-255.

Lee, E.R., Marshall, J., Siegel, C.S., Jiang, C., Yew, N.S., Nichols, M.R., Nietupski, J.B., Ziegler, R.J., Lane, M.B., Wang, K.X., Wan, N.C., Scheule, R.K., Harris, D.J., Smith, A.E., and Cheng, S.H. (1996) Detailed analysis of structures and formulations of cationic lipids for efficient gene transfer to the lung. Human Gene Therapy **7**, 1701-1717.

Legendre, J.Y. and Szoka, F.C. (1992) Delivery of plamsmid DNA into mammalian cell lines using pH senstive liposomes: comparison with cationic liposomes. Pharm. Res. **9**, 1235-1242.

Liu, Y., Mounkes, L.C., Liggitt, H.D., Brown, C.S., Solodin, I., Heath, T.D., and Debs, R.J. (1997) Factors influencing the efficiency of cationic liposome-mediated intravenous gene delivery. Nature Biotechnology **15**, 167-173.

Marcel, T. and Grausz, J.D. (1997) The TMC worldwide gene therapy enrollment report, end 1996. Human Gene Therapy **8**, 775-800.

Matsui, H., Johnson, L.G., Randell, S.H., and Boucher, R.C. (1997) Loss of binding and entry of liposome-DNA complexes decreases trasnfection efficiency in differentiated airway epithelial cells. Jour. Biol. Chem. **272**, 1117-1126.

McLachlan, G., Davidson, D.J., Stevenson, B.J., P.Dickinson, Davidson-Smith, H., Dorin, J.R., and Porteous, D.J. (1995) Evaluation *in vitro* and *in vivo* of cationic liposome-expression construct complexes for cystic fibrosis gene therapy. Gene Therapy **2**, 614-622.

Nabel, G.J., Nabel, E.G., Yang, Z.-Y., Fox, B.A., Plutz, G.E., Gao, X., Huang, L., Shu, S., Gordon, D., and Chang, A.E. (1993) Direct gene transfer with DNA-liposome complexes in melanoma: Expression, biologic activity, and lack of toxicity in humans. Proc. Natl. Acad. Sci. (USA) **90**, 11307-11311.

Nabel, G.J., Gordon, D., Bishop, D.K., J.Nickoloff, B., Yang, Z.-Y., Aruga, A., Cameron, M.J., Nabel, E.G., and Chang, A.E. (1996) Immune response in human melanoma after transfer of an allogenic class I major histocompatibility complex gene with DNA-liposome complexes. Proc. Natl. Acad. Sci. (USA)**93**, 15388-15393.

Plautz, G.E., Yang, Z.Y., Wu, B., Gao, X., Huang, L., and Nabel, G.J. (1993) Immunotherapy of malignacy by *in vivo* gene transfer into tumors. Proc. Natl. Acad. Sci. (USA) **90**, 4645-4649.

Porteous, D., Dorin, J.R., McLachlan, G., Davidson-Smith, H., Davidson, H., Stevenson, B.J., Carothers, A.D., Wallace, W.A., Moralee, S., Hoenes, C., Kallmeyer, G., Michaelis, U., Naujoks, K., Ho, L.P., Samways, J.M., Imrie, M., Greening, A.P., and Innes, J.A. (1997) Evidence for safety and efficacy of DOTAP cationic liposome-mediated CFTR gene transfer to the nasal epithelium of patients with cystic fibrosis. Gene Therapy **4**, 210-218.

Rädler, J., Koltover, I., Salditt, T., and Safinya, C.R. (1997) Structure of DNA-cationic liposome complexes: DNA intercalation in multilamellar membranes in distinct interhelical packing regimes. Science **275**, 810-814.

Roessler, B.J. and Davidson, B.L. (1994) Direct plasmid mediated transfection of adult murine brain cells *in vivo* using cationic liposomes. Neurosci. Lett. **167**, 5-10.

Smith, J.G., Walzem, R.L., and German, J.B. (1993) Liposomes as agents of DNA transfer. Biochim. Biophys. Acta **1154**, 327-340.

Smyth Templeton, N., Lasic, D.D., Frederik, P.M., Strey, H.H., Roberts, D.D., and Pavlakis, G.N. (1997) Improved DNA:liposome complexes for increased systemic delivery and gene expression. Nature Biotechnology **15**, 647-652.

Spector, M.S. and Schnur, J.M. (1997) DNA ordering on a lipid membrane. Science **275**, 791-792.

Stephan, D.J., Yang, Z.-Y., San, H., Simari, R.D., Wheeler, C.J., Felgner, P.L., Gordon, D., Nabel, G.J., and Nabel, E.G. (1996) A new cationic liposome DNA complex enhances the efficiency of arterial gene transfer *in vivo*. Human Gene Therapy **7**, 1803-1812.

Sternberg, B., Sorgi, F.L., and Huang, L. (1994) New structures in complex formation between DNA and cationic liposomes visualised by freeze-fracture electron microscopy. FEBS Letters **356**, 361-366.

Stewart, M.J., Plautz, G.E., Del Buono, L., Yang, Z.Y., Xu, L., Gao, X., Huang, L., Nabel, E.G., and Nabel, G.J. (1992) Gene transfer *in vivo* with DNA-liposome complexes: safety and acute toxicity in mice. Human Gene Therapy **3**, 267-275.

Stribling, R., Brunette, E., Liggitt, D., Gaensler, K., and Debs, R. (1992) Aerosol gene delivery *in vivo*. Proc. Natl. Acad. Sci. (USA) **89**, 11277-11281.

Thierry, A.R., Lunardi-Iskander, Y., Bryant, J.L., Rabinovich, P., Gallo, R.C., and Mahans, L.C. (1995) Systemic gene therapy: biodistribution and long-term expression of a transgene in mice. Proc. Natl. Acad. Sci. (USA) **92**, 9742-9746.

Wheeler, C., Felgner, P.L., Tsai, Y.J., Marshall, J., Sukhu, L., Doh, S.G., Hartikka, J., Nietupski, J., Manthorpe, M., Nichols, M., Plewe, M., Liang, X., Norman, J., Smith, A., and Cheng, S.H. (1996) A novel cationic lipid greatly enhances plasmid DNA delivery and expression in mouse lung. Proc. Natl. Acad. Sci. (USA) **93**, 11454-11459.

Wrobel, I. and Collins, D. (1995) Fusion of cationic liposomes with mammalian cells occurs after endocytosis. Biochim. Biophys. Acta **1235**, 296-304.

Xing, X., Liu, V., Xia, W., Stephens, L.C., Huang, L., Lopez-Berestein, G., and Hung, M.-C. (1997) Safety studies of the intraperitoneal injection of E1A-liposome complex in mice. Gene Therapy **4**, 238-243.

Xu, Y. and Szoka, F.C. (1996) Mechanism of DNA release from cationic liposome/DNA complexes used in cell transfection. Biochemistry **35**, 5616-5623.

Yoshimura, K., Rosenfield, M.A., Nakamura, H., Scherer, E.M., Pavirani, A., Lecocq, J.-P., and Crystal, R.G. (1992) Expression of the human cystic fibrosis transmembrane conductance regulator gene in the mouse lung after *in vivo* intratracheal plasmid-mediated gene transfer. Nuc. Acids Res. **20**, 3233-3240.

Zabner, J., Fasbender, A.J., Moninger, T., Poellinger, K.A., and Welsh, M.J. (1995) Cellular and molecular barriers to gene transfer by a cationic lipid. Jour. Biol. Chem. **270**, 18997-19007.

Zhou, X., Kilbanov, A.L., and Huang, L. (1991) Lipophilic polylysines mediate efficient DNA transfection in mammalian cells. Biochim. Biophys. Acta **1065**, 8-14.

Zhou, X. and Huang, L. (1994) DNA transfection mediated by cationic liposomes containing lipopolysine: characterization and mechanism of action. Biochim. Biophys. Acta **1189**, 195-203.

Zhu, J., Zhang, L., Hanisch, U.K., Fegner, P.L., and R.Reszka. (1996) A continuous intracereberal gene delivery system for *in vivo* liposome-mediated gene therapy. Gene Therapy **3**, 472-476.

Zhu, N., Liggitt, D., Liu, Y., and Debs, R. (1993) Systemic gene expression after intravenous DNA delivery into adult mice. Science **261**, 209-211.

Theory and Practice of Using Polycationic Amphiphiles and polymers for In Vitro and In Vivo Gene Transfer

B. A. Demeneix, D. Goula, C. Benoist, J.S. Rémy* and J. P. Behr*.

Laboratoire de Physiologie Générale et Comparée, U.R.A.90 CNRS, Museum National d'Histoire Naturelle, F-75231, Paris Cedex 5, France

* Laboratoire de Chimie Génétique associé au CNRS, Faculté de Pharmacie, Université Louis Pasteur de Strasbourg, F-67401 Illkirch.

Abstract.

The mechanisms underlying the actions of polycationic (as opposed to monocationic) gene transfer vectors is described. Two main types of vectors are examined, polycationic amphiphiles such as DOGS (Transfectam™) and Lipofectamine on the one hand and cationic polymers such as polyethyleneimine on the other hand. The gene transfer performances of these molecules is a function of their DNA condensing capacity, their interactions with anionic proteoglycans of the cell membrane and their capactiy to induce endosome swelling and rupture. The importance of taking into account the overall charge ratio of complexes when carrying out *in vitro* or *in vivo*gene transfer is emphasized.

A central problem remaining to be solved in the field of gene therapy is that of effective gene transfer. Three broad classes of approaches can be used to administer DNA in a therapeutic setting: genetically engineered cells, viruses and plasmid-based delivery systems. Each field is undergoing intensive research and effective therapies could well involve more than one approach. This review considers how plasmid DNA can be transferred into cells by the use of polycationic (as opposed to monocationic) vectors. Indeed, although in the early period of research on gene therapy little interest was shown in non-viral or plasmid-based methods, of late there has been a clear increase in activity and more attention is being paid to their potential for gene transfer and therapy.

However, getting DNA into the cell is not all the story. First, in an in vivo setting the DNA/vector complex must reach the target cell, and this is no mean feat, with circulating and extracellular matrix proteins creating major obstacles to complex diffusion and stability. Second, if the target cell is reached, then transfer of complexes into the cell must be followed by a cascade of events including release from the endosome, intracellular trafficking and crossing of the nuclear membrane, all of this culminating in the synthesis of a large number of effector protein molecules. As will be emphasised below, the

NATO ASI Series, Vol. H 105
Gene Therapy
Edited by Kleanthis G. Xanthopoulos

chemical nature of the vector can influence each of the delivery steps and in particular that of membrane rupture. This step could, in principle, occur directly at the cell surface, but all current ultrastructural studies suggest that complexes enter cells by endocytosis and that membrane rupture probably occurs after endocytosis.

Mechanism of polycationic amphiphile- and polymer-based gene transfer

Before dealing with the different polycationic molecules that have been developed and reviewing some of their uses in in vitro and in vivo settings, we will deal with some of the hypotheses behind their mode of action at the levels of DNA compaction, interaction with the cell surface and membrane rupture. First, complex formation between DNA and cationic lipids or polymers is a process which is still largely empirical and uncontrolled. Basically, the anionic plasmid and cationic vector will collapse into 0.05-0.3 micrometer particles. It is important to note that with polycationic (as distinct from monocationic) molecules, like charges borne on the molecule will repel each other, thus extending the molecule and optimising counterion collapse on interaction with the polyanionic DNA. Such properties will also affect interaction with the cell membrane, as high charge density will favour strong interactions with the cell surface (unfortunately also with the extracellular matrix and with the complement system). Indeed, provided the net charge of the complex formed is cationic (with an excess ratio of cationic charges to nucleic acid phosphates) cooperative ionic interaction will again enable the complexes to bind to polyanionic glycosaminoglycans of the cell membrane (1; 2). In vitro, studies with both cationic lipids and cationic polymers, have shown that electrostatic interactions between the negatively charged cell membranes and the positively charged DNA/vector complexes are enhanced by increasing the overall charge of the complexes, which in turn is achieved by increasing the ratio of vector to DNA (3;4;). Moreover, electron microscopy has been used to follow these interactions in vitro, and such studies show that on adherent cells, interactions of positively charged complexes with the cell membrane will lead to endocytosis (1).

As to the process of endosomal membrane rupture, again, if an extended polycation (such as spermine) is used as head group for an amphiphile, it will have a favorable effect as its overall molecular shape will be that of a wedge, which upon self-assembling leads to tubular cell membrane-perturbing phases. Another membrane rupturing property of polyamines could come from their buffering potential at physiological pH. This hypothesis is supported by the observation that transfection with polyamines can not be improved by the addition of fusogenic lipids or lysosomotropic bases (5). Moreover, the potentiometric protonation states of the amines shows that at physiological pH only three of the four nitrogens in the spermine head are cationic (Fig. 1). The pKa of the last amine is 5.5 (6), half way between the extracellular and intralysosomal pH values. This feature contrasts with the headgroups of monovalent cationic lipids which contain a constitutively charged quaternary ammonium that cannot provide any

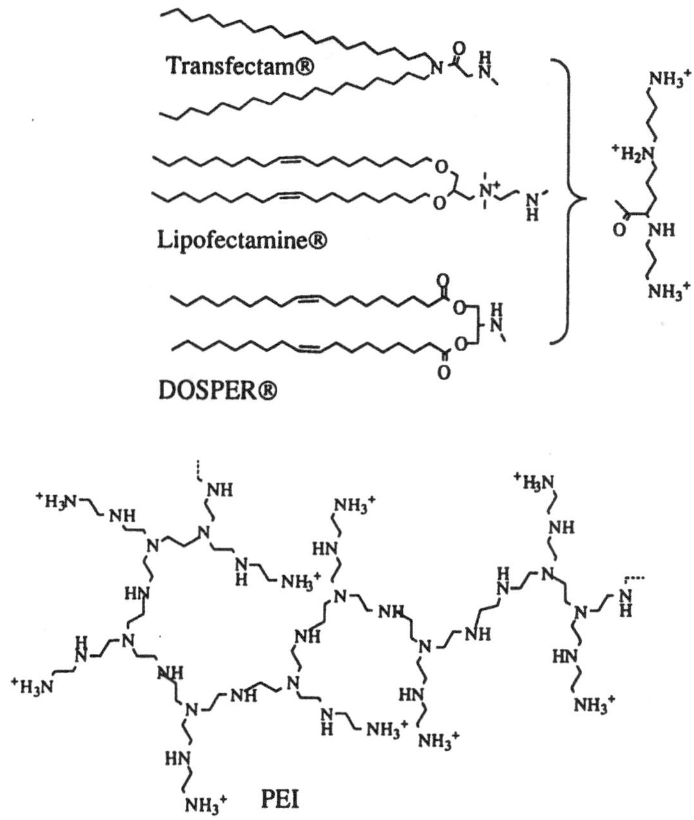

Figure 1 Structure of some of the commercially available polycationic vectors described in the text: Transfectam™, Lipofectamine™, DOSPER ™and PEI.

buffering capacity in endosomes. The buffering hypothesis is bolstered by results obtained with some cationic polymers. On the one hand, there are the polyamidoamine dendrimers (7;8); these quasispherical macromolecules bear a large number of amine groups and again, as for the lipopolyamines, not all of these amines are protonated at physiological pH (Fig. 1). On the other hand, we have polyethyleneimine (PEI, Fig. 1; ref 5), a commercially available polymer in which one in every third atom is an amino nitrogen that can be protonated. In fact, PEI is the cationic polymer having the highest charge density potential and the overall protonation level of PEI increases from 20 to 45% between pH 7 and 5 (9).

Therefore, given that certain polycationic vectors can provide a substrate for a protonation process (i.e. cationic charge=protonation of nitrogen atom), the more the process continues the more it will result in a distribution of decreasing pK's. This in turn may result in buffering capacity around pH 5 to 6. This will not only tend to inhibit the action of the lysosomal nucleases that have an acidic optimal pH, but will also alter the osmolarity of the vesicle.

Indeed, the accumulation of protons brought in by the endosomal ATPase is coupled to an influx of chloride anions (10). In the presence of a protonatable polycation such as PEI there will be a large increase in the ionic concentration within the endosome, resulting in a swelling of the polymer by internal charge repulsion and osmotic swelling of the endosome, due to water entry. As the two phenomena occur simultaneously it is likely that endosomal life expectancy is rapidily reduced! How the subsequent steps of complex dissociation and intracellular trafficking, nuclear membrane crossing occur is unclear, but it seems it is inefficient.

Polycationic amphiphiles.

The DNA condensing capacity of lipopolyamines was first described over a decade ago (11). The demonstration that the DNA/lipid complexes so formed could be used for transfection followed three years later (3). The commercialization in 1990 of one of the lipopolyamines, dioctadecylamidoglyclspermine (DOGS, Transfectam™), described in these early papers was followed by that of two others, Lipofectamine™ and DOSPER™, in 1992 and 1996 respectively. All contain the same spermine headgroup chemically derived from bis-aminopropylation of ornithine (Figure 1). Lipofectamine is a 3/1 mixture of 2,3-dioleyoxy-N-{2(sperminecarboxamido)ethyl}-N,N-dimethyl-1-propanaminium trifluoroacetate (DOSPA) and a neutral lipid dioleoylphosphatidylethanolamine (DOPE), whereas DOSPER (1,3-dioleoyloxy-2-(6-carboxyspermyl)-propylamide) is supplied alone .

These three lipopolyamines have been widely used for transfection of a very large variety of primary cultures and cell lines in vitro. They have been shown to be consistently more efficient than monocationic lipids (see introduction for current hypotheses as to why). Furthermore, a systematic study of the transfection performances of a large series of cationic lipids showed the polycationic nature to be essential for transfection efficiency (12). An independent line of argument comes from the work of Wheeler and co-workers (13), who added a protonatable amine function to a mono-cationic lipid to produce ßaminoethyl-DMRIE. This compound, unlike the parent lipid DMRIE, provided high levels of transfection without the addition of a helper lipid such as DOPE. Plasmid DNA, antisense oligonucleotides, mRNA and even protein can be delivered to eukaryotic cells by these polycationic cytofectins. Such techniques have been applied either for studying gene regulation per se or for testing *in vitro* models of experimental therapies. Also, they have been combined with replication deficient viruses (14) to obtain higher levels of transfection than with either component alone.

Each of the lipopolyamines described above has been tested for their capacities to deliver DNA in vivo. Most often, those conditions which were defined as optimal (high charge ratio) for *in vitro* work proved to be unsatisfactory for *in vivo* use (Fig. 2), as complexes bearing strong positive charge interact preferentially with the extracellular matrix and the

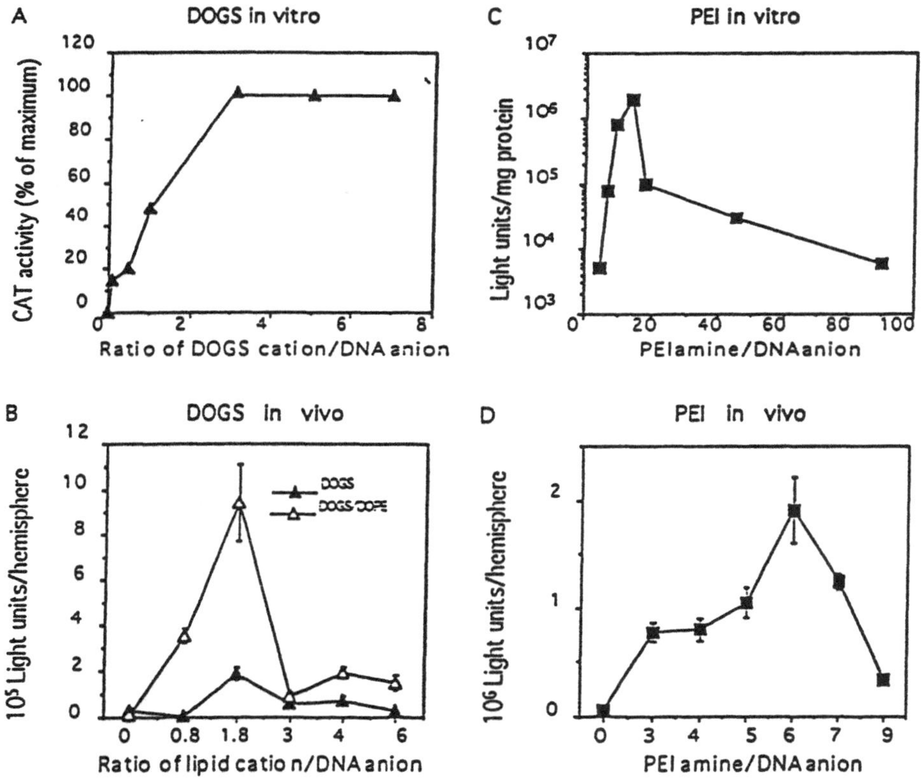

Figure 2.
Comparaison of transfection efficencies obtained with different amine/phosphate ratios (R) when using either DOGS *in vitro* (A) and *in vivo* (B) or PEI *in vitro* (C) and *in vivo* (D) . A) Transfection of primary neurons from new born rat brain, transfected with increasing amounts of DOGS and a constant amount (5µg) of a chloramphenicol acetyl transferase (CAT) expression vector. B) Levels of firefly luciferase expression resulting from *in vivo* transfection of newborn mouse brains with 1µg CMV-luc complexed with increasing amounts of DOGS alone or DOGS used with a two-molar excess of DOPE. Note, comparing A and B, that maximal transfection in vivo is obtained at R1.8, whereas *in vitro*, maximal transfection can be obtained at R 3-R6, conditions which are ineffective *in vivo*. C) Expression of firefly luciferase following in vitro transfection of 3T3 fibroblasts with CMV-luc and different PEI amine/phosphate ratios (Equivalents, eq). Note that as only one in five of the protonatable PEI amines is charged at physiological pH, 20 eq gives a charge ratio of 4. D) Levels of firefly luciferase (expressed as light units per brain hemisphere) resulting from *in vivo*stereotaxic transfection of adult mouse brains with 5µg CMV-luc complexed with increasing PEI amine/phosphate ratios. Comparing C and D we note again that maximum transfection *in vitro* is obtained at 20 eq (R4) whilst *in vivo* transfection is optimal at 6eq (R1.2).

complement system (15). However, DOGS has been used alone in slightly positively charged particles, both for transfecting chick embryos by depositing complexes directly on the embryo (16) and also, somewhat surprisingly, for intravenous delivery to foetuses in gestating mice (17). However, if good distribution is to be obtained on injection, then very low amounts of cationic lipid must be used, so that the complexes do not bear an excessive cationic charge. Usually the low amounts of DOGS will require the addition of a two fold molar excess of DOPE to provide stable complexes and more efficient transfection. DOGS/DOPE complexes have been used successfully for introducing genes into the newborn mouse brain by direct injection (18) and for intravenous delivery (19).

Cationic polymers.
Cationic polymers have long been used for gene transfer and we can currently view the perspective on three generations. The first, DEAEdextran and polybrene have effectively been surpassed by the second and third generations. The main player in the second generation was polylysine (PLL). However, to approach the efficiencies of the lipopolyamines, PLL, whether conjugated to a ligand or not, requires numerous additives such as chloroquine (see for instance, 20) fusogenic peptides (21) or inactivated adenovirus (22). Each of these partners act at the same level: that of endosome release. This, again, underlines the intrinsic endolytic potential of the polyamines that do not require such additions.
Other members of this second generation of cationic polymeric vectors that appeared between the late-eighties and mid-nineties were nuclear proteins such histones and high mobility group proteins (HMG). Bottger et al. (23) showed that the chromosomal non-histone protein HMG1 produced condensed monodisperse complexes with plasmid DNA, and that the complexes could be used successfully for in vitro transfection. Use of histones has been more frequent and often involve coupling to ligands . More recently, a histone protein (H1) has been used with anionic lipids to make transfection competent complexes in which the DNA is present on the outside of the DNA/lipid particles (24).
Overall, most cationic polymers are not very good artificial vectors per se, with two notable exceptions however. Using the firefly luciferase reporter gene, Remy and co-workers (25) compared the transfection activities of polyethylenimines (PEIs) of various molecular weight, whether branched or linear, PEIs modified with epichlorhydrine or ethoxylated, 57kDa polyallylamine, 32kDa polyhistidine, chitosan, (3,10)n-polyamines, polyamidoamine and polypropylenimines dendrimers of different sizes, and 50kDa polylysine on 3T3 cells. The results showed that only PEIs of molecular weight above 2kDa and dendrimers formed of either polyamidoamine or polypropylenimine perform well in gene transfer. Each of these compounds is able to buffer endosomes, producing the proton sponge effect (26) with swelling at acidic pH which causes lysis of the endosome . The spherical

cationic polyamidoamines (PAMAM) polymers or starburst dendrimers consist of primary amines on the surface and tertiary amines inside. Their transfection activity is greatly improved by heat treatment which induces significant degradation and « fracturation » of the polymer. Indeed, Tang et al. (8) show that there is a correlation between transfection efficiency and the degree of flexibility of the molecules. It is proposed that this flexibility enables the fractured dendrimer to be compact when complexed with DNA and yet swell (and producing an endolytic effect) when released from the nucleic acid. The transfection efficiency of PEI has been compared to that of lipopolyamines on a large variety of cell lines and primary cultures. The results show efficiencies at least as high as the best currently available synthetic vectors (4) indicating an entirely new function for this simple molecule. Moreover a number of reports show PEI to be suitable for *in vivo* gene transfer. In both the adult and the newborn mouse brain PEI/ DNA complexes provide levels of transfection equal to those found in vitro for the same amount of DNA applied to primary neuronal cultures (up to 10^6 RLU per µg DNA injected). The best levels of expression in both models are obtained with the branched 25kDa polymer (27) or the linear 22Kd molecule (Goula et al., unpublished observations). Double immunostaining with antibodies against cell specific markers and transgene products show that both neurons and glia can be transduced by PEI transfection in vivo (27). Moreover, toxicity is low, no mortality being observed in injected animals and no necrosis at the site of injection. Also of interest is that when transfecting neuronal cells in culture with PEI, no interference with membrane excitability is seen (28).
As for when working with polycationic lipids, different amine/phosphate ratios were required for optimal transfection efficencies in vitro and in vivo (Fig. 2 C and D). In both the newborn and adult central nervous system, complexes with low overall charges provide the best transfection efficiencies. Ratios of 6 or 9 amines per DNA phosphate were found to be optimal in the adult and immature brain (27; Goula et al. unpublished observations). Theoretically, these ratios produce complexes bearing only low net cationic charges. Such complexes would be expected to diffuse in the extracellular matrix (whereas highly charged cationic complexes such as used in the in vitro situations would likely stick to substrates around the injection site).
A final point about PEI is that it is a simple molecule that can be chemically modified (i.e. ethoxylated or modified by epichlorhydrine) without loss of activity. This latter result shows that PEIs will provide ideal starting points for chemical modifications aimed at acheiving improved distribution in vivo or cell specific targeting.

References

1. Labat-Moleur F, Steffan AM, Brisson C, Perron H, Feugeas O, Furstenberger P, Oberling F, Brambilla E, Behr JP : **An electron microscopy study into the mechanism of gene transfer with lipopolyamines.** Gene Ther. (1996). 3 : 1010-1017

2. Mislick KA, Baldeschwieler JD : **Evidence for the role of proteoglycans in cation-mediated gene transfer.** Proc. Natl. Acad. Sci. USA (1996). **93** : 12349-12354

3. Behr JP, Demeneix BA, Loeffler J, Perez-Mutul J : **Efficient gene transfer into mammalian primary endocrine cells with lipopolyamine-coated DNA.** Proc. Natl. Acad. Sci. USA (1989). **86** : 6982-6986

4. Boussif O, Lezoualc'h F, Zanta MA, Mergny M, Scherman D, Demeneix BA, Behr J-P : **A novel, versatile vector for gene and oligonucleotide transfer into cells in culture and in vivo: polyethylenimine.** Proc. Natl. Acad. Sci. USA. (1995). **92** : 7297-7303

5. Remy JS, Kichler A, Mordinov V, Schuber F, Behr JP : **Targeted gene transfer into hepatoma cells with lipopolyamine-condensed DNA particles presenting galactose ligands: a stage toward artificial viruses.** Proc. Natl. Acad. Sci. USA (1995). **92** : 1744-1748

6. Behr JP : **Gene transfer with synthetic cationic amphiphiles: prospects for gene therapy.** Bioconjugate Chem. (1994). **5** : 382-389

7. Haensler J, Szoka FC : **Polyamidoamine cascade polymers mediate efficient transfection of cells in culture.** Bioconjugate Chem. (1993). **4** : 32-39.

8. Tang MX, Redemann CT, Szoka FC : ***In vitro* gene delivery by degraded polyamidoamine dendrimers.** Bioconjugate Chemistry (1996). **7** : 703-714

9. Suh J, Paik H-J, Hwang BK : **Ionization of polyethylenimine and polyallylamine at various pH's.** Bioorg. Chem. (1994). **22** : 318-327

10. Nelson N : **Structure and pharmacology of the proton ATPases.** Trends Pharmacol. Sci. (1991). **12** : 71-75

11. Behr JP : **DNA strongly binds to micelles and vesicles containing lipopolyamines or lipointercalants.** Tetrahedron Lett. (1986). **27** : 5861-5864

12. Remy JS, Sirlin CS, Vierling P, Behr JP : **Gene transfer with a series of lipophilic DNA-binding molecules.** Bioconjugate Chem. (1994). **5** : 647-654

13. Wheeler CJ, Sukhu L, G. Y, Tsai YJ, C. B, Felgner PL, J. N, Marshall J, Manthorpe M : **Converting an alcohol to an amine in a cationic lipid dramatically alters the co-lipid requirement, cellular transfection activity and the ultrastructure of DNA-cytofectin complexes.** Biochim. Biophys. Acta (1996). **1280** : 1-11

24. Meunier-Durmont C, Grimal H, Sachs L, Demeneix BA, Forest C : **Adenovirus enhancement of polyethyeleimine-mediated transfer of regulated genes in differentiated cells.** Gene Ther. (1997). In press

15. Plank C, Mechtler K, Szoka FC, Wagner E : **Activation of the complement system by synthetic DNA complexes: A potential barrier for intravenous gene delivery.** Hum. Gene Ther. (1996). **7** : 1437-1446

16. Demeneix BA, Abdel-Taweb H, Benoist C, Seugnet I, Behr JP : **Temporal and spatial expression of lipospermine-compacted genes transferred into chick embryos *in vivo*.** BioTechniques (1994). **16** : 496-501

17. Tsukamoto M, Ochiya T, Yoshida S, Sugimura T, Terada M : **Gene transfer and expression in progeny after intravenous DNA injection into pregnant mice.** Nature Genet. (1995). **9** : 243-248

18. Schwartz B, Benoist C, Abdallah B, Scherman D, Behr JP, Demeneix BA : **Lipospermine-based gene transfer into the newborn mouse brain is optimized by a low lipospermine/DNA charge ratio.** Hum. Gene Ther. (1995). **6** : 1515-1524

19. Thierry AR, Lunardy-Iskandar Y, Bryant JL, Rabinovich P, Gallo RC, Mahan LC : **Systematic gene therapy: Biodistribution and long-term expressionof a transgene in mice.** Proc. Natl. Acad. Sci. USA (1995). **92** : 9742-9746

20. Stewart AJ, Pichon C, Meunier L, Midoux P, Monsigny M, Roche AC : **Enhanced biological activity of oligonucleotides complexed with glycosylated poly-L-lysine.** Mol. Pharmacol. (1996). **50** : 1487-1494

21. Wagner E, Plank C, Zatloukal K, Cotten M, Birnstiel ML : **Influenza virus hemagglutinin HA-2 N-terminal fusogenic peptides augment gene transfer by transferrin-polylysine-DNA complexes: toward a synthetic virus-like gene transfer vehicle.** Proc. Natl. Acad. Sci. U.S.A. (1992). **89** : 7934-7938

22. Cook DR, Maxwell IH, Glode LM, Maxwell F, J.O. S, Purner MB, Wagner E, Curiel DT, Curiel TJ : **Gene therapy for B-cell lymphoma in a SCID mouse model using an immunoglobulin-regulated diphteria toxin gene delivery by a novel adenovirus-polylysine conjugate.** Cancer Biother. (1994).9:131-141

23. Bottger M, Vogel F, Platzer M, Kiessling U, Grade K, Strauss M : **Condensation of vector DNA by the chromosomal protein HMG1 results in efficient transfection.** Biochim Biophys Acta (1988). **950** : 221-228

24. Hagstrom JE, Sebestyen MG, Budker V, Ludtke JJ, Fritz JD, Wolff JA : **Complexes of non-cationic liposomes and histone H1 mediate efficient transfection of DNA without encapsulation.** Biochim. Biophys. Acta (1996). **1284** : 47-55

25. Remy JS, Abdallah B, Zanta MA, Boussif O, Behr JP, Demeneix BA : **Gene transfer with lipospermines and polyethylenimines.** Advanced Drug Delivery Reviews (1997). In press.

26. Demeneix BA, Behr JP. **The proton sponge: a trick the viruses did not exploit.** Artificial Self-Assembling Systems for Gene Delivery (1996). 146-151

27. Abdallah B, Hassan A, Benoist C, Goula D, Behr JP, Demeneix BA : **A powerful non-viral vector for in-vivo gene transfer into the adult mammalian brain : polyethylenimine.** Hum. Gene Ther. (1996).

28. Lambert RC, Maulet Y, Dupont JL, Mykita S, Craig P, Volsen S, Feltz A : **Polyethylenimine-mediated DNA Transfection of Peripheral and Central Neurons in Primary Culture: Probing Ca2+ Channel Structure and Function with Antisens Oligonucleotides.** Mol. Cell. Neurosci. (1996). 7 : 239-246

Lipid Gene Transfer and Clinical Gene Therapy

Natasha J. Caplen.
Clinical Gene Therapy Branch, National Human Genome Research Institute, National Institutes of Health, Bethesda, MD 20892, USA.

Keywords. Liposome-mediated gene transfer, cystic fibrosis, clinical trials

1 Introduction

There are several features of liposome-mediated DNA transfer that make it a viable delivery system for clinical gene therapy. 1) In contrast to viral delivery systems, contamination with exogenous protein or other nucleic acid is minimal, so the potential for immunity is significantly reduced, making repeated administration more feasible. 2) The components can be easily produced on a large scale. 3) The components can be easily manipulated and there is theoretically no size restriction on the DNA that can be delivered.

At present four cationic liposomes have been, or are being assessed in clinical gene therapy trials, namely, DC-Chol:DOPE (Gao and Huang 1991), DMRIE (Vical Inc., San Diego, California USA) (Felgner *et al.*, 1994), DOTAP (Boehringer Mannheim, Mannheim, Germany) and GL67 (Genzyme Inc., Framingham, Mass., USA) (Lee *et al.*, 1996). Clinical trials have include studies in patients with both genetic and acquired diseases, principally cystic fibrosis and melanoma. These trials have on the whole been successful. There have been no overt safety problems, in the majority of subjects there has been evidence of DNA transfer and in some cases there have been indications of efficacy. However, in transferring this technology from the laboratory to testing first in animals and then in human subjects some interesting issues have been brought to light which are now having to be actively addressed.

The most significant issues in applying lipid gene transfer clinically have been that of: (1) transfection efficiency, and the level of transgene expression thus obtained; (2) the persistence of transgene expression; (3) the determination of formulations which can be practically administered to human subjects, and (4) the minimization of acute cytotoxicity. This review will discuss these issues using examples from the pre-clinical and clinical studies that have been conducted over the last 5 years with the aim of developing cationic liposome-mediated gene therapy for patients with cystic fibrosis.

2 Cystic Fibrosis

Cystic fibrosis (CF) is the most common lethal inherited disease in Europe and North America. CF is inherited as an autosomal recessive character, with approximately 1 in 20 people heterozygous for the abnormal gene. About 1 in 2000 live births in

NATO ASI Series, Vol. H 105
Gene Therapy
Edited by Kleanthis G. Xanthopoulos

Europe and North America are children with CF, which is usually diagnosed during the first 5 years of life. Approximately 10% of cases are diagnosed at birth as infants suffering with intestinal blockage; the remainder usually present with an increased susceptibility to respiratory infection. The principal pathology centers around the respiratory and intestinal tracts. In both cases, these hollow, epithelial-lined organs become filled with thickened, tenacious secretions. The intestinal symptoms are generally milder and are of two types; obstruction of the small intestine, which can normally be corrected by surgery, and reduced secretion of the pancreatic enzymes needed for absorption of gut contents. Pancreatic insufficiency, as this is termed eventually occurs in 80-90% of patients requiring enzyme treatment and other dietary supplements.

The principal clinical problem of CF is lung damage and respiratory failure as a result of bacterial colonization and recurrent chest infections. The main early bacterial infections are due to *Staphylococcus aureus* and *Haemophilus influenzae*, with *Pseudomonas aeruginosa* and *Burkholderia cepacia* subsequently becoming the principal pathogens. Colonization with these bacteria contributes significantly to the cycles of inflammation which lead to bronchietasis and respiratory failure in CF patients. Other clinical features of CF include the development of liver disease as a result of blockage of the intrahepatic bile ducts, and azo-spermia as a result of absence or obstruction of vas deferens which renders CF males sterile (Boat *et al.*, 1989). Fifty years ago the mean life expectancy for an individual with CF was approximately 1 year; this has now risen to around 30 years. This increase in life expectancy is mainly due to improved antibiotics, the use of intensive physiotherapy and (in rare cases) heart and lung transplantation. However, CF remains a disease associated with significant morbidity and mortality and a huge social and financial cost; hence, the urgent need for the development of new therapies.

2.1 CF genetics and physiology

The gene effected in individuals with CF was identified and isolated in 1989 and is termed the cystic fibrosis transmembrane conductance regulator (*CFTR*) gene (Riordan *et al.*, 1989; Rommens *et al.*, 1989). The gene is located on the long arm of chromosome 7, contains 27 exons, covers approximately 250 kb of genomic DNA and encodes an mRNA of about 6.5 kb. The CFTR protein is a cAMP-regulated chloride channel found on the apical membrane of epithelial cells (Welsh *et al.*, 1992). The CFTR protein consists of two transmembrane regions, two nucleotide-binding domains and a regulatory region containing a number of potential phosphorylation sites (Riordan *et al.*, 1989). The epithelial cells lining the secretory ducts in CF individuals exhibit absent or significantly reduced cAMP-mediated chloride transport across the apical membrane of those cells as a result of dysfunctional CFTR. Over 400 mutations in the *CFTR* gene have been described (Tsui 1995). The most common mutation associated with CF is a 3 bp deletion in exon 10 of the *CFTR* gene (ΔF508). This mutation accounts for approximately 70% of CF chromosomes and removes a phenylalanine residue within the first nucleotide binding fold of CFTR, this results in a mutant protein which is never trafficked to the cell membrane (Cheng *et al.*, 1990). Other mutations result in a 'null phenotype' as no full length *CFTR* mRNA is produced or in reduced or unregulated chloride transport (Welsh and Smith 1993).

Whilst our understanding of the genetics and the basic cellular defect underlying CF have significantly improved in recent years, the relationship between alterations in the function of CFTR and the pathology of CF is only poorly understood. One theory suggests that the abnormal ion transport characteristic of CF epithelia leads to disrupted water movement which in turn results in dehydration of the epithelial cell surface and associated fluids. In the lung this may have the additional consequence of disturbing mucocilliary clearance, the process which ensures removal of inhaled particles and bacteria through the synchronized beating of cilia. However, more recent studies have began to suggest that there may be a more direct link between CFTR and bacterial adherence and/or colonization of the CF airway. One study has shown that transfer of the human *CFTR* cDNA to primary CF nasal cells significantly reduces the number of bound *P. aeruginosa* associated with these cells compared to un-transfected or mock transfected CF cells (Davies *et al.*, 1997). A second study has demonstrated that normal airway surface fluid contains a bactericidal activity not present in fluid obtained from individuals with CF. Transfer of the *CFTR* cDNA to CF cells corrected this defect in bacterial killing (Smith *et al.*, 1996).

2.2 Gene therapy for cystic fibrosis

2.2.1 Background

The first studies to suggest that transfer of the *CFTR* gene might have an effect on the basic cellular defect *in vitro* where conducted using retroviral-mediated transfer (Drumm *et al.*, 1990; Rich *et al.*, 1990). In theses studies CFTR gene transfer was shown to restore normal ion transport to cells derived from CF patients. Subsequently many groups observed similar findings using adenoviral based vectors, adeno-associated viral vectors and cationic liposomes (Egan *et al.*, 1992; Mittereder *et al.*, 1994; Rosenfeld *et al.*, 1994a; McLachlan *et al.*, 1995; Stern *et al.*, 1995). *In vivo*, the normal human *CFTR* gene has been successfully transferred to a variety of animals including, mice, rats, rabbits and non-human primates (Rosenfeld *et al.*, 1992; Yoshimura *et al.*, 1992; Flotte *et al.*, 1993; Yei *et al.*, 1994; Zabner *et al.*, 1994b; Logan *et al.*, 1995). In proceeding to clinical studies, research has had to address several issues, including: (1) what is the minimum level of CFTR expression that will be required for therapy; (2) are there any risks associated with over or inappropriate expression of CFTR; (3) where to target gene transfer and, (4) what vector delivery system should be used?

2.2.2 *CFTR* transgene expression and therapeutic benefit

An important consideration in the development of a gene therapy strategy for CF is the need to establish the minimum level of gene transfer which could still modulate the progression of the disease. As the normal level of expression of CFTR in the lung is low (Crawford *et al.*, 1991; Trapnell *et al.*, 1991; Kartner *et al.*, 1992) and

CF carriers have only half that and yet have no clinical phenotype the required level gene transfer may not be that high. *In vitro* retroviral transduction of between 5 and 20% of CF cells grown as a complete monolayer results in total restoration of chloride conductance (Johnson *et al.*, 1992; Zabner *et al.*, 1994a; Goldman *et al.*, 1995) (Johnson *et al.*, 1992; Zabner *et al.*, 1994a; Goldman *et al.*, 1995). *In vivo*, CF transgenic mice expressing only 5% of the normal level of *Cftr* show disproportionately more chloride secretion (approximately 50% of wild-type) than if the relationship were linear (Dorin *et al.*, 1996). However, it should be stressed that it is unclear how these findings relate to any likely clinical benefit, though in CF transgenic mice expression of only 5% of wild-type *Cftr* a significantly reduces intestinal disease (Dorin *et al.*, 1996). It should also be noted that correction of sodium hyperabsorption, another ion transport abnormality characteristic of CF epithelia, appears to require much higher levels of gene transfer, may be up to 100% (Goldman *et al.*, 1995).

2.2.3 Over-expression of CFTR

The potential problem of over expression of CFTR or expression in sites where CFTR is not normally expressed has been addressed in only a limited number of studies, however, the data from these is broadly encouraging. *In vitro*, Rosenfeld and co-workers have shown that increasing levels of CFTR protein does not induce levels of cAMP chloride secretion over a certain level (Rosenfeld *et al.*, 1994b). *In vivo*, transgenic mice expressing both human and mouse CFTR in the lung had normal lung weight, morphology and overall growth (Whitsett *et al.*, 1992). As yet there have been no reports of antibodies to the CFTR protein following CFTR transgene expression though this remains a theoretical problem, particularly in patients with a 'null' phenotype.

2.2.4. Spatial expression of CFTR

Though CF is a multi-organ disorder, it is the respiratory aspects of the disease which are associated with the greatest morbidity and mortality and thus the airway has been the principal target for CF gene therapy. It is unclear from present data examining the natural distribution of CFTR expression in the airway as to exactly which cell type should be the target of gene therapy for CF. The highest level of CFTR expression is in the submucosal glands of the proximal airways, but sub-populations of more peripheral epithelial cells, both ciliated and non-ciliated, also express CFTR (Engelhardt *et al.*, 1992; Engelhardt *et al.*, 1994). Interestingly, though the lower respiratory tract remains the most medically relevant target, several investigators have chosen to start their clinical studies by looking at gene transfer to the nasal mucosa. The nasal cavity represents a useful clinical test area for initial assessment of the safety and efficacy of *CFT*R gene transfer and has several advantages in comparison with the lung as there is easier access for both for gene transfer and measurements of safety and a reduced risk in case of the occurrence of side-effects. In addition, well defined measurements of ion transport can be performed on the nasal epithelium that clearly differentiate CF and non-CF (Middleton *et al.*,

1994b; Knowles *et al.*, 1995b) and can be used to measure the efficacy of *CFTR* gene transfer.

2.2.5. Choice of vector delivery system

The first reports of *in vitro* correction of the CF chloride channel defect used retroviral-mediated transfer. However, *in vivo* retroviruses have been considered to be unsuitable for mediation of gene transfer to the airway. This is because retroviral vectors require cell proliferation for proviral integration and gene expression and airway surface epithelial cells are terminally differentiated and thus non- or slowly dividing. Only in the presence of a damaged epithelium which will stimulate proliferation and allow access to the more rapidly dividing underlying progenitor cells has retroviral-mediated gene transfer been obtained in the airway (Halbert *et al.*, 1996).

Adenovirus (Ad) is a particularly attractive vector for gene transfer to the airway as it naturally infects the respiratory epithelia. Currently there are five published clinical trials assessing Ad-CFTR transfer (Zabner *et al.*, 1993; Crystal *et al.*, 1994; Hay *et al.*, 1995; Knowles *et al.*, 1995a; Zabner *et al.*, 1996; Bellon *et al.*, 1997), while several others are on-going. In summary, these studies have shown DNA transfer and transgene expression. However, evidence for functional correction is slightly more variable but, where assessed, some individuals have shown alterations in their electrophysiological characteristics consistent with restoration in CFTR function. Whilst this data is encouraging certain safety issues have been observed during the course of these and other studies which may have an impact on the long-term clinical use of adenoviral vectors in CF patients. In both pre-clinical and clinical studies a dose and time-dependent inflammation in the airways treated with adenoviral vector has been observed (Engelhardt *et al.*, 1993; Simon *et al.*, 1993; Brody *et al.*, 1994; Crystal *et al.*, 1994; Yei *et al.*, 1994; Knowles *et al.*, 1995a). The inflammation seen in these studies appears to be related to the antigenicity of the adenoviral coat proteins leading to a cytotoxic T-lymphocyte response (Yang *et al.*, 1995a). In addition, there is accumulating data that an immune response to adenovirus can be invoked that results in the production of neutralizing antibodies at a sufficient level to reduce gene transfer following a second administration of the vector (Yang *et al.*, 1995a; Yang *et al.*, 1995b). In the one published clinical study assessing repeated administration of an Ad-CFTR vector a reduction in CFTR functional correction was seen following the second and third administrations of the vector. This reduction in correction was paralleled by evidence of an immune response against the adenoviral vector (Zabner *et al.*, 1996). It is for these reasons, and others, that several groups have considered cationic liposomes as a viable alternative gene transfer system for cystic fibrosis.

3. Cationic liposome-mediated gene transfer and cystic fibrosis

3.1 Pre-clinical studies

The first demonstration of *in vivo*, delivery of *CFTR* plasmid DNA complexed with a cationic liposome was demonstrated by Yoshimura and co-workers (Yoshimura *et al.*, 1992). Using the cationic liposome DOTMA:DOPE, complexes were introduced into the lungs of mice by intratracheal administration. Human specific *CFTR* mRNA transcripts were detected in the lung for up to 4 weeks following administration. Subsequently, three different cationic liposomes have been used to deliver *CFTR* containing plasmids to CF transgenic mice. Hyde and co-workers (Hyde *et al.*, 1993) instilled a cationic liposome (DOTMA:DOPE) complexed with a *CFTR* cDNA into the trachea of the CF transgenic mice and showed restoration of cAMP-stimulated chloride secretion. Nebulized DC-Chol:DOPE *CFTR* cDNA complexes delivered to the CF transgenic mice showed correction of the CF chloride defect in the trachea and to a lesser extent to the nose (Alton *et al.*, 1993) as did direct application to the nose and tracheal instillation of the same transgenic mice with *CFTR* plasmid DNA complexed with DOTAP (McLachlan *et al.*, 1996). In a similar study using the cationic liposome DMRIE-DOPE human *CFTR* mRNA was detected in rats instilled three days previously (Logan *et al.*, 1995). In most case the presence of endogenous CFTR function cannot be distinguished from functional expression of the *CFTR* transgene, however, in the rat cAMP mediated chloride conductance is minimal. Thus, Logan and co-workers also measured the ion transport characteristics of excised rat tracheas. Bioelectrical responses consistent with *CFTR* transfer were observed including cAMP mediated chloride secretion 3 days after instillation.

3.2 Clinical studies

The success of these and many other pre-clinical studies have enabled several groups to initiate clinical protocols testing cationic liposome-mediated *CFTR* gene transfer in individuals with CF. Three clinical trials of administration of DNA-liposome complexes to the nasal epithelium of CF subjects have now been completed (Caplen *et al.*, 1995; Gill *et al.*, 1997; Porteous *et al.*, 1997). In the first 15 ΔF508 homozygous subjects with CF were enrolled in a double-blind, placebo-controlled trial (nine *CFTR* treated, receiving either 20, 200 or 600 μg DNA plus lipid, and six DC-Chol:DOPE alone) (Caplen *et al.*, 1995). No safety problems were encountered, either in the routine clinical assessment or by a blinded, semiquantative analysis of nasal biopsies. Both plasmid DNA and *CFTR* mRNA were detected from the nasal biopsies in five of eight samples available from *CFTR* treated subjects. Sodium-related measurements, were significantly reduced (approximately 20% toward values seen in patients without CF). However, it is important to note that these changes fell within the coefficient of variation of these measurements. More importantly, chloride secretion, assessed by perfusion with a low chloride solution, also showed a significant 20% increase toward normal values, a change well outside the a variation in these measurements. In two subjects, these chloride responses reached values within the range seen in subjects without CF. These changes in the sodium- and chloride-related measurements paralleled each other, and were no longer present at 7 days.

The trial performed by Gill and colleagues also used the cationic liposome DC-Chol:DOPE, but in a slightly different formulation and method of administration to the nasal mucosa. In addition, the *CFTR* cDNA was driven by the RSV-LTR

(Gill *et al.*, 1997). In this study 12 CF subjects (of mixed genotype) were randomized to either one of two doses of *CFTR* cDNA (40 and 400 μg) or placebo (buffer alone or lipid plus plasmid carrying a non-functional CFTR cDNA). No safety problems were encountered. Analysis of the electrophysiological characteristics of isolated nasal cells taken 5 days after DNA/liposome administration and *in vivo* ion transport measurements taken at regular intervals, showed changes consistent with restoration of chloride secretion in 6 out of the 8 *CFTR* treated subjects. In most cases the level of correction was partial, however, in 2 patients (1 from each dosing group) the chloride responses were within the non-CF range. These changes were more sustained than in the previously reported trial, lasting for approximately 7 days. This difference may reflect the effect of using the RSV-LTR promoter as against the SV40 early promoter which was used to drive *CFTR* expression in the first trial.

In the third trial the cationic lipid used was DOTAP and the *CFTR* cDNA was driven by the CMV promoter (Porteous *et al.*, 1997). Sixteen CF subjects (of mixed genpotype) were studied. Eight subjects received 400 μg DNA plus lipid, and 8 recived placebo, buffer alone. Again no safety problems were encountered. Vector DNA was detected in 7 out of 8 patients at day 3 and day 7 and at day 28 in 2 patients. Vector derived mRNA was detected in 2 patients at day 3 and day 7. Whilst there was much variability in the ion transport measurements observed two patients showed changes in chloride conductance consistent with correction, with a mean change towards normal values of 20%.

3.3 On-going studies

Recently new cationic liposomes specifically developed for gene therapy have been described (Lee *et al.*, 1996; Wheeler *et al.*, 1996). One of the most promising of these, GL67 (Genzyme Corp., Framingham, MA, USA) is currently being assessed in two clinical trials in CF patients. In a nasal study a *CFTR* cDNA complexed with GL67 has been administered to several CF patients, however, as yet no clear effect has been seen on ion transport parameters significantly different from administration of DNA alone (D. Meeker, personnel communication). In the second trial, a collaboration between the National Heart and Lung Institute/Royal Brompton Hospital, London, UK and Genzyme Corp. *CFTR* plasmid DNA/GL67 complexes are being applied to both the nasal mucosa and to the lung by aerosolization, the results of this trial should be available shortly (E. Alton personnel communication).

The effect of repeated administration with DNA-liposome complexes has been assessed in CF transgenic mice. Preliminary data shows that the degree of correction of chloride secretion seen after two doses of CFTR/DC-Chol:DOPE, 10 days apart, is not significantly different from that seen after a single dose (Goddard *et al.*, 1996). The same group are currently conducting a clinical trial assessing three doses of DNA-liposome complexes in CF patients, administered 28 days apart (D.Gill and S.Hyde, personnel communication).

3.4 Safety studies

Early safety studies showed DNA-liposome complexes to have little toxicity (Nabel *et al.*, 1992; Stewart *et al.*, 1992) and insufflation of the cationic liposome DC-Chol:DOPE into the human nose did not alter ion transport measurements, themselves a sensitive index of cellular function and integrity (Middleton *et al.*, 1994a). More recently though a dose-dependent pulmonary inflammation characterized by infiltrates of neutrophils and to a lesser extent macrophages and lymphocytes was observed following the intra-nasal administration of the cationic liposome GL67 to mice (Scheule *et al.*, 1997), however, aerosolized GL67 administered alone to normal volunteer subjects showed no safety issues (E. Alton personnel communication).

4. Future perspectives

The principal problem associated with cationic liposome-mediated DNA gene transfer to the airway is that of efficiency. Currently, on a per molecule of DNA basis, viral vectors are several orders of magnitude more efficient than liposomes. However, it is unclear at what point in the transfer process the majority of DNA is lost following liposome delivery and much research is still required for us to obtain a better understanding of the basic mechanism(s) of cationic liposome-mediated DNA transfer. Given that gene expression using this system is transient repeated administration will be required; this appears to be feasible but it is unclear what will be the long term effect of multiple exposures of large amounts of lipid on the delicate architecture of the lung. In spite of these problems, the fact that measurable changes in the basic cellular defect have been observed in several CF patients following CFTR cDNA/liposome transfer must justify continued research and development of this technology.

5. References

Alton, E.W.F.W., Middleton, P.G., Caplen, N.J., Smith, S.N., Steel, D.M., Munkonge, F.M., Jeffery, P.K., Geddes, D.M., Hart, S.L., Williamson, R., Fasold, K.I., Miller, A.D., Dickinson, P., Stevenson, B.J., McLachlan, G., Dorin, J.R., and Porteous, D.J. (1993) Non-invasive liposome-mediated gene delivery can correct the ion transport defect in cystic fibrosis mutant mice. Nature Genetics **5**, 135-142.

Boat, T.F., Welsh, M.J., and Beaudet, A.L. (1989) Cystic Fibrosis. In *The Metabolic Basis of Inherited Diseases*, eds. C. R. Scriver, A. L. Beaudet, W. S. Sly, and D. Valle. 6th ed., Vol. New York: McGraw-Hill, 2649-2680.

Brody, S.L., Metzger, M., Danel, C., Rosenfeld, M.A., and Crystal, R.G. (1994) Acute response of non-human primates to airway delivery of an adenovirus vector containing the human cystic fibrosis transmembrane conductance regulator cDNA. Human Gene Therapy **5**, 8821-836.

Caplen, N.J., Alton, E.W.F.W., Middleton, P.G., Dorin, J.R., Stevenson, B.J., Gao, X., Durham, S., Jeffery, P.K., Hodson, M.E., Coutelle, C., Huang, L., Porteous, D.J., Williamson, R., and Geddes, D.M. (1995) Liposome-mediated

CFTR gene transfer to the nasal epithelium of patients with cystic fibrosis. Nature Medicine **1**, 39-46. Addendum 1 (3) 272.

Cheng, S.H., Gregory, R.J., Marshall, J., Paul, S., Souza, D.W., White, G.A., O'Riordan, C.R., and Smith, A.E. (1990) Defective intracellular transport and processing of CFTR is the molecular basis of most cystic fibrosis. Cell **63**, 827-834.

Crawford, I., Moloney, P.C., Zeitlin, P.L., Guggino, W.B., Hyde, S.C., Turley, H., Gatter, K.C., Harris, A., and Higgins, C. (1991) Immunocytochemical localization of the cystic fibrosis gene product CFTR. Proc. Natl. Acad. Sci. USA **88**, 9262-9266.

Crystal, R.G., McElvaney, N.G., Rosenfeld, M.A., Chu, C.-S., Mastrangeli, A., Hay, J.G., Brody, S.L., Jaffe, H.A., Eissa, N.T., and Danel, C. (1994) Administration of an adenovirus containing the human CFTR cDNA to the respiratory tract of individuals with cystic fibrosis. Nature Genetics **8**, 42-51.

Davies, J.C., Stern, M., Dewer, A., Caplen, N.J., Munkonge, F.M., Pitt, T., Sorgi, F., Huang, L., Bush, A., Geddes, D.M., and Alton, E.W.F.W. (1997) *CFTR* gene transfer reduces the binding of *Pseudomonas aeruginosa* to cystic fibrosis respiratory epithelium. Am. J. Respir. Cell Mol. Biol. **16**, 657-663.

Dorin, J., Farley, R., Webb, S., Smith, S.N., Farini, E., Delaney, S.J., Wainwright, B.J., Alton, E.W.F.W., and Porteous, D.J. (1996) A demonstration using mouse models that successful gene therapy for cystic fibrosis requires only partial gene correction. Gene Therapy **3**, 797-801.

Drumm, M.L., Pope, H.A., Cliff, W.H., Rommens, J.M., Marvin, S.A., Tsui, L.-C., Collins, F.S., Frizzel, R.A., and Wilson, J.M. (1990) Correction of the cystic fibrosis defect *in vitro* by retrovirus-mediated gene transfer. Cell **62**, 1227-1233.

Egan, M., Flotte, T., Afione, S., Solow, R., Zeitlin, P.L., Carter, B.J., and Guggino, W.B. (1992) Defective regulation of outwardly rectifying Cl- channels by protein kinase A corrected by insertion of CFTR. Nature **358**, 581-584.

Engelhardt, J.F., Simon, R.H., Yang, Y., Zepeda, M., Weber-Pendleton, S., Doranz, B., Grossman, M., and Wilson, J.M. (1993) Adenovirus-mediated transfer of the CFTR gene to lung of non-human primates: biological efficiacy study. Human Gene Therapy **4**, 759-769.

Engelhardt, J.F., Yankaskas, J., Ernst, S.A., Yang, Y., Marino, C.R., Boucher, R.C., Cohn, J.A., and Wilson, J.M. (1992) Submucosal glands are the predominant site of CFTR expression in the human bronchus. Nature Genetics **2**, 240-247.

Engelhardt, J.F., Zepeda, M., Cohn, J.A., Yankaskas, J.R., and Wilson, J.M. (1994) Expression of the cystic fibrosis gene in adult human lung. J. Clin. Invest. **93**, 737-749.

Felgner, J.H., Kumar, R., Sridhar, C.N., Wheeler, C.J., Tsai, Y.J., Border, R., Ramsey, P., Martin, M., and Felgner, P.L. (1994) Enhanced gene delivery and mechanism studies with a novel series of cationic lipid formulations. Jour. of Biol. Chem. **269**, 2550-2561.

Flotte, T.R., Afione, S.A., Conrad, C., McGath, S.A., Solow, R., Oka, H., Zeitlin, P.L., Guggino, W.B., and Carter, B.J. (1993) Stable *in vivo* expression of the cystic fibrosis transmembrane conductance regulator with an adeno-associated virus vector. Proc. Natl. Acad. Sci (USA) **90**, 10613-10617.

Gao, X. and Huang, L. (1991) A novel cationic liposome reagent for efficient transfection of mammalian cells. Biochem. Biophys. Res. Comm. **179**, 280-285.

Gill, D., Southern, K.W., Mofford, K.A., Seddon, T., Huang, L., Sorgi, F., Thomson, A., MacVinish, L.J., Ratcliff, R., Bilton, D., Lane, D.J., Littlewood, J.M., Webb, A.K., Middleton, P.G., Colledge, W.H., Cuthbert, A.W., Evans, M.J., Higgins, C.F., and Hyde, S.C. (1997) A placebo-controlled study of liposome-mediated gene transfer to the nasal epithelium of patients with cystic fibrosis. Gene Therapy **4**, 199-209.

Goddard, C.A., Ratcliff, R., Gill, D., Hyde, S., cuthbert, A.w., Anderson, J.R., Evans, M.J., Colledge, W.H., and MacVinish, L.J. (1996) A second dose delivery of a CFTR/liposome complex is as effective as the first dose in restoring Cl- channel function in null mice. Pedatric Pulmonology **Suppl. 13**, 265.

Goldman, M.J., Yang, Y., and Wilson, J.M. (1995) Gene therapy in a xenograft model of cystic fibrosis lung corrects chloride transport more effectively than the sodium defect. Nature Genetics **9**, 121-131.

Halbert, C.L., Aitken, M.L., and Miller, A.D. (1996) Retroviral vectors efficiently transduce basal and secretory airway epithelial cells *in vitro* resulting in persistent gene expression in organotypic culture. Human Gene Therapy **7**, 1871-1881.

Hyde, S., Gill, D., Higgins, C., Treizie, A., MacVinish, L.J., Cuthburt, A., Ratclife, R., Evans, M., and Colledge, W. (1993) Correction of the ion transport defect in cystic fibrosis transgenic mice by gene therapy. Nature **362**, 250-255.

Johnson, L.G., Olsen, J.C., Sarkadi, B., Moore, K.L., Swanstrom, R., and Boucher, R.C. (1992) Efficiency of gene transfer for restoration of normal airway epithelial function in cystic fibrosis. Nature Genetics **2**, 21-25.

Kartner, N., Augustinas, O., Jensen, T.J., Naismith, A.L., and Riordan, J.R. (1992) Mislocalizaion of ΔF508 CFTR in cystic fibrosis sweat gland. Nature Genetics **1**, 321-327.

Knowles, M.R., Hohneker, K.W., Zhou, Z., Olsen, J.C., Noah, T.L., Hu, P.-C., Leigh, M.W., Engelhardt, J.F., Edwards, L.J., Jones, K.R., Grossman, M., Wilson, J.M., Johnson, L.G., and Boucher, R.C. (1995a) A controlled study of adenoviral-vector-mediated gene transfer in the nasal epithelium of patients with cystic fibrosis. N. Engl. J. Med. **333**, 823-831.

Knowles, M.R., Paradiso, A.M., and Boucher, R.C. (1995b) *In vivo* nasal potential difference: techniques and protocols for assessing efficiacy of gene transfer in cystic fibrosis. Human Gene Therapy **6**, 445-455.

Lee, E.R., Marshall, J., Siegel, C.S., Jiang, C., Yew, N.S., Nichols, M.R., Nietupski, J.B., Ziegler, R.J., Lane, M.B., Wang, K.X., Wan, N.C., Scheule, R.K., Harris, D.J., Smith, A.E., and Cheng, S.H. (1996) Detailed analysis of structures and formulations of cationic lipids for efficient gene transfer to the lung. Human Gene Therapy **7**, 1701-1717.

Logan, J.J., Bebok, Z., Walker, L.C., Peng, S., Felgner, P.L., Siegal, G.P., Frizzell, R.A., Dong, J., Howard, M., Matalon, S., Lindsey, J.R., DuVall, M., and Sorscher, E.J. (1995) Cationic lipids for reporter gene and CFTR transfer to rat pulmonary epithelium. Gene Therapy **2**, 38-49.

McLachlan, G., Davidson, D.J., Stevenson, B.J., P.Dickinson, Davidson-Smith, H., Dorin, J.R., and Porteous, D.J. (1995) Evaluation *in vitro* and *in vivo* of cationic liposome-expression construct complexes for cystic fibrosis gene therapy. Gene Therapy **2**, 614-622.

McLachlan, G., Ho, L.P., Davidson-Smith, H., Samways, J., Davidson, H., Stevenson, B.J., Carothers, A.D., Alton, E.W., Middleton, P.G., Smith, S.N., Kallmeyer, G., Michaelis, U., Seeber, S., Naujoks, K., Greening, A.P., Innes, J.A., Dorin, J.R., and Porteous, D.J. (1996) Laboratory and clinical studies in support of cystic fibrosis gene therapy using pCMV-CFTR-DOTAP. Gene Therapy **3**, 1113-1123.

Middleton, P.G., Caplen, N.J., Gao, X., Haung, L., Gaya, H., Geddes, D.M., and Alton, E.W.F.W. (1994a) Nasal application of the cationic liposome DC-Chol:DOPE does not alter ion transport, lung function or bacterial growth. Eur. Resp. J. **7**, 442-445.

Middleton, P.G., Geddes, D.M., and Alton, E.W. (1994b) Protocols for *in vivo* measurement of ion transport defects in cystic fibrosis nasal epithelium. Eur. Respir. J. **7**, 2050-2056.

Mittereder, N., Yei, S., Bachurski, C., Cuppoletti, J., Whitsett, J.A., Tolstoshev, P., and Trapnell, B.C. (1994) Evaluation of the efficacy and safety of *in vitro*, adenovirus-mediated transfer of the human cystic fibrosis transmembrane conductance regulator cDNA. Human Gene Therapy **5**, 719-731.

Nabel, E.G., Gordon, D., Yang, Z.-Y., Xu, L., San, H., Plutz, G.E., Wu, B.-Y., Gao, X., Huang, L., and Nabel, G.J. (1992) Gene transfer *in vivo* with DNA-Liposome complexes: Lack of autoimmunity and gonadal localization. Human Gene Therapy **3**, 649-656.

Porteous, D., Dorin, J.R., McLachlan, G., Davidson-Smith, H., Davidson, H., Stevenson, B.J., Carothers, A.D., Wallace, W.A., Moralee, S., Hoenes, C., Kallmeyer, G., Michaelis, U., Naujoks, K., Ho, L.P., Samways, J.M., Imrie, M., Greening, A.P., and JA, J.A.I. (1997) Evidence for safety and efficacy of DOTAP cationic liposome-mediated CFTR gene transfer to the nasal epithelium of patients with cystic fibrosis. Gene Therapy **4**, 210-218.

Rich, D.P., Anderson, M.P., Gregory, R.J., Cheng, S.H., Paul, S., Jefferson, D.M., McCann, J.D., Klinger, K.W., Smith, A.E., and Welsh, M.J. (1990) Expression of cystic fibrosis transmembrane conductance regulator corrects defective chloride channel regulation in cystic fibrosis airway epithelial cells. Nature **347**, 358-363.

Riordan, J.R., Rommens, J.M., Kerem, B.-S., Alon, N., Rozmahel, R., Grzelczak, Z., Zielenski, J., Lok, S., Plavsic, N., Chou, J.-L., Drumm, M.L., Iannuzzi, M.C., Collins, F.S., and Tsui, L.-C. (1989) Identification of the cystic fibrosis gene: cloning and characterization of complementary DNA. Science **245**, 1066-1073.

Rommens, J.M., Iannuzzi, M.C., Kerem, B.-S., Drumm, M.L., Melmer, G., Dean, M., Rozmahel, R., Cole, J.L., Kennedy, D., Hidaka, N., Zsiga, M., Buchwald, M., Riordan, J.R., Tsui, L.-C., and Collins, F.S. (1989) Identification of the cystic fibrosis gene: chromosome walking and jumping. Science **245**, 1059-1065.

Rosenfeld, M.A., Chu, C.-S., Seth, P., Danel, C., Banks, T., Yoneyama, K., Yoshimura, K., and Crystal, R.G. (1994a) Gene transfer to freshly isolated human respiratory epithelial cells in vitro using a replication-deficient adenovirus containing the human cystic fibrosis transmembrane conductance regulator cDNA. Human Gene Therapy **5**, 331-342.

Rosenfeld, M.A., Rosenfeld, S.J., Danel, C., Banks, T.C., and Crystal, R.G. (1994b) Increasing expression of the normal human CFTR cDNA in cystic fibrosis epithelial cells results in a progressive increase in the level of CFTR protein expression, but a limit on the level of cAMP-stimulated chloride secretion. Human Gene Therapy **5**, 1121-1129.

Rosenfeld, M.A., Yoshimura, K., Trapnell, B.C., Yoneyama, K., Rosenthal, E.R., Dalemans, W., Fukayama, M., Bargon, J., Stier, L.E., Stratford-Perricaudet, L., Perricaudet, M., Guggino, W.B., Pavirani, A., Lecocq, J.-P., and Crystal, R.G. (1992) *In vivo* transfer of the human cystic fibrosis transmembrane conductance regulator gene to the airway epithelium. Cell **68**, 143-155.

Scheule, R.K., George, J.A.S., Bagley, R.G., Marshall, J., Marshall, J.M., Kaplan, J.M., Akita, G.Y., Wang, K.X., Lee, E.R., Harris, D.J., Jiang, C.W., Yew, N.S., Smith, A.E., and Cheng, S.H. (1997) Basis of pulmonary toxicity associated with cationic lipid-mediated genetransfer to the mammalian lung. Human Gene Therapy **8**, 689-707.

Simon, R.H., Engelhardt, J.F., Y.Yang, Zepeda, M., Weber-Pendleton, S., Grossman, M., and Wilson, J.M. (1993) Adenovirus-mediated transfer of the CFTR gene to lung of nonhuman primates: toxicity study. Human Gene Therapy **4**, 771-780.

Smith, J.J., Travis, S.M., Greenberh, E.P., and Welsh, M.J. (1996) Cystic fibrosis airway epithelia fails to kill bacteria because of abnormal airway surface fluid. Cell **85**, 229-236.

Stern, M., Munkonge, F.M., Caplen, N.J., Sorgi, F., Huang, L., Geddes, D., and Alton, E.W.F.W. (1995) Quantitative fluoresecence measurement of chloride secretion in native airway epithelium from CF and non-CF subjects. Gene Therapy **2**, 766-774.

Stewart, M.J., Plautz, G.E., Del Buono, L., Yang, Z.Y., Xu, L., Gao, X., Huang, L., Nabel, E.G., and Nabel, G.J. (1992) Gene transfer *in vivo* with DNA-liposome complexes: safety and acute toxicity in mice. Human Gene Therapy **3**, 267-275.

Trapnell, B.C., Chu, C.-S., Paako, P.K., Banks, T.C., Yoshimura, K., Ferrans, V.J., Chernick, M.S., and Crystal, R. (1991) Expression of the cystic fibrosis transmembrane regulator gene in the respiratory tract of normal individuals and individuals with cystic fibrosis. Proc. Natl. Acad. Sci. USA **88**, 6565-6569.

Tsui, L.-C. (1995) The cystic fibrosis transmembrane conductance regulator gene. Am. J. Respir. Crit. Care **151**, S47-S53.

Welsh, M.J., Anderson, M.P., Rich, D.P., Berger, H.A., Denning, G.M., Ostedgaard, L.S., Sheppard, D.N., Cheng, S.H., Gregory, R.J., and Smith, A.E. (1992) Cystic Fibrosis Transmembrane Conductance Regulator: A chloride channel with novel regulation. Neuron **8**, 821-829.

Welsh, M.J. and Smith, A.E. (1993) Molecular mechanisms of CFTR chloride channel dysfunction in cystic fibrosis. Cell **73**, 1251-1254.

Wheeler, C., Felgner, P.L., Tsai, Y.J., Marshall, J., Sukhu, L., Doh, S.G., Hartikka, J., Nietupski, J., Manthorpe, M., Nichols, M., Plewe, M., Liang, X., Norman, J., Smith, A., and Cheng, S.H. (1996) A novel cationic lipid greatly enhances plasmid DNA delivery and expression in mouse lung. Proc Natl Acad Sci U S A **93**, 11454-11459.

Whitsett, J.A., Dey, C.R., Stripp, B.R., Wikenheiser, K.A., Clark, J.C., Wert, S.E., Gregory, R.J., Smith, A.E., Cohn, J.A., Wilson, J.M., and Engelhardt, J. (1992) Human cystic fibrosis transmembrane conductance regulator directed to respiratory epithelial cells of transgenic mice. Nature Genetics **2**, 13-20.

Yang, Y., Li, Q., Ertl, H.C., and Wilson, J.M. (1995a) Cellular and humoral immune responses to viral antigens create barriers to lung-directed gene therapy with recombinant adenoviruses. Journal of Virology **69**, 2004-2015.

Yang, Y., Trinchieri, G., and Wilson, J.M. (1995b) Recombinant IL-12 prevents formation of blocking IgA antibodies to recombinant adenovirus and allows repeated gene therapy to mouse lung. Nature Medicine **1**, 890-893.

Yei, S., Mittereder, N., Wert, S., Whitsett, J.A., Wilmott, R.W., and Trapnell, B. (1994) *In vivo* evaluation of the safety of adenovirus-mediated transfer of the human cystic fibrosis transmembrane conductance regulator cDNA to the lung. Human Gene Therapy **5**, 731-744.

Yoshimura, K., Rosenfield, M.A., Nakamura, H., Scherer, E.M., Pavirani, A., Lecocq, J.-P., and Crystal, R.G. (1992) Expression of the human cystic fibrosis transmembrane conductance regulator gene in the mouse lung after *in vivo* intratracheal plasmid-mediated gene transfer. Nucleic Acids Res. **20**, 3233-3240.

Zabner, J., Couture, L.A., Smith, A.E., and Welsh, M.J. (1994a) Correction of cAMP-stimulated fluid secretion in cystic fibrosis airway epithelia: efficiency of adenovirus-mediated gene transfer in vitro. Human Gene Therapy **5**, 585-593.

Zabner, J., Petersen, D.M., Puga, A.P., Graham, S.M., Couture, L.A., Keyes, L.D., Lukason, M.J., George, J.A.S., Gregory, R.J., Smith, A.E., and Welsh, M.J. (1994b) Safety and efficacy of repetitive adenovirus-mediated transfer of CFTR cDNA to airway epithelia of primates and cotton rats. Nature Genetics **6**, 75-83.

Zabner, J., Ramsey, B.W., Meeker, D.P., Aitken, M.L., Balfour, R.P., Gibson, R.L., Launspach, J., Moscicki, R.A., Richards, S.M., Standaert, T.A., Williams-Warren, J., Wadsworth, S.C., Smith, A.E., and Welsh, M.J. (1996) Repeat administration of an adenovirus vector encoding cystic fibrosis transmembrane conductance regulator to the nasal epithelium of patients with cystic fibrosis. J. Clin. Invest. **97**, 1504-1511.

Human Dystrophin Gene Expression in mdx Muscles After In Vivo Ballistic Transfection, Application of Synthetic Oligopeptide Complexes and Cationic Liposomes

V. Baranov, A. Zelenin, Engelgard's Institute of Molecular Biology, Russian Academy of Sciences, Moscow, Russia; O.Tarasenko, V. Kolesnikov, Engelgard's Institute of Molecular Biology, Russian Academy of Sciences, Moscow, Russia; V. Mikhailov, Institute of Cytology, Russian Academy of Sciences, St. Petersburg, Russia; T. Ivaschenko, A. Kiselev, O.Artemyeva, O. Evgrafov, Research Center for Medical Genetics, Russian Academy of Medical Sciences, Moscow, Russia; I. Zelenina, State University, Moscow, Russia, R.Shafei, State University, Moscow, Russia T. Kascheeva, G. Dickson, Royal Holloway University of London, UK, A. Baranov. Ott's Institute of Obstetrics & Gynecology, Russian Academy of Medical Sciences, St. Petersburg, Russia;

SUMMARY

Immunocytochemical analysis of cryostat sections of the quadriceps muscle was carried out after in vivo gene transfection of mdx mice with full length human dystrophin gene cDNA in plasmid constructions pHSADy or pMLVDy directed by ballistic particles (ballistic transfection -BT), by liposome (lipofectamine-LFA) or by synthetic oligopeptide complex (SOC). The number of dystrophin positive myofibres (DPM) in experimental limbs varied from 3 to 17% after BT (17 days of exposure), remained within control limits (0.2-0.6%) after LFA and reached 17% on day 20th after injection of SOC + pHSADy mixture compared to control mdx mice (0.2%), injection of naked plasmid (2.2%) or BT with naked metallic particles (1%). Unexpectedly high levels of DPM were recorded in the contralateral limbs of mdx mice 60 days after BT (2.8%), 20 days after SOC + pHSADY injection (12%), as well as after a single intracardial injection of naked pHSADy (4.9%).

INTRODUCTION

Duchenne muscle dystrophy (DMD) is an X-linked, lethal disorder of skeletal muscle caused by a defect in the dystrophin gene. The dystrophin gene, being the largest identified gene, so far in the human genome (2.4 megabase), encodes a 14 kb cDNA responsible for the synthesis of 427 kDa muscle membrane protein - dystrophin. Mutations of the dystrophin gene are predominantly deletions of different length and result in the absence or in truncated forms of the protein dystrophin at the internal side of the sarcolemma membrane with subsequent instability of the muscle membrane and progressive muscle fibre degeneration [1,2,3]. The absence of positive pharmaceutical treatment and very poor results with direct normal myoblast transplantation into affected muscles of DMD subjects [4,5] stimulated an intesive search for its efficient gene therapy via genetic complimentation strategies [6,7]. Meanwhile numerous attempts in this direction involving naked plasmids, liposomes, retroviral and adenoviral vectors as summarized by Fassati et al. 1997 [8] as well as more recent original findings [9,10] are still far from ultimate solution of this complex problem.

NATO ASI Series, Vol. H 105
Gene Therapy
Edited by Kleanthis G. Xanthopoulos

The principal goal of our study is to elucidate the efficiency of human dystrophin gene expression in affected muscle of mdx mice - a biological model of DMD, after ballistic transfection with full length human dystrophin cDNA or after its delivery into mdx muscles by plasmid vectors with liposome or syntethic olygopeptide complexes as vehicles.

RESULTS AND DISCUSSION

BT as well as all other studies were carried out in 2-6 weeks old male C57BI/6j or C57BI/ mdx-mice (courtesy of prof. T. Partridge, UK). For BT suprafacial gluteus muscle and quadriceps muscle of experimental hind limb were opened surgically under anesthesia and bombarded by means of an original gen-gun [II] with the mixture of golden or tungsten microparticles (1-4 micron in diameter) coated with plasmid DNA pHSADy or pMLVDy by the Ca-Pi precipitation method as described elsewhere [II]. Both plasmids carried full length dystrophin cDNA (12 Kb) in eukaryotic expression vector pJ40 supplemented with the promoter (MLV-LTR) and mRNA processing signal (SV-40 splice/poly A) for plasmid pMLVDy, or by human alpha-skeletal promotor for pHSADY [12,13].

Experimental and contralateral quadriceps muscle were surgically removed 14-60 days after BT transfection with 10-13 micrograms of plasmid DNA and processed for crosscryosection and immunostaining analysis with dystrophin specific rabbit polyclonal antibodies P6 (courtesy of Dr. Caroline Sewry and Dr. Peter Strong, UK). Different mdx tissue samples were used for analysis of human dystrophin cDNA by amplification of exons 52-53 in one PCR reaction as well as for dystrophin protein detection by immunoblotting.

The proportion of dystrophin-positive myofibres (DPM) was at average 2% on day 17 and increased to 3% by day 60 after BT with pMLVDy compared to only 0.2% of revertant DPM in mdx mice of relevant age. Dystrophin staining was highly variable between DPM after BT with pHSADy with the maximum score (17%) of DPM on the 20th day after BT compared to 0.5% DPM after BT with naked metallic particles and to 1% of DPM after BT with pCMVIacZ (Fig 1).

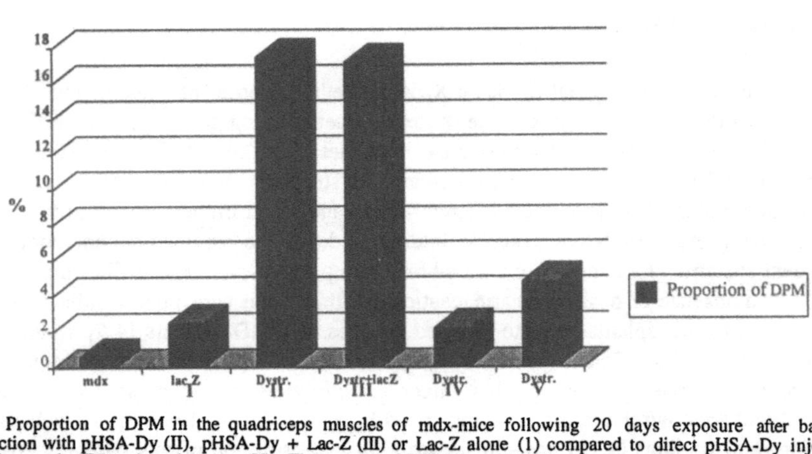

Fig 1. Proportion of DPM in the quadriceps muscles of mdx-mice following 20 days exposure after ballistic transfection with pHSA-Dy (II), pHSA-Dy + Lac-Z (III) or Lac-Z alone (1) compared to direct pHSA-Dy injection into this muscle (IV) or into the heart (V). Three types of DPM shown in Fig. 4 were registered. Note substantial increase of DPM (2.5%) after ballistic transfection with Lac-Z (1), highly significant transfection efficiency (17%) after pHSA-Dy alone (II) or in combination with Lac-Z (III) compared to injection of naked plasmid in the same muscle (2%)-IV or into the heart (4%) of mdx mice.

An unusual finding of both experimental groups with BT was a statistically significant increase of DPM counts in non-bombarded contralateral hind limb of experimental mice with 2.8% of DPM on Day 60 and 12% on day 20th (Fig. 1) after BT with pMLVDy and pHSADy respectively.

PCR analysis of human dystrophin cDNA proved the presence of amplification product corresponding to exons 52-53 (325 bp) in DNA samples from all mdx tissues studied (quadriceps muscle of both limbs, intestine, heart, tongue, brain), but not in human genome DNA or in mdx muscles of control mice.

Immunoblot analysis substantiated PCR findings and proved the presence of specific 427 kDa dystrophin gene protein product in the muscles of the bombarded limb.

Thus BT in vivo with the human dystrophin gene seems to be more efficient compared to injection of naked cDNA plasmids [14] or with retroviral vectors carrying the mini-dystrophin gene [12]; it is quite comparable with efficiency of muscle transfection with adenovirus vectors [10] or HVJ-liposomes [9].

Dystrophin cDNA transfection with cationic liposomes (lipofectamine) resulted in a very low number of DPM being only slightly higher than the control level (0.6%) in the muscle injected with 10 micrograms of plasmid pHSADy DNA (0.6%) Fig.2. DPM count in the relevant muscle of the contralateral limb was almost 4 times more than that in the experimental muscle of the same mdx mouse (Fig.2). Thus being quite efficient for myoblasts in vitro lipofection does not seem to be a valuable approach for dystrophin gene delivery into myofibres in vivo.

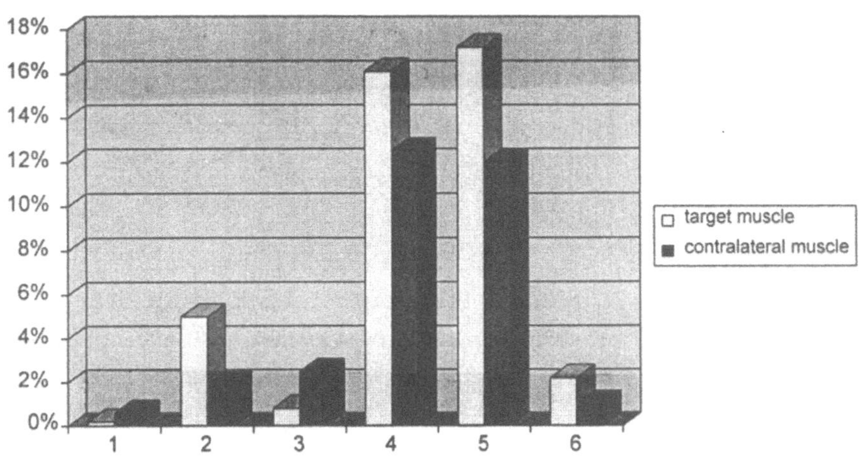

Fig. 2 Proportion of DPM in the quadriceps muscle of mdx-mice in control mdx (1), compared to different types of pHSA-Dy plasmid transfection: direct intracardial injection of naked plasmid (2), liposome transfection (LFA) (3), synthetic oligopeptides (K-8+JTS1) single (4) and triple injections (5), intramuscular injection of naked plasmid (6).

SOC has been recently designed and tested in vitro by Gottschalk e.a. [15]. It consists of two short synthetic peptides which emulate viral functions - a DNA condensing agent YKAK8WK (K8) and an amphipathic, pH-dependent endosomal releasing agent (JTS-1) which mimics virus lytic peptides. Both peptides were synthesized by the solid phase method. In preliminary studies, different ratios of DNA, K8 and JTS-1 were tested and the proportion of DNA/K8 1:3 found as the most efficient pose was used in our subsequent studies. Self assembled SOC loaded with 15-20 µg of pHSADy resulted in substantial (16%) increase of the DPM count 5 days after its intramuscular injection. The same amount of DPM was recorded after 20 days exposure (Fig 2). Maximal DPM score (12%) in the contralateral limb was registered on the 20th day after triple injection of SOC (Fig.2). It was also found that packaging peptide K8 mixed with plasmid pHSADy DNA completely abolishes PCR amplification of exons 52-53 both in vitro and in vivo. These findings stimulated us to trace pHSADy + SOC in injected mdx muscles. After RNAase treatment cryosections of experimental muscles were processed for in situ hybridization with biotin-labelled pHSADy as DNA probe. Clusters of biotin positive DNA granules were found both in myoplasm and nuclei of myofibers 3 hours after SOC injection. More specific labelling of microsatellite and myofibers nuclei was recorded 14 weeks after SOC injection with the groups of specific DNA granules inside or close to the nuclear membrane.

More studies are needed to find out the most appropriate conditions (DNA doses, exposure time, mode of delivery, etc.) for dystrophin gene delivery in vivo by BT and SOC. Meanwhile both of these vehicles seem to be rather promising for the efficient transfection of skeletal muscles in mdx mice. A substantial increase in the proportion of DPM in the contralateral mdx muscles after BT, SOC, LFA or direct intracardial injection of the naked plasmid is of special interest. To our knowledge such clearcut "distant" effect of dystrophin cDNA has not been reported so far. Taken with caution this unexpected finding is now under thorough experimental investigation in our lab.

Acknowledgements

We would like to thank Prof. Terry Partridge (Hammersmith Hospital, London) for the stock of mdx-mice and Dr. Caroline Swery and Dr. Peter Strong (Hammersmith Hospital, London) for the samples of P6 antibodies.

References

1. Emery A.E.H. Duchenne Muscular Dystrophy (2nd edt.) Oxford Monograph on Medical Genetics. Oxford University Press, Oxford, UK 1993, 257 p.
2. Love D.R., Byth B.C., Tinsley J.M., Blake D.J., Davies K.E. Dystrophin and dystrophin related proteins: a review of protein and RNA studies. Neuromusc. Disorder 1993, v. 3, 5-21.
3. Dunckley M.G., Piper T.A., G. Dickson "Towards a gene therapy for duchenne muscular dystrophy". Mental Rtard. & Develop. Disabilities. Res. Review 1995, 1, 71-78.
4. Stelnman L., Blau HM. Normal dystrophin transcripts detected in Duchenne muscular dystrophy patients after myoblast transplantation. Nature 1992, 356, 435-438.
5 Karpati G., Ajdukovic D., Amold D., Gledhill R. B., et al. Myoblast transfer in Duchenne muscular dystrophy. Ann. Neurol 1993, 34, 8-17.
6. Huser M. A., Hamberlain J. S. Progress towards gene therapy for Duchenne muscular dystrophy. J. Endocrinol. 1996, 149, 373-378.
7. Dickson G. Gene transfer to muscle. Biochem. Soc. Transactions 1996, 24, 514-518.
8. Fassati A., St. Murphy, G. Dickson Gene therapy of Duchenne muscular dystrophy. Adv. Genet. 1997, 35, 117-153.
9. Yanagihara I., Inui K., Dickson G., Turner G., Piper T., Kaneda Y., Okada. Expression of full-length human dystrophin CDNA in mdx mouse muscle by HVJ-liposome injection. Gene Therapy 1996, 3, 549-553.
10. Clemens P.R., Kochanek S., Sumada Y. S. Chan, H. H. Chen, K. P. Campbell, C. T. Caskey. In vivo muscle gene transfer of full-length dystrophin with adenovirus vector that lacks all viral genes. Gene Therapy 1996, 3, 11, 965-972.
11. Zelenin A., O. Tarasenko, V. A. Kolesnikov, R. Shafei et al. Bacterial beta-galactosidase and human dystrophin genes are expressed in mouse skeletal muscles fibres after ballistic transfection. FEBs Letters 1997, in press.

12. Dickson G., Love D., Davies K., Wells K. E., Piper T. A., Walsh F. S. Human dystrophin gene transfer: production and expression of a functional recombinant DNA-based gene. Hum. Genet. 1991, 88, 53-58.
13. Wells D. J., Wells K. E., Asante E. A. et al. Expression of human full length and minidystrophin in transgenic mdx mice: implication for gene therapy of Duchenne muscular dystrophy. Hum. Mol. Genet. 1995, 4, N8, 1245-1250
14. Ascadi G., Dickson G., Lovee D.R., Jami A. et al. Human dystrophin expression in mdx mice after intramuscular injection of DNA constracts. Nature 1991, 352, 815-818.
15. Gottschalk S., Sparrow J. T., Hauer J., Mims MP., Leland F. E., Woo S. L. C., Smith L. C. A novel DNA-peptide complex for efficient gene transfer and epression in mammalian cells. Gene Therapy 1996, 3, 448-457.

NATO ASI Series H

Vol. 1: **Biology and Molecular Biology of Plant-Pathogen Interactions.** Edited by J.A. Bailey. 415 pages. 1986.

Vol. 2: **Glial-Neuronal Communication in Development and Regeneration.** Edited by H.H. Althaus and W. Seifert. 865 pages. 1987.

Vol. 3: **Nicotinic Acetylcholine Receptor: Structure and Function.** Edited by A. Maelicke. 489 pages. 1986.

Vol. 4: **Recognition in Microbe-Plant Symbiotic and Pathogenic Interactions.** Edited by B. Lugtenberg. 449 pages. 1986.

Vol. 5: **Mesenchymal-Epithelial Interactions in Neural Development.** Edited by J. R. Wolff, J. Sievers, and M. Berry. 428 pages. 1987.

Vol. 6: **Molecular Mechanisms of Desensitization to Signal Molecules.** Edited by T M. Konijn, P J. M. Van Haastert, H. Van der Starre, H. Van der Wel, and M.D. Houslay. 336 pages. 1987.

Vol. 7: **Gangliosides and Modulation of Neuronal Functions.** Edited by H. Rahmann. 647 pages. 1987.

Vol. 8: **Molecular and Cellular Aspects of Erythropoietin and Erythropoiesis.** Edited by I.N. Rich. 460 pages. 1987.

Vol. 9: **Modification of Cell to Cell Signals During Normal and Pathological Aging.** Edited by S. Govoni and F. Battaini. 297 pages. 1987.

Vol. 10: **Plant Hormone Receptors.** Edited by D. Klämbt. 319 pages. 1987.

Vol. 11: **Host-Parasite Cellular and Molecular Interactions in Protozoal Infections.** Edited by K.-P. Chang and D. Snary. 425 pages. 1987.

Vol. 12: **The Cell Surface in Signal Transduction.** Edited by E. Wagner, H. Greppin, and B. Millet. 243 pages. 1987.

Vol. 13: **Toxicology of Pesticides: Experimental, Clinical and Regulatory Perspectives.** Edited by L.G. Costa, C.L. Galli, and S.D. Murphy. 320 pages. 1987.

Vol. 14: **Genetics of Translation. New Approaches.** Edited by M.F. Tuite, M. Picard, and M. Bolotin-Fukuhara. 524 pages. 1988.

Vol. 15: **Photosensitisation. Molecular, Cellular and Medical Aspects.** Edited by G. Moreno, R. H. Pottier, and T. G. Truscott. 521 pages. 1988.

Vol. 16: **Membrane Biogenesis.** Edited by J.A.F Op den Kamp. 477 pages. 1988.

Vol. 17: **Cell to Cell Signals in Plant, Animal and Microbial Symbiosis.** Edited by S. Scannerini, D. Smith, P. Bonfante-Fasolo, and V. Gianinazzi-Pearson. 414 pages. 1988.

Vol. 18: **Plant Cell Biotechnology.** Edited by M.S.S. Pais, F. Mavituna, and J. M. Novais. 500 pages. 1988.

NATO ASI Series H

Vol. 19: **Modulation of Synaptic Transmission and Plasticity in Nervous Sys tems.**
Edited by G. Hertting and H.-C. Spatz. 457 pages. 1988.

Vol. 20: **Amino Acid Availability and Brain Function in Health and Disease.**
Edited by G. Huether. 487 pages. 1988.

Vol. 21: **Cellular and Molecular Basis of Synaptic Transmission.**
Edited by H. Zimmermann. 547 pages. 1988.

Vol. 22: **Neural Development and Regeneration. Cellular and Molecular Aspects.**
Edited by A. Gorio, J. R. Perez-Polo, J. de Vellis, and B. Haber. 711 pages. 1988.

Vol. 23: **The Semiotics of Cellular Communication in the Immune System.**
Edited by E.E. Sercarz, F. Celada, N.A. Mitchison, and T. Tada. 326 pages. 1988.

Vol. 24: **Bacteria, Complement and the Phagocytic Cell.**
Edited by F. C. Cabello und C. Pruzzo. 372 pages. 1988.

Vol. 25: **Nicotinic Acetylcholine Receptors in the Nervous System.**
Edited by F. Clementi, C. Gotti, and E. Sher. 424 pages. 1988.

Vol. 26: **Cell to Cell Signals in Mammalian Development.**
Edited by S.W. de Laat, J.G. Bluemink, and C.L. Mummery. 322 pages. 1989.

Vol. 27: **Phytotoxins and Plant Pathogenesis.**
Edited by A. Graniti, R. D. Durbin, and A. Ballio. 508 pages. 1989.

Vol. 28: **Vascular Wilt Diseases of Plants. Basic Studies and Control.**
Edited by E. C. Tjamos and C. H. Beckman. 590 pages. 1989.

Vol. 29: **Receptors, Membrane Transport and Signal Transduction.**
Edited by A. E. Evangelopoulos, J. P. Changeux, L. Packer, T. G. Sotiroudis, and K.W.A. Wirtz. 387 pages. 1989.

Vol. 30: **Effects of Mineral Dusts on Cells.**
Edited by B.T. Mossman and R.O. Begin. 470 pages. 1989.

Vol. 31: **Neurobiology of the Inner Retina.**
Edited by R. Weiler and N.N. Osborne. 529 pages. 1989.

Vol. 32: **Molecular Biology of Neuroreceptors and Ion Channels.**
Edited by A. Maelicke. 675 pages. 1989.

Vol. 33: **Regulatory Mechanisms of Neuron to Vessel Communication in Brain.**
Edited by F. Battaini, S. Govoni, M.S. Magnoni, and M. Trabucchi. 416 pages. 1989.

Vol. 34: **Vectors asTools for the Study of Normal and Abnormal Growth and Differentiation.**
Edited by H. Lother, R. Dernick, and W. Ostertag. 477 pages. 1989.

Vol. 35: **Cell Separation in Plants: Physiology, Biochemistry and Molecular Biology.** Edited by D. J. Osborne and M. B. Jackson. 449 pages. 1989.

NATO ASI Series H

Vol. 36: **Signal Molecules in Plants and Plant-Microbe Interactions.** Edited by B.J.J. Lugtenberg. 425 pages. 1989.

Vol. 37: **Tin-Based Antitumour Drugs.** Edited by M. Gielen. 226 pages. 1990.

Vol. 38: **The Molecular Biology of Autoimmune Disease.** Edited by A.G. Demaine, J-P. Banga, and A.M. McGregor. 404 pages. 1990.

Vol. 39: **Chemosensory Information Processing.** Edited by D. Schild. 403 pages. 1990.

Vol. 40: **Dynamics and Biogenesis of Membranes.** Edited by J. A. F. Op den Kamp. 367 pages. 1990.

Vol. 41: **Recognition and Response in Plant-Virus Interactions.** Edited by R. S. S. Fraser. 467 pages. 1990.

Vol. 42: **Biomechanics of Active Movement and Deformation of Cells.** Edited by N. Akkas. 524 pages. 1990.

Vol. 43: **Cellular and Molecular Biology of Myelination.** Edited by G. Jeserich, H. H. Althaus, and T. V. Waehneldt. 565 pages. 1990.

Vol. 44: **Activation and Desensitization of Transducing Pathways.** Edited by T. M. Konijn, M. D. Houslay, and P. J. M. Van Haastert. 336 pages. 1990.

Vol. 45: **Mechanism of Fertilization: Plants to Humans.** Edited by B. Dale. 710 pages. 1990.

Vol .46: **Parallels in Cell to Cell Junctions in Plants and Animals.** Edited by A. W Robards, W. J . Lucas, J . D. Pitts, H . J . Jongsma, and D. C. Spray. 296 pages. 1990.

Vol. 47: **Signal Perception and Transduction in Higher Plants.** Edited by R. Ranjeva and A. M. Boudet. 357 pages. 1990.

Vol. 48: **Calcium Transport and Intracellular Calcium Homeostasis.** Edited by D. Pansu and F. Bronner. 456 pages. 1990.

Vol. 49: **Post-Transcriptional Control of Gene Expression.** Edited by J. E. G. McCarthy and M. F. Tuite. 671 pages. 1990.

Vol. 50: **Phytochrome Properties and Biological Action.** Edited by B. Thomas and C. B. Johnson. 337 pages. 1991.

Vol. 51: **Cell to Cell Signals in Plants and Animals.** Edited by V. Neuhoff and J. Friend. 404 pages. 1991.

Vol. 52: **Biological Signal Transduction.** Edited by E. M . Ross and K . W. A. Wirtz. 560 pages. 1991.

Vol. 53: **Fungal Cell Wall and Immune Response.** Edited by J. P. Latge and D. Boucias. 472 pages. 1991.

NATO ASI Series H

Vol. 54: **The Early Effects of Radiation on DNA.**
Edited by E. M. Fielden and P. O'Neill. 448 pages. 1991.

Vol. 55: **The Translational Apparatus of Photosynthetic Organelles.**
Edited by R. Mache, E. Stutz, and A. R. Subramanian. 260 pages. 1991.

Vol. 56: **Cellular Regulation by Protein Phosphorylation.**
Edited by L. M. G. Heilmeyer, Jr. 520 pages. 1991.

Vol. 57: **Molecular Techniques in Taxonomy.**
Edited by G . M . Hewitt, A. W. B. Johnston, and J. P. W. Young . 420 pages. 1991.

Vol. 58: **Neurocytochemical Methods.**
Edited by A. Calas and D. Eugene. 352 pages. 1991.

Vol. 59: **Molecular Evolution of the Major Histocompatibility Complex.**
Edited by J. Klein and D. Klein. 522 pages. 1991.

Vol. 60: **Intracellular Regulation of Ion Channels.**
Edited by M. Morad and Z. Agus. 261 pages. 1992.

Vol. 61: **Prader-Willi Syndrome and Other Chromosome 15q Deletion Disorders.**
Edited by S. B. Cassidy. 277 pages. 1992.

Vol. 62: **Endocytosis. From Cell Biology to Health, Disease and Therapie.**
Edited by P. J. Courtoy. 547 pages. 1992.

Vol. 63: **Dynamics of Membrane Assembly.**
Edited by J. A. F. Op den Kamp. 402 pages. 1992.

Vol. 64: **Mechanics of Swelling. From Clays to Living Cells and Tissues.**
Edited by T. K. Karalis. 802 pages. 1992.

Vol. 65: **Bacteriocins, Microcins and Lantibiotics.**
Edited by R. James, C. Lazdunski, and F. Pattus. 530 pages. 1992.

Vol. 66: **Theoretical and Experimental Insights into Immunology.**
Edited by A. S. Perelson and G. Weisbuch. 497 pages. 1992.

Vol. 67: **Flow Cytometry. New Developments.**
Edited by A. Jacquemin-Sablon. 1993.

Vol. 68: **Biomarkers. Research and Application in the Assessment of Environmental Health.** Edited by D. B. Peakall and L. R. Shugart. 138 pages. 1993.

Vol. 69: **Molecular Biology and its Application to Medical Mycology.**
Edited by B. Maresca, G. S. Kobayashi, and H. Yamaguchi. 271 pages. 1993.

Vol. 70: **Phospholipids and Signal Transmission.**
Edited by R. Massarelli, L. A. Horrocks, J. N. Kanfer, and K. Löffelholz. 448 pages. 1993.

Vol. 71: **Protein Synthesis and Targeting in Yeast.**
Edited by A. J. P. Brown, M. F. Tuite, and J. E. G. McCarthy. 425 pages. 1993.

NATO ASI Series H

Vol. 72: **Chromosome Segregation and Aneuploidy.** Edited by B. K. Vig. 425 pages. 1993.

Vol. 73: **Human Apolipoprotein Mutants III. In Diagnosis and Treatment.** Edited by C. R. Sirtori, G. Franceschini, B. H. Brewer Jr. 302 pages. 1993.

Vol. 74: **Molecular Mechanisms of Membrane Traffic.** Edited by D. J. Morré, K. E. Howell, and J. J. M. Bergeron. 429 pages. 1993.

Vol. 75: **Cancer Therapy. Differentiation, Immunomodulation and Angiogenesis.** Edited by N. D'Alessandro, E. Mihich, L. Rausa, H. Tapiero, and T. R.Tritton. 299 pages. 1993.

Vol. 76: **Tyrosine Phosphorylation/Dephosphorylation and Downstream Signalling.** Edited by L. M. G. Heilmeyer Jr. 388 pages. 1993.

Vol. 77: **Ataxia-Telangiectasia.** Edited by R. A. Gatti, R. B. Painter. 306 pages. 1993.

Vol. 78: **Toxoplasmosis.** Edited by J. E. Smith. 272 pages. 1993.

Vol. 79: **Cellular Mechanisms of Sensory Processing. The Somatosensory System.** Edited by L. Urban. 514 pages. 1994.

Vol. 80: **Autoimmunity: Experimental Aspects.** Edited by M. Zouali. 318 pages. 1994.

Vol. 81: **Plant Molecular Biology. Molecular Genetic Analysis of Plant Development and Metabolism.** Edited by G. Coruzzi, P. Puigdomènech. 579 pages. 1994.

Vol. 82: **Biological Membranes: Structure, Biogenesis and Dynamics.** Edited by Jos A. F. Op den Kamp. 367 pages. 1994.

Vol. 83: **Molecular Biology of Mitochondrial Transport Systems.** Edited by M. Forte, M. Colombini. 420 pages. 1994.

Vol. 84: **Biomechanics of Active Movement and Division of Cells.** Edited by N. Akkaş. 587 pages. 1994.

Vol. 85: **Cellular and Molecular Effects of Mineral and Synthetic Dusts and Fibres.** Edited by J. M. G. Davis, M.-C. Jaurand. 448 pages. 1994.

Vol. 86: **Biochemical and Cellular Mechanisms of Stress Tolerance in Plants.** Edited by J. H. Cherry. 616 pages. 1994.

Vol. 87: **NMR of Biological Macromolecules.** Edited by C. I. Stassinopoulou. 616 pages. 1994.

Vol. 88: **Liver Carcinogenesis. The Molecular Pathways.** Edited by G. G. Skouteris. 502 pages. 1994.

NATO ASI Series H

Vol. 89: **Molecular and Cellular Mechanisms of H^+ Transport.**
Edited by B. H. Hirst. 504 pages. 1994.

Vol. 90: **Molecular Aspects of Oxidative Drug Metabolizing Enzymes: Their Significance in Environmental Toxicology, Chemical Carcinogenesis and Health.**
Edited by E. Arınç, J. B. Schenkman, E. Hodgson. 623 pages. 1994.

Vol. 91: **Trafficking of Intracellular Membranes: From Molecular Sorting to Membrane Fusion.**
Edited by M. C. Pedroso de Lima, N. Düzgüneş, D. Hoekstra. 371 pages. 1995.

Vol. 92: **Signalling Mechanisms – from Transcription Factors to Oxidative Stress.**
Edited by L. Packer, K. Wirtz. 466 pages. 1995.

Vol. 93: **Modulation of Cellular Responses in Toxicity.**
Edited by C. L. Galli, A. M. Goldberg, M. Marinovich. 379 pages. 1995.

Vol. 94: **Gene Technology.**
Edited by A. R. Zander, W. Ostertag, B. V. Afanasiev, F. Grosveld. 556 pages. 1996.

Vol. 95: **Flow and Image Cytometry.**
Edited by A. Jacquemin-Sablon. 241 pages. 1996.

Vol. 96: **Molecular Dynamics of Biomembranes.**
Edited by J. A. F. Op den Kamp. 424 pages. 1996.

Vol. 97: **Post-transcriptional Control of Gene Expression.**
Edited by O. Resnekov, A. von Gabain. 283 pages. 1996.

Vol. 98: **Lactic Acid Bacteria: Current Advances in Metabolism, Genetics and Applications.**
Edited by T. F. Bozoglu, B. Ray. 412 pages. 1996.

Vol. 99: **Tumor Biology Regulation of Cell Growth, Differentiation and Genetics in Cancer.**
Edited by A. S. Tsiftsoglou, A. C. Sartorelli, D. E. Housman, T. M. Dexter. 342 pages. 1996.

Vol. 100: **Neurotransmitter Release and Uptake.**
Edited by Ş. Pöğün. 344 pages. 1997.

Vol. 101: **Molecular Mechanisms of Signalling and Membrane Transport.**
Edited by K. W. A. Wirtz. 344 pages. 1997.

Vol. 102: **Interacting Protein Domains**
Edited by L. Heilmeyer. 296 pages. 1997.

Vol. 103: **Molecular Microbiology.**
Edited by S. J. W. Busby, C. M. Thomas, N. L. Brown. 334 pages. 1998.

NATO ASI Series H

Vol. 104: **Cellular Integration of Signalling Pathways in Plant Development.**
Edited by F. Lo Schiavo, R. L. Last, G. Morelli, N. V. Raikhel.
330 pages. 1998.

Vol. 105: **Gene Therapy.**
Edited by Kleanthis G. Xanthopoulos.
238 pages. 1998.

Vol. 106: **Lipid and Protein Traffic.**
Pathways and Molecular Mechanisms.
Edited by J. A. F. Op den Kamp.
374 pages. 1998.

MIX
Papier | Fördert gute Waldnutzung
FSC® C083411

Kolibri 360 GmbH
Ferdinand-Jühlke-Straße 7
99095 Erfurt, Deutschland
produktsicherheit@kolibri360.de